Toxins as Tools in Neurochemistry

Toxins as Tools in Neurochemistry

Proceedings of the Symposium
Berlin (West), March 22–24, 1983

Editors
Ferdinand Hucho
Yuri A. Ovchinnikov

Walter de Gruyter · Berlin · New York 1983

Editors
Ferdinand Hucho, Professor, Dr. rer. nat.
Free University Berlin
Department of Chemistry
Institute of Biochemistry
Fabeckstrasse 34
D-1000 Berlin 33
Germany

Yuri A. Ovchinnikov, Professor, D. Sc.
Shemyakin Institute of Bioorganic Chemistry
USSR Academy of Sciences
Moscow
USSR

CIP-Kurztitelaufnahme der Deutschen Bibliothek

Toxins as tools in neurochemistry: proceedings of the symposium Berlin (West), March 22–24, 1983/
ed. Ferdinand Hucho; Yuri A. Ovchinnikov. –
Berlin; New York: de Gruyter, 1983.
ISBN 3-11-009593 9
NE: Hucho, Ferdinand [Hrsg.]

Library of Congress Cataloging in Publication Data

Toxins as tools in neurochemistry.
Organized by members of the USSR Academy of Sciences and the Free University Berlin.
Bibliography: p.
Includes index.
1. Neurochemistry -- Congresses. 2. Neurotoxic agents -- Physiological effect -- Congresses. 3. Ion channels -- Congresses.
I. Hucho, Ferdinand, 1939 – II. Ovchinnikov, IU. A. (IUriĭ Anatol'evich), 1934 – III. Akademiia nauk SSSR.
IV. Freie Universität Berlin. [DNLM: 1. Neurotoxins -- Congresses. QW 630 T7555]
QP356.3.T67 1983 591.1'88 83-14479
ISBN 3-11-009593-9

 Printing: Gerike GmbH, Berlin. – Binding: Dieter Mikolai, Berlin. – Printed in Germany.

Preface

This symposium took place in March 1983 in Berlin (West), organized jointly by members of the USSR Academy of Sciences and the Free University of Berlin. The aim of the meeting was to review the present state of toxin application in neurochemistry.

The proceedings of the symposium contain reviews and reports on new and established toxins used for the identification and molecular characterization of key functions of the membranes of excitable cells (nerve and muscle). Like monoclonal antibodies, toxins have specificity and high affinity for their site of action. But they are by definition superior to the former as far as their effectiveness in modulating (activating or inhibiting) the respective nerve function is concerned.

Though many people question the usefulness of proceedings of this kind, we feel they are helpful and acceptable provided they appear shortly after the respective meeting. No other type of publication offers such an up-to-date and yet comprehensive account of scientific methods and data. To avoid delay in publication we chose not to spend a large amount of time polishing the texts, but to accept the individual manuscripts largely as submitted. We ask the reader to take into account the fact that none of the participants has English as his mother tongue.

A meeting of this sort can be successful only with the help of many people. We would like to thank all who gave a helping hand. We owe special thanks to Frau Greiner who kept minor catastrophes to a minimum and who also dealt with the manuscripts and technical organization of this volume with great competence.

Grateful acknowledgement is also due to several sponsors, including the Senator für Wirtschaft und Verkehr der Stadt Berlin (West), the Gesellschaft für Biologische Chemie, Schering AG, Boehringer Mannheim GmbH and Kontron Analytic for their financial support.

Berlin, July 1983

Y.A. Ovchinnikov
Ferdinand Hucho

Contents

List of participants

Bandini, G.
Institut für Biochemie, Freie Universität Berlin, Fabeckstr.34, 1ooo Berlin 33

Baumann, H.
Institut für Biochemie, Freie Universität Berlin, Fabeckstr.34, 1ooo Berlin 33

Béress, L.
Institut für Toxikologie der Universität Kiel, Hospitalstr.4-6, 23oo Kiel

Betz, H.
Max-Planck-Institut für Psychiatrie, Abt. Neurochemie
Am Klopferspitz 18a, 8o33 Martinsried b.München

Breer, H.
Universität Osnabrück, Abt. Zoophysiologie, Seminarstr.2o, 45oo Osnabrück

Bystrov, V.F.
Shemyakin Institute of Bioorganic Chemistry, USSR Academy of Sciences, Moscow

Ebeling, M.
Institut für Biochemie der Freien Universität Berlin, Fabeckstr.34, 1ooo Berlin 33

Yelin, E.A.
Shemyakin Institute of Bioorganic Chemistry, USSR Academy of Sciences, Moscow

Fels, G.
Max-Planck-Institut für Ernährungsphysiologie, Rheinlanddamm 2o1, 46oo Dortmund

Grünhagen, H.-H.
BASF AG, Hauptlabor B9, 67oo Ludwigshafen

Habermehl, G.
Chemisches Institut der Tierärztlichen Hochschule Hannover, Bischofsholer Damm 15, 3ooo Hannover

Hahn,U.
Institut für Kristallographie der Freien Universität Berlin, Takustr.6, 1ooo Berlin 33

Hammann,P.
Chemisches Institut der Tierärztlichen Hochschule Hannover, Bischofsholer Damm 15, 3ooo Hannover

Hucho,F.
Institut für Biochemie der Freien Universität Berlin,Fabeck-str.34-36, 1ooo Berlin 33

Jakob,R.
Institut für Kristallographie der Freien Universität Berlin, Takustr.6, 1ooo Berlin 33

Jaweed, S.
Institut für Biochemie der Freien Universität Berlin, Fabeck-str.34-36, 1ooo Berlin 33

Khodorov,B.I.
Vishnevsky Surgery Institute, USSR Academy of Medical Sciences, Moscow

Kirsch, J.
Chemisches Institut der Tierärztlichen Hochschule Hannover, Bischofsholer Damm 15, 3ooo Hannover

Koschel, K.
Institut für Virologie und Immunbiologie der Universität Würzburg, Versbacherstr.7, 87oo Würzburg

Kostyuk,P.G.
A.A.Bogomoletz Institute of Physiology, Academy of Sciences of the Ukrainian SSR, Kiev

Lauffer, L.
Institut für Biochemie der Freien Universität Berlin, Fabeck-str.34-36, 1ooo Berlin 33

Lishko,V.K.
Palladin Institute of Biochemistry, Ukrainian SSR Academy of Sciences, Kiev

Maelicke,A.
Max-Planck-Institut für Ernährungsphysiologie, Rheinland-Damm 2o1, 46oo Dortmund

Mebs,D.
Zentrum der Rechtsmedizin der Universität Frankfurt, Kennedyallee 1o7, 6ooo Frankfurt

Mewes,H.
I.Physiologisches Institut der Universität des Saarlandes, 665o Homburg/Saar

Meyers,H.-W.
Max-Planck-Institut für Ernährungsphysiologie, Rheinland-Damm 2o1, 46oo Dortmund

Muhn,P.
Institut für Biochemie der Freien Universität Berlin, Fabeckstr.34-36, 1ooo Berlin 33

Naumov,A.P.
Institute of Cytology, Academy of Sciences of the USSR, Leningrad, USSR

Nuske,J.
Institut für Biochemie der Freien Universität Berlin, Fabeckstr.34-36, 3ooo Berlin 33

Pluzhnikov,K.A.
Shemyakin Institute of Bioorganic Chemistry, USSR Academy of Sciences, Moscow

Rack,M.
Institut für Physiologische Chemie der Universität des Saarlandes, 665o Homburg/Saar

Ravens,U.
Abt.Pharmakologie, Klinisch-Theoretische Medizin II der Christian-Albrecht-Universität, Hospitalstr.4-6, 23oo Kiel

Rehm,H.
Max-Planck-Institut für Psychiatrie, Abt.Neurochemie, Am Klopferspitz 18a, 8o33 Martinsried b.München

Reulecke,M.
Institut für Biochemie der Freien Universität Berlin, Fabeckstr.34-36, 1ooo Berlin 33

Saenger,W.
Institut für Kristallographie der Freien Universität Berlin, Takustr.6, 1ooo Berlin 33

Salikhov,I.
Institute of Bioorganic Chemistry, Uzbeck SSR Academy of Sciences, Tashkent

Schmidtmayer,J.
Physiologisches Institut der Christian-Albrecht-Universität, Olshausenstr.4o-6o, 23oo Kiel

Schneider,M.
Institut für Kristallographie der Freien Universität Berlin, Takustr.6, 1ooo Berlin 33

Soldatov,N.M.
Shemyakin Institute of Bioorganic Chemistry, USSR Academy of Sciences, Moscow

Surin,A.
Shemyakin Institute of Bioorganic Chemistry, USSR Academy of Sciences, Moscow

Tesseraux,I.
Institut für Toxikologie der Universität Kiel, Hospitalstr. 4-6, 23oo Kiel

Ulbricht,W.
Physiologisches Institut der Christian-Albrecht-Universität, Olshausenstr.4o-6o, 23oo Kiel

Utkin,Yu.N.
Shemyakin Institute of Bioorganic Chemistry, USSR Academy of Sciences, Moscow

Verdenhalven,J.
Institut für Biochemie der Freien Universität Berlin, Fabeckstr.34-36, 1ooo Berlin 33

I. Na-Channels

KINETICS OF NEUROTOXIN ACTION AT THE NODAL MEMBRANE

Werner Ulbricht
Physiologisches Institut der Universität Kiel, D-2300 Kiel

Introduction

Neurotoxins and other agents that specifically act on excitable membranes are increasingly used to explore the channels through which ions move during excitation. For the electrophysiologist such agents serve to alter details of channel function with the idea to deduce structural details complementary to the structure of the agent. Extension of such studies to the kinetics of drug-channel reactions may yield additional information about accessibility (and hence location) of sites and about possible interactions of two drugs at this site. The present paper reviews relevant experiments on sodium channels of Ranvier nodes, done in our laboratory.

Methods

The experiments (mostly in the voltage clamp) were done on myelinated nerve fibres of the frog, *Rana esculenta*, in two setups. One was the standard Nonner clamp (3) with a chamber that permitted an exchange of Na^+ ions at the membrane with a half time of ca. 0.45 s (8). In the other setup, described in Ref. (5) exchange was ca. 20 times faster (10) but the speed of voltage clamping was limited, hence only slower current phases were recorded. In either setup sodium current, I_{Na}, was recorded after almost complete suppression of potassium current by 10 mM TEA. Capacity and leakage currents were subtracted either by an analogue device (12) or with the aid of a signal

Toxins as Tools in Neurochemistry

averager. Records obtained in the fast-exchange setup were corrected only for leakage. Potentials are given as deviations, V, from the normal resting potential, depolarizations being positive. - Mean values are given ± S.E.M.

Results

Potential-Insensitive Block

The ideal neuroactive agent for kinetic studies should reversibly react with the channel, the reaction should have immediately measurable consequences and be insensitive to secondary conditions such as membrane potential or channel gating. The low-molecular neurotoxins tetrodotoxin, TTX, and saxitoxin, STX, seem to come close to the ideal; they block sodium channels possibly by plugging them (2) in a one-to-one reaction as suggested by the Hill coefficients $n_H \approx 1$ of either block (8, 12). These reactions can be written

$$A \quad + \quad R \underset{k_{2A}}{\overset{k_{1A}}{\rightleftharpoons}} A.R$$

$$B \quad + \quad R \underset{k_{2B}}{\overset{k_{1B}}{\rightleftharpoons}} B.R$$

where A and B are the toxin molecules, R is the receptor, A.R and B.R are toxin-receptor complexes whose formation are thought to induce immediate complete block of a channel. k_{1A} and k_{1B} are the rate constants of association, k_{2A} and k_{2B} those of dissociation. Let us first consider the single toxin (e.g. TTX) case. At equilibrium with [A] the fraction $y_A(\infty)$ of channels blocked by A is given by

$$y_A(\infty) = [A.R]\ /([A.R] + [R]) = c_A/(c_A + 1) \qquad (1)$$

where $c_A = k_{1A}\ [A]/k_{2A} = [A]/K_A$ with K_A = equilibrium dissociation constant. On a step change in [A], y_A will change exponentially with a time constant

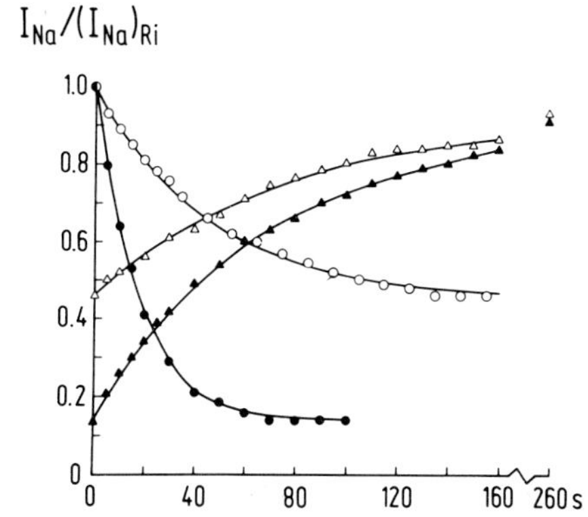

Fig. 1. Onset (circles) and offset (triangles) of TTX action on normalized peak I_{Na} vs. time after solution change. Open symbols refer to 3.1 nM (τ_{on} = 45 s, τ_{off} = 77 s), filled symbols to 15.5 nM TTX (τ_{on} = 17 s, τ_{off} = 78 s). Curves fitted with $k_{1A} = 2.9 \times 10^6 M^{-1}s^{-1}$, $k_{2A} = 1.3 \times 10^{-2}s^{-1}$. From (8) by permission.

$$\tau_A = 1/k_{2A}(c_A + 1). \qquad (2)$$

From equ. (2) it follows that τ_A decreases with increasing c_A but becomes independent on washout when c_A drops to zero: $\tau_A = 1/k_{2A}$. Fig.1 shows that this is indeed observed and that the time course with which peak I_{Na} decreases on applying two TTX concentrations and recovers on washing can be described by a single set of rate constants. The figure also shows that block by TTX is slow enough to be studied in the standard chamber.

We now have to ask whether k_{1A} and k_{2A} extracted from experiments as in Fig. 1 are the true (limiting) rates of reaction. The answer is affirmative because of, among other points, the high Q_{10} values corresponding to apparent activation energies of 41 and 86 kJ/mol for k_{1A} and k_{2A} (8). Also, diffusional access cannot be rate limiting since there is no appreciable delay in the onset of action.

STX ("B") acts very similarly but the rates are higher. In a comparative study on the same preparations we found k_{1B} = 2.3

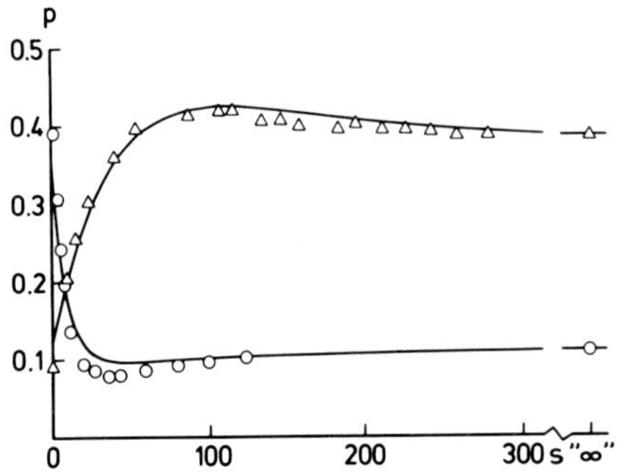

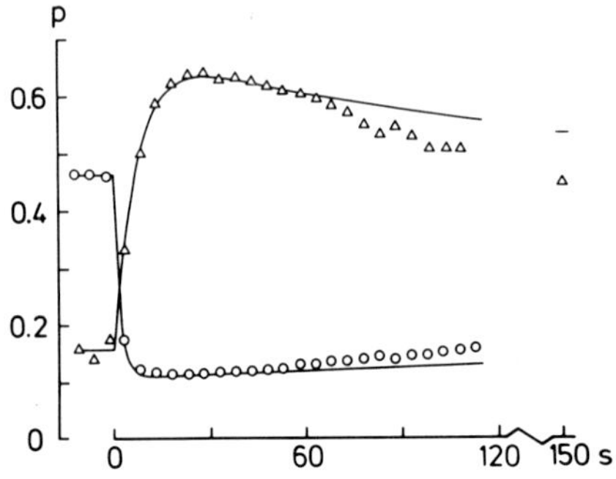

Fig. 2 (left). Time course of p, I_{Na} normalized to control, on adding 9 nM STX after equilibration in 3.1 nM TTX (○) and on taking out STX after equilibration in STX-TTX (△). Curves computed with c_A = 1.6, c_B = 0 or 6.3, $k_{2A} = 0.81 \times 10^{-2} s^{-1}$, $k_{2B} = 1.38 \times 10^{-2} s^{-1}$. 14.8°C. From (12) by permission.

Fig. 3 (right). Time course of p on adding 0.5 mM benzocaine (B) to 3 μM ervatamine (A;○) and on taking out B from the mixture (△). Curves computed with c_A = 1.16, c_B = 0 or 4.29, $k_{2A} = 1.8 \times 10^{-2} s^{-1}$, $k_{2B} = 1.05 \times 10^{-1} s^{-1}$. 14.7°C. From (4) by permission.

k_{1A}, $k_{2B} = 1.7\ k_{2A}$, K_B = 1.4 nM and K_A = 2.0 nM with this batch of TTX. The fact that $k_{2B} > k_{2A}$ has an interesting kinetic consequence if A (=TTX) and B (=STX) compete for the same receptor i.e. if "R" in the two reaction schemes belongs to the same pool: after equilibration in a STX-TTX mixture, taking out STX leads to a non-monotonic recovery as illustrated by the triangles in Fig. 2. The reason is that in the mixture of [A] + [B] the fraction of channels blocked by the respective toxin is $y'_A(\infty) + y'_B(\infty)$ where $y'_A(\infty) = c_A/(c_A + c_B + 1)$. At equilibrium with [A] alone the fraction of channels blocked is $y_A(\infty) = c_A/(c_A + 1)$ i.e. $y_A(\infty) > y'_A(\infty)$. On taking out STX ($c_B$= 0) channels formerly occupied by this toxin become free faster than TTX can claim its share which leads to a transiently overshooting recovery whose time course can be calculated (1, 12). Computations predict that the overshoot increases with increasing ratio k_{2B}/k_{2A}. Clearly the occurence of a non-monotonic recovery is proof of a common binding site or at least of mutually exclusive binding.

Potential-Sensitive Block

Block of sodium channels by local anaesthetics is, in contrast to that by TTX, modulated by membrane potential which much complicates the description in terms of drug-receptor reaction. Hence only two types of experiments shall be described here. One is the application of the principle illustrated by Fig. 2 for STX-TTX to a mixture of the very fast acting benzocaine and the slowly blocking indole alkaloid ervatamine. In Fig. 3 we see again an enormously overshooting recovery from block when benzocaine is omitted after equilibrating the node in a benzocaine-ervatamine mixture (4). This result suggests that benzocaine and the chemically unrelated ervatamine interact in blocking sodium channels. Possibly the local anaesthetic binding site admits quite diverse molecules. It certainly does not distinguish between permanently uncharged anaesthetics such as benzocaine and, at neutral pH, partially dissociated anaesthetics such as lidocaine as established experimentally by the same basic argument (6).

The other type of experiments was designed to test whether local anaesthetics and STX block by binding to different sites, one each per channel. Let, in a mixture of A (procaine) and B (STX), fractions y_A and y_B of pertinent <u>sites</u> be occupied. A channel is blocked if either site is occupied so that the fraction, Y, of <u>channels</u> blocked will be given by (13)

$$Y = y_A + y_B - y_A y_B. \qquad (3)$$

This is also true for non-equilibrium cases. For y_B = const. (after equilibration in STX) we have

$$dY/dt = dy_A/dt\ (1 - y_B). \qquad (4)$$

This is indeed observed in Fig. 4 which shows that the rate of use-dependent recovery from procaine block is unchanged by the presence of STX. In this series of experiments the mean $y_A(\infty)$ was 0.81 (corresponding to c_A = 4.26), $y_B(\infty)$ was 0.90

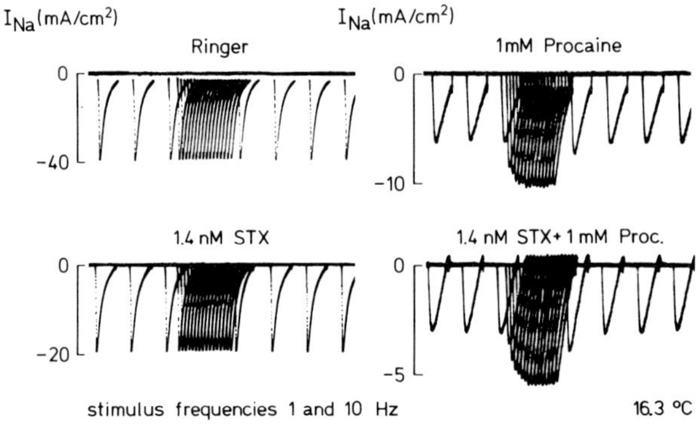

Fig. 4. Effect of increasing the rate of pulses (50-ms prepulse, V = -40 mV, test pulse, V = 60 mV) from 1 to 10 Hz on peak I_{Na}, standing pictures recorded on moving film: in Ringer solution and 1.3 nM STX no effect, in 1 mM procaine with and without STX use-dependent recovery. Note that 1.4 nM STX reduces I_{Na} by ca. 50% in either case. No TEA in solutions. From (13) with permission.

(corresponding to $c_B = 0.96$) and $Y(\infty)$ was 0.90, exactly as predicted by equ. (3). Note that in the case of competition we would expect a lower value $Y'(\infty) = (c_A + c_B)/(c_A + c_B + 1) = 0.84$. Thus the kinetic and equilibrium results point to separate and independently accessible receptors for the two agents. STX most likely binds to the external channel entrance, benzocaine somewhere inside the channel where it supposedly gets directly from the lipid phase or, after penetrating it, through the axoplasmic entrance (2).

Modified Sodium Channel Inactivation

If the drug-receptor reaction leads to blockage the maximum effect is easily defined as total block i.e. complete disappearance of I_{Na}. With drugs that modify channel gating the situation is more complicated as illustrated by Fig. 5 which shows equilibrium effects of 3 different concentrations of *Anemonia sulcata* toxin II, ATX II. The prominent reversible

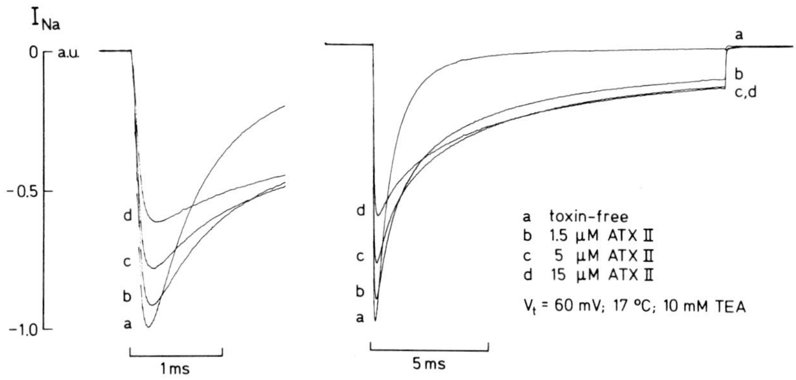

Fig. 5. Equilibrium effects of three [ATX II] on $I_{Na}(t)$, normalized to peak in the control. Constant test pulse (15 ms, V = 60 mV), left-hand family 4 × expanded in time. From (11) by permission of MASSON S. A. Paris.

effect of this toxin is to slow inactivation i.e. the decline of I_{Na} on sustained depolarization. Inactivation is moreover incomplete so that a sizeable current, I_{15ms}, is flowing at the end of a 15-ms pulse. Obviously I_{15ms} saturates for [ATX II] ≧ 5 μM whereas peak I_{Na} is further reduced (11). Thus complete loss of inactivation (with I_{15ms} ≈ control peak I_{Na}) cannot be achieved with ATX II. The reason is unclear but a few clues may be obtained from kinetic studies which were done to learn about the accessibility of the toxin site. With I_{15ms} as a measure, onset and offset of ATX II action were recorded in the fast exchange setup (5). An example is shown in Fig. 6A for 5 μM (○), for a two-step application, 1.25 and 5 μM (△) and for a very short (●, 1.3 s; 5 μM) application which was followed by a clearly overshooting reaction. Onset and offset were sigmoid, reversibility was complete. Cooling definitely slowed the rates to an extent suggesting an almost limiting toxin-channel reaction. The best fit was obtained with a two-step reaction

$$A + R \underset{k_{2A}}{\overset{k_{1A}}{\rightleftharpoons}} A.R_1 \underset{k_{4A}}{\overset{k_{3A}}{\rightleftharpoons}} A.R_2$$

where $A.R_1$ is a silent complex that on conversion to $A.R_2$ ex-

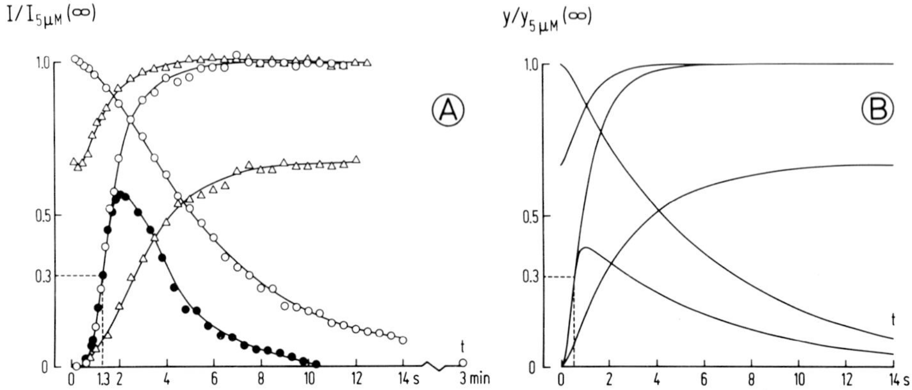

Fig. 6. A. Time course of normalized I_{15ms}, $I/I_{5\mu M}(\infty)$, on applying and washing out ATX II as described in text vs. time after solution change corrected for arrival of Na^+ at the membrane. At the end of the short (1.3 s) application the signal was 0.3 but continued to rise to ca. 0.6 on washing. 15.4°C.
B. Simulation of A, computation based on two-step reaction with $k_{1A} = 1.48 \times 10^5\ M^{-1}s^{-1}$, $k_{2A} = 0.296\ s^{-1}$, $k_{3A} = k_{4A} = 2.083\ s^{-1}$, $y_{1.25\mu M}(\infty) = 0.28$, $y_{5\mu M}(\infty) = 0.42$. Short (here 0.65 s) application until signal was 0.3. From (5) with permission.

presses neurotoxicity. Fig. 6B gives computed fractions of sites in the active configuration, $y = [A.R_2]/([R]+[A.R_1]+[A.R_2])$, normalized to the value after equilibration in 5 μM ATX II. If $y/y_{5\mu M}(\infty)$ can be equated with the normalized I_{15ms}, the scheme could explain why I_{15ms} saturates at a relatively low level since at high [A], $[A.R_2]$ is solely determined by k_{4A}/k_{3A} and could be much less than unity.

Delay of onset and overshoot are more pronounced in the real experiment than in curves computed on the assumption of a step change in toxin concentration at the site which may not be valid for the diffusion of ATX II (MWt ca. 5000) through the unstirred layer covering the membrane [see discussion in Ref. (5)]. At any rate the fast onset of toxin action suggests an easily accessible site which also seems to admit freely Ca^{2+}. Calcium not only shifts (as in unpoisoned fibres) the peak

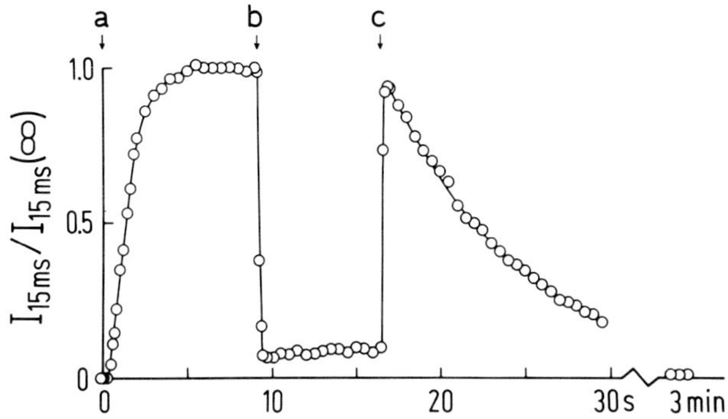

Fig. 7. Effect of Ca^{2+} on normalized I_{15ms} as described in text. 17.0°C.

I_{Na}-V relation to more positive potentials but it also nearly suppresses I_{15ms} for $60 < V < 80$ mV. This modulating Ca^{2+} effect is very fast as illustrated by Fig. 7 which shows the normalized I_{15ms} during onset of 5 μM ATX II, 2 mM Ca^{2+} (a), on changing to toxin + 10 mM Ca^{2+} (b) and on washing with toxin-free solution containing 2 mM Ca^{2+} (c) which leads to a prompt recovery of the signal (Ca^{2+} effect) before it declines (wash-out). The mechanism of this Ca^{2+} action is unknown but it is likely due to a direct interference with the toxin-site reaction and not, indirectly, due to a change in surface potential commonly thought to be the reason for the shift in peak $I_{Na}(V)$. In line with this idea, the sizeable I_{15ms} observed on internal iodate treatment (9) is not affected by external Ca^{2+} whereas peak I_{Na} is shifted in these circumstances (7). These results point to a superficial receptor for ATX II and it remains to be found out how its occupation can affect the inactivation "gate" which is held to be located at the axoplasmic mouth of the channel.

In summary, the binding sites for ATX II and for TTX-STX appear to be externally located and at least the latter is separate from that for local anaesthetics and similarly acting but chemically quite diverse agents as the alkaloid ervatamine.

Acknowledgement. I should like to thank the Deutsche Forschungsgemeinschaft for support and Ms. E. Dieter for continuing help.

References

1. Colquhoun, D.: Proc. R. Soc. (Lond.) B 170, 135-154 (1968)
2. Hille, B.: J. Gen. Physiol. 69, 497-515 (1977)
3. Nonner, W.: Pflügers Arch. 309, 176-192 (1969)
4. Pichon, Y., Schmidtmayer, J., Ulbricht, W.: Neurosci. Lett. 22, 325-330 (1981)
5. Schmidtmayer, J., Stoye-Herzog, M., Ulbricht, W.: Pflügers Arch. 394, 313-319 (1982)
6. Schmidtmayer, J., Ulbricht, W.: Pflügers Arch. 387, 47-55 (1980)
7. Schmidtmayer, J., Ulbricht, W.: Pflügers Arch. 394, R 45 (1982)
8. Schwarz, J.R., Ulbricht, W., Wagner, H.-H.: J. Physiol. 233, 167-194 (1973)
9. Stämpfli, R.: Experientia 30, 505-508 (1974)
10. Ulbricht, W.: Physiol. Rev. 61, 785-828 (1981)
11. Ulbricht, W., Schmidtmayer, J.: J. Physiol. (Paris) 77, 1103-1111 (1981)
12. Wagner, H.-H., Ulbricht, W.: Pflügers Arch. 359, 297-315 (1975)
13. Wagner, H.-H., Ulbricht, W.: Pflügers Arch. 364, 65-70 (1976)

MODIFICATION OF SODIUM CHANNELS WITH SCORPION TOXINS AND ALKALOIDS

A.P.Naumov

Institute of Cytology, Academy of Sciences of the USSR,
Leningrad , USSR

Introduction

Sodium channels of excitable cells can be defined as water-filled pores through the membrane, controlled by some electrosensitive gates. Normally when membrane potential is negative enough (inside minus outside) most of the channels are closed; during depolarization they open within fractions of millisecond (activation) and then become nonconducting again (inactivation). Sodium channels are primary targets for a number of neurotoxins each of which causes specific alterations in channel functions.

The present communication summarizes some results of our investigations into effects of the polypeptide scorpion toxins and the alkaloid neurotoxins aconitine and batrachotoxin on sodium channels.

Results

Fig.1 shows the effect of one of the toxins (ScTX) from Buthus eupeus scorpion venom on Na current in the neuroblastoma membrane. It is seen that the toxin slows down Na inactivation and makes it incomplete. Firstly such an effect was demonstrated by Koppenhöfer and Schmidt (1) with the whole venom of the scorpion Leiurus quinquistriatus. The Figure shows also that the toxin is effective only being added to the external solution. Internal application of much higher concentrations of the same toxin has no effect. So

Toxins as Tools in Neurochemistry

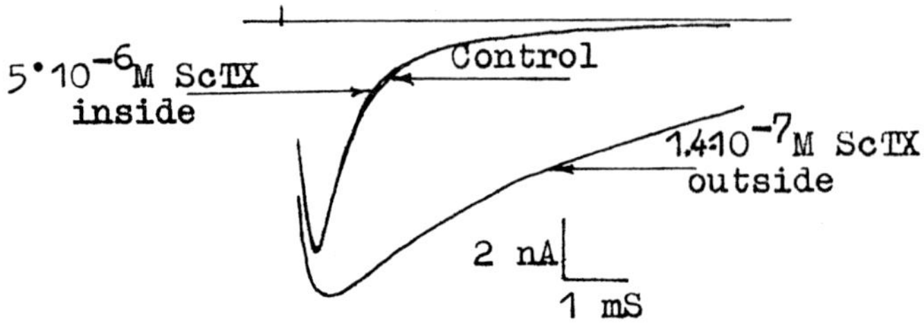

Fig.1 Effect of ScTX (conditional name 2oo1) on Na current in neuroblastoma cell. Shown are three current traces: Control, ScTX added to internal solution and ScTX added to external solution. Test potential -2o mV. From (2).

receptors for ScTX are located on the external side of the membrane. An interesting feature of the ScTX binding to Na channels is its potential dependence: Toxin binding decreases when depolarizing the membrane (3). This phenomenon deserves attention because it may reflect some potential induced changes in channel conformation, for example associated with opening or (and) inactivation. We have investigated ScTX binding to Na channels under voltage clamp conditions with frog myelinated nerve (4). This method enables to vary membrane voltage (E) over a wide range. The estimate of the fraction of poisoned channels at given toxin concentrations and potentials was based on the difference in inactivation kinetics for normal and poisoned channels. Also a fraction of noninactivated channels (h_∞) and the normalized peak conductance (g_p) were determined as functions of E. When analysing the data we used the simplest model, according to which the Na channel can assume, at least, three functional states: closed (1), open (2) and inactivated (3). According to the model h_∞ and steady-state conductance (g_∞) are given by the following equations:

$$h_\infty = \frac{1 + \exp a_2 (E - E_2)}{1 + \exp a_2(E - E_2) + \exp a_3(E - E_3)} \qquad (1)$$

$$g_{\infty} = \frac{\exp a_2 (E - E_2)}{1 + \exp a_2(E - E_2) + \exp a_3(E - E_3)} \qquad (2)$$

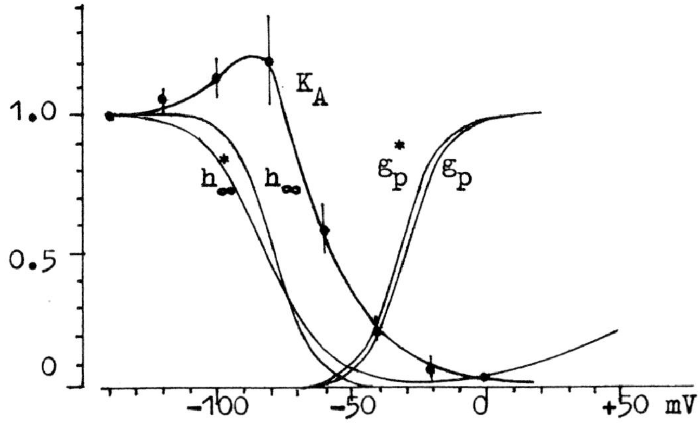

Fig.2 Apparent binding constant (K_A) of ScTX (conditional name M_7), h_∞ and g_p as functions of E for nodal membrane. Asterisks denote curves relevant to the channels poisoned by toxin. Circles with vertical bars represent mean K_A values ± S.D. Smooth lines are calculated according to the three-state model with mean values of parameters. Adopted from (4).

where a_2 and a_3 (in mV^{-1}) are the factors determining the steepness of voltage dependence of free energies of open and inactivated states, respectively. E_2 and E_3 are E at which energies of open and inactivated states are equal to the energy of the closed state. The closed state is taken as reference. All these parameters can be easily determined from experimental $h_\infty(E)$ and $g_p(E)$ curves. For example the $g_p(E)$ curve can be fitted by equation 2 after omitting the term relevant for the inactivated state. Fig.2 shows h_∞ and g_p curves calculated according to the model with mean parameter values determined in ref. (4). The main effect of ScTX is the decrease of steepness parameter for the inactivated state (a_3). For this reason the term $\exp a_3(E - E_3)$ for poisoned channels

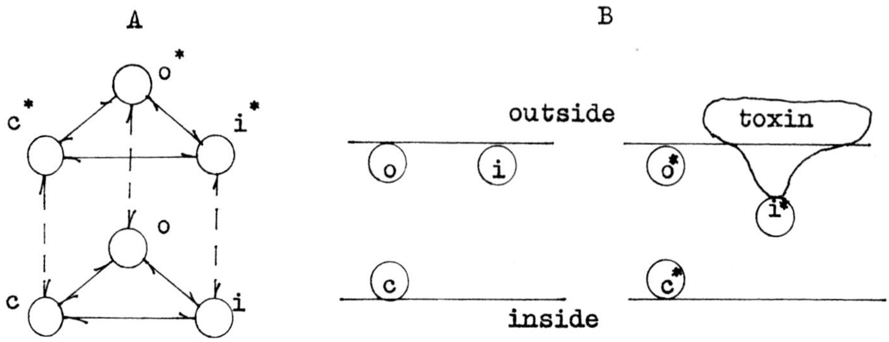

Fig.3 State diagram (A) and mechanical representation (B) of the three-state model. c, o and i are closed, open and inactivated states, respectively. Asterisks denote states of the toxin-poisoned channel. See text.

grows less steeply with depolarization than exp $a_2(E - E_2)$ and as a result the $h_\infty(E)$ curve rises at $E > -40$ mV. In other words, energies and consequently probabilities of open and inactivated states in the poisoned channel become closer to one another as E becomes more positive. In normal channels the inactivated state is much more favorable energetically than the open state at all potentials and therefore g_∞ in untreated membranes is very low. As the channel can be either bound or free of toxin the number of states is doubled (Fig.3). Experiments show that the rates of interaction of the toxin with the channel are much slower than those of activation and inactivation. In this case the apparent binding constant (K_A) is given by:

$$K_A = K_1 \frac{1 + \exp a_2^*(E - E_2^*) + \exp a_3^*(E - E_3^*)}{1 + \exp a_2(E - E_2) + \exp a_3(E - E_3)} \qquad (3)$$

where K_1 is the binding constant for the channel in state 1, asterisks indicating parameters related to poisoned channels. Binding constants for open and inactivated channels are as follows:

$$K_2 = K_1 \exp \; a_2^*(E - E_2^*) - a_2(E - E_2) \qquad (4)$$

$$K_3 = K_1 \exp \; a_3(E - E_3) - a_3(E - E_3) \qquad (5)$$

K_1 does not depend on the potential, because K_A tends to a steady level at high negative E and on the other hand, K_1 according to equation 3 tends to K_1 over this potential range. ScTX does not alter the parameters of the open state, so K_2 is nearly the same as K_1. Equation 5 shows that K_3 changes with E because $a_3^* < a_3$. The more negative is E, the larger is K_3. At high negative E K_3 becomes larger than K_1, but the number of inactivated channels decreases, that is why the $K_A(E)$ curve has a maximum. Fig.2 shows that the model gives a rather good quantitative description of the voltage-dependency of ScTX binding. The mode of toxin-channel interaction may be illustrated by the simple mechanical model (Fig.3 B). The circles show positions of some generalized gating charge (it is assumed as a positive one, for the sake of simplicity) corresponding to closed, open and inactivated states. The toxin molecule reduces the effective charge of the inactivated state by decreasing the respective distance travelled by the charged particle in the electric field. For this reason ScTX decreases the slope of the $h_\infty(E)$ curve. Further, on depolarization, this charged particle will tend to occupy its normal position, thereby displacing the toxin molecule. That is why K_A decreases with depolarization. Whatever the actual mechanism of inactivation may be, it is clear that this process involves conformational changes of the channel structure from the internal to external surfaces of the membrane.

Quite another chemical nature have the alkaloid neurotoxins aconitine (Ac) and batrachotoxin (BTX). The most obvious effects of these alkaloids on Na channel gating are the shift of voltage range of activation to more negative potentials and the increase of steady-state Na conductance. Analysis of these results with the three-state model reveals that the second effect is a consequence of the first one. Fig.4 shows currents through Ac-modified Na channels, $g_p(E)$ and $h_\infty(E)$ curves for

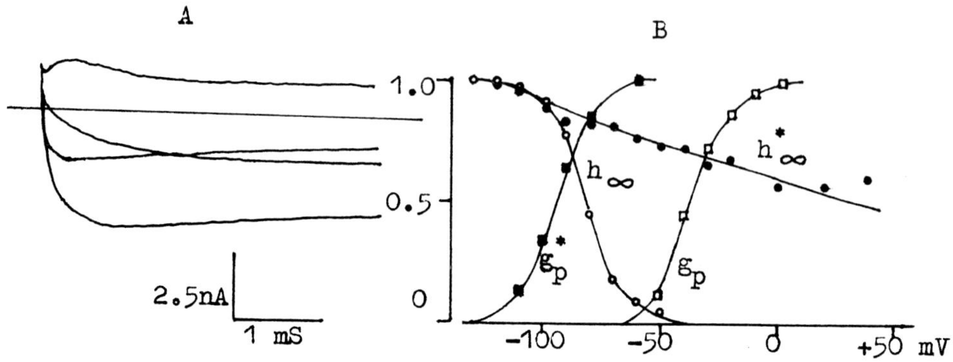

Fig.4 Accnitine-modified Na channels of the frog node. A. Currents at -9o, -5o, -1o and +3o mV (from (5)); B. Activation (g_p and g_p^*) and inactivation (h_∞ and h_∞^*) curves for normal and Ac-modified (asterisks) channels. Smooth curves are calculated according to the three-state model with a_2 = o.13o mV^{-1}, a_2^* = o.12o mV^{-1}, E_2 = -38 mV, E_2^* = -94.5 mV, a_3 = -o.133 mV^{-1}, a_3^* = o.133 mV^{-1}, E_3 = -82 mV, E_3^* = -82 mV. From (6).

normal and modified channels in frog node. It is seen that currents in modified channels inactivate only partially and $h_\infty(E)$ values do not fall lower than o.5. The $g_p(E)$ curve for modified channels is shifted to the left on the E-axis by about 5o mV. In terms of the model Ac makes E_2 more negative. This difference in E_2 alone proved to be quite enough to account for the observed change in $h_\infty(E)$ curve in Ac-treated membrane. The reason for this can be explained as follows. The negative shift of E_2 means that the open state becomes energetically more favorable and consequently more probable. Because the sum of probabilities of all states is equal to unity, the increase of probability of the open state makes the decrease in probability of the inactivated state inevitable. The larger the negative shift of E_2 the smaller is the fraction of the channels inactivated on depolarization. This rule is confirmed by two observations. Unlike with nodal mem- Ac-modified Na channels in neuroblastoma cells inactivate nearly completely (Fig.5). This difference can be accounted

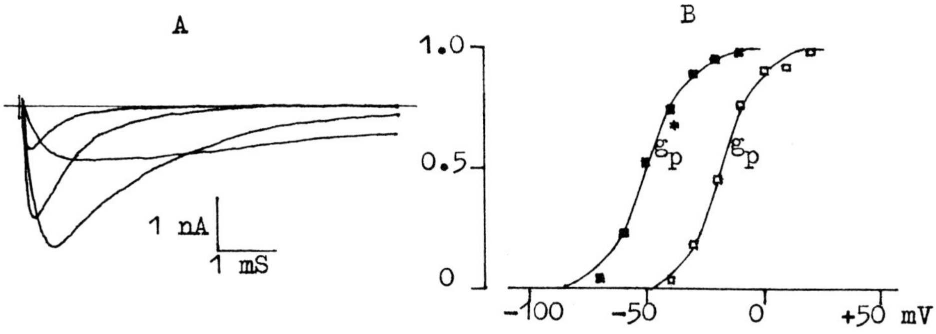

Fig.5 Aconitine-modified Na channels in neuroblastoma cells. A. Currents at -6o, -4o, -2o and 0 mV; B. Activation curves for normal and Ac-modified (asterisks) channels. Smooth curves are calculated according to the three-state model (see text) with a_2 = o.13o mV^{-1}, E_2 = -17mV, a_2^* = o.12o mV^{-1}, E_2^* = -5o mV. From (7).

for by the fact that in this cells Ac induces a relatively small (about 2o - 3o mV) negative shift of the g_p(E) curve. BTX exerts effects qualitatively similar to those of Ac (9),but BTX induces larger E_2 shifts both in the node and in neuroblastoma cells; respectively, it induces a larger steady-state Na conductance. From what is stated above it is clear that the drug-induced conversion of normally inactivating channels into partly- or noninactivating ones is not necessarily due to changes in parameters of the inactivated state(s).

The substitution of the parameters of the open state for normal and modified channels in equation 4 shows that binding constantsfor Ac and BTX should increase by more than two orders when opening the channels. Indeed, repetitive depolarization of the membrane sufficient to open channels, dramatically accelerates the modification of the channels by Ac and BTX (8,9). FurthermoreScTX increases the alkaloid binding to the channels at low potentials (1o). This phenomenon seems to be due to the fact that ScTX increases the probability of the open state at these potentials by means of changing parameters of the inactivated state (see above). Hence, one can expect that the higher the level of h_∞ at low E for ScTX-poisoned channels the stronger is the alkaloid binding. We have investigated two toxins from Buthus eupeus venom (conditional names S-1 and E-3). Their effects on the $h_\infty(E)$ curve were qualitatively analogous, but S-1 induced a higher h_∞ level at low E (Fig.6). Accordingly, binding experiments show that S-1 is more potent in strengthening BTX-binding (11). According to the model the one type of toxin influences the binding of the other, that is alkaloids are expected to increase ScTX binding too.

Of course the three-state model is too simple to give a detailed quantitative description of the behaviour of the

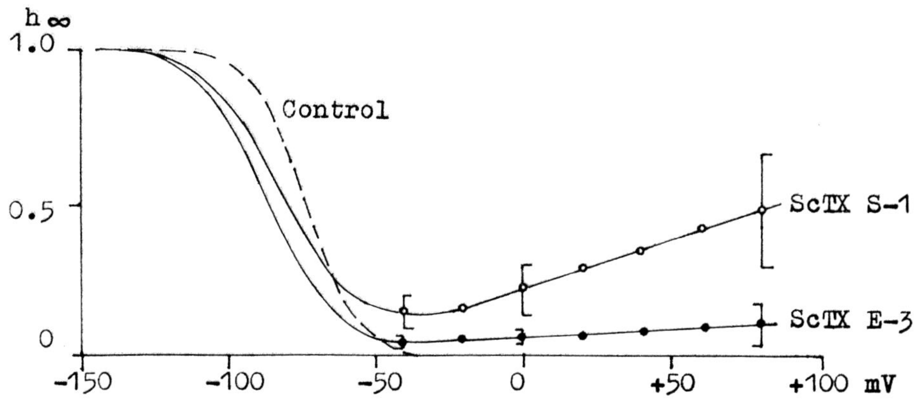

Fig.6 Effect of two toxins (S-1 and E-3) from Buthus eupeus scorpion venom on the h_∞ (E) curve of Na channels of the frog nerve .ymbols with vertical bars are mean ± S.D. (n = 7 for S-1 and n = 5 for E-3). Smooth lines at $E < -40$ mV are calculated according to equation 1 with the following mean values: $a = 0.125\ mV^{-1}$, $E = -73.5$ mV (control), $a_3^* = 0.074$ mV, $E_3^* = -79.2$ mV (S-1) and $a_3^* = 0.087\ mV^{-1}$, $E_3^* = -84.8$ mV (E-3). Peak conductance curves were not measured in these experiments, so calculations were limited to $E < -40$ mV, where terms relevant to the open state are negligible (Mozhayeva, Naumov & Nosyreva, to be published).

channels and their interaction with neurotoxins. Thus, the number of possible states for channel is, obviously, more than three. Nevertheless it enables to describe qualitatively and in some cases quantitatively a number of experimental data. So it seems to reflect essential traits of the relationships between the most important functional states or types of states of the Na channel.

Alkaloid neurotoxins do not only alter the gating but also the selectivity of Na channels. This was demonstrated for Ac (8,12). In the nodal membrane the order of permeability for Ac-modified channels determined from reversal potential measurements is: NH_4:Na:K:Rb:Cs = 1.36:1:0.36:0.19:0.12 (8). BTX and veratridine also alter the channel selectivity but to a lesser degree (9,13,14). Thus the order of permeability for BTX-modified Na channels in nodal membrane is Na:NH_4:K = 1:0.49:0.19 (15). Similar figures for Ac- and BTX-modified channels were obtained in experiments on neuroblastoma cells (7,16). Ac and BTX increase proton-to-sodium permeability ratio from 25o to 12oo and 53o respectively (17,18). In spite of the high proton permeability the channel conductance in acid solutions is low. This fact can be accounted for by binding of H^+ to acid group(s) in the Na channel. There are two or more acid groups in the Na channel (19,2o) one of them being localized presumably in the selectivity filter (21). H^+ block experiments show that Ac and BTX make the pK_a of this group lower by about o.8 and o.4 units, respectively (15,22). Thus one can think that the selectivity of the channel depends to a large extent on properties of this acidic group.

Ac and BTX do not change the sensitivity of the Na channel to tetrodotoxin (TTX) (12). Taking into account the effect of these alkaloids on the selectivity one can conclude that structures responsible for TTX binding and those for selectivity are separate ones. The TTX receptor seems to be localized just at the external mouth of the pore, whreras the "selectivity filter" within the pore.

The fact that alkaloid neurotoxins alter simultaneously both gating and selectivity suggests that the receptor for these alkaloids is situated somewhere between structures responsible for both functions. BTX (23) and Ac (7) exert their effects being applied from both sides of the membrane. It is natural to suppose that alkaloid molecules reach their receptors via the lipophilic core of the membrane. This supposition is consistent with the fact that these alkaloids are lipid soluble substances. If this supposition is correct one can furthermore suggest that pore walls and part of the gating machinery are in contact with membrane lipids. It is quite possible that some quantitative differences in properties of Na channels from membrane of various cells are due to differences in lipid composition of the membranes.

References

1) Koppenhöfer,E., Schmidt,H.: Pflügers Arch. 3o3, 133-149 (1968)

2) Zubov, A.N.: Tsitologia 22, 12o7-1213 (198o)

3) Catterall,W.A.: J.Biol.Chem.23, 866o-8668 (1977)

4) Mozhayeva,G.N., Naumov,A.P., Nosyreva,E.D., Grishin, E.V.: Biochim.Biophys.Acta 597, 587-6o2 (198o)

5) Mozhayeva,G.N., Naumov,A.P.: Neurophysiology (Kiev) 14, 5o8-516 (1982)

6) Mozhayeva,G.N., Naumov,A.P., Nosyreva,E.D.: Neurophysiology (Kiev) 12, 612-618 (198o)

7) Grishchenko,I.I., Naumov,A.P., Zubov,A.N.: Tsitologia 24, 658-666 (1982)

8) Mozhayeva,G.N., Naumov,A.P., Negulyaev,Yu.A., Nosyreva,E.D.: Biochim.Biophys. Acta 466, 461-473 (1977)

9) Khodorov,B.I.: Revenko,S.V.: Neuroscience 4, 1315-133o (1979)

1o) Catterall,W.A.: J.Biol. Chem. 252, 8669-8676 (1977)

11) Soldatov,N.M.: this edition

12) Mozhayeva,G.N., Naumov,A.P., Negulyaev,Yu.A.: Neurophysiology (Kiev) 8, 152-16o (1976)

13) Naumov,A.P., Negulyaev,Yu.A., Nosyreva,E.D.: Tsitologia 21, 692-696 (1979)

14) Lazdunski,M., Balerna,M., Barhanin,J., Chicheportiche,R., Fosset,M., Frelin,C., Jacques,Y., Lombet,A., Pouyssegur,J., Renaud,J., Romey,G., Schweitz,H., Vincent,J.P.: Ann.New-York Acad. Sci.,358, 169-182 (198o)

15) Mozhayeva,G.N., Naumov,A.P., Khodorov,B.I.: Gen.Physiol. Biophys. 1, 453-455 (1982)

16) Zubov,A.N., Naumov,A.P., Khodorov,B.I.: Gen.Physiol.Biophys. 2, 75-77 (1983)

17) Mozhayeva G.N., Naumov,A.P.: Pflügers Arch.396, 163-173 (1983)

18) Mozhayeva,G.N., Naumov,A.P., Khodorov,B.I.: Gen. Physiol. Biophys. 1, 463-464 (1982)

19) Spalding,B.C.: J.Physiol.3o5, 485-5oo (198o)

2o) Mozhayeva,G.N., Naumov,A.P., Negulyaev,Yu.A.: Gen. Physiol. Biophys.1, 5-19 (1982)

21) Hille,B.: Fed.Proc.34, 1318-1321 (1975)

22) Mozhayeva,G.N., Naumov,A.P., Negulyaev,Yu.A.: Gen Physiol. Biophys.1, 21-35 (1982)

23) Albuquerque,E.X., Daly,J.W., Witkop,B.: Science 172, 995-1oo2 (1971)

SYNTHESIS OF STEROID ALKALOIDS ACTIVE ON SODIUM CHANNELS

E.Yelin, V. Leonov, O.Tikhomirova, I. Torgov
Shemyakin Institute of Bioorganic Chemistry, USSR Academy of Sciences, Moscow, USSR

Introduction

Batrachotoxin (BTX) (I) a steroid alkaloid contained in skin extracts (1) of the Columbian frogs Phyllobates aurotaenia, attracted great attention because of its ability to affect excitable membranes (2) to a high degree. A batrachotoxin molecule consists of a steroid moiety called batrachotoxinin A (II) and a dimethylpyrrolcarboxylic acid residue connected with (II) by an ester link. After elucidation of BTX (I) structure (3,4) efforts were made to synthesize it. In the beginning of the 7o-s the syntheses of (II) (5) and its dihydroanalog (III) (6) were completed by Wehrli and coworkers in a series of brilliant works. Necessity to have batrachotoxin or its closely related analog as a tool for investigations of excitable membranes prompted us to start a synthesis of 7.8-dihydrobatrachotoxinin A (III). We believed that the lack of Δ^7-double bond would not have a critical influence on the biological activity of (III), which was proved by further investigations.

Results

The synthesis of 7.8-dihydrobatrachotoxinin A (III) challenges chemists to solve a number of problems. The main tasks are construction of a hemiacetal grouping in A and C rings and functionalization of the inaccessible and unreactive 18-methyl group. The first problem was principally solved by Wehrli and we doubt whether a better solution of this problem could be found. We would like to mention that the ability of the $\Delta^{9(II)}$-

Toxins as Tools in Neurochemistry

double bond in steroid molecules with cis-conjunction of A/B rings to be oxidized with osmium tetraoxide has already been known in our laboratory by the time of publication of the Wehrli work. As to functionalization of the 18-methyl group we were interested in the possibility of applying the Barton reaction (7) to solve this problem. As a starting material we choose progesterone (IY) transformed to II α-hydroxyprogesterone (Y) by microbiological hydroxylation with the culture Rhizopus nigricans. The function of this compound was to provide stereospecific hydrogenation of the Δ^4-double bond, functionalization of the 18-methyl group and further introduction of the $\Delta^{9(II)}$-double bond necessary to construct a hemiacetal grouping.

Hydrogenation of II α-hydroxyprogesterone (Y), oxidation of the IIα-hydroxy group, ketalization and reduction of the II-keto group with sodium borohydride resulted in formation of bisethyleneketal pregnan-5β-o1-IIβ-dione-3,2o (YI). After treating bisethyleneketal (YI) with nitrosil chloride in pyridine the resultant nitrite ester was exposed to the Barton reaction (7) - photolysis in xylene solution using a high - pressure mercury lamp. The resulting reaction mixture consisted of three isomeric products (YII), (YIII), (IX) each of them containing a nitrogen atom, and IIβ, 19-cyclic ether (X).

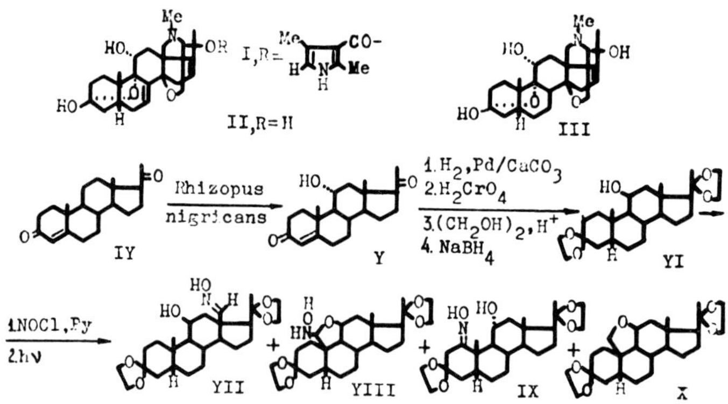

The fact that one of angular methyl groups disappeared in the course of photolysis and a new AB-system arose according to the NMR-spectrum of (X) as well as to the IR- and mass-spectra points to formation of an oxide ring. The choice between two possible ways of ring formation, i.e. between 18- and 19-methyl group was made after removing the ethyleneketal groups. Upfield shift of the 18-methyl group in the NMR-spectrum of the saponificated product as a result of the shielding influence of the 2o-keto group allowed to ascribe the IIβ, 19-oxide structure to compound (X). The structure of compound (YIII) was established on the basis of its transformation to IIβ, 19-oxide (X). The absence of the C = N double bond absorption from the IR-spectrum of (YIII) as well as the absence of the azomethine proton from the NMR-spectrum suggested intramolecular addition of the IIβ-hydroxy group to 19-aldoxime with formation of the hydrolamine derivative.

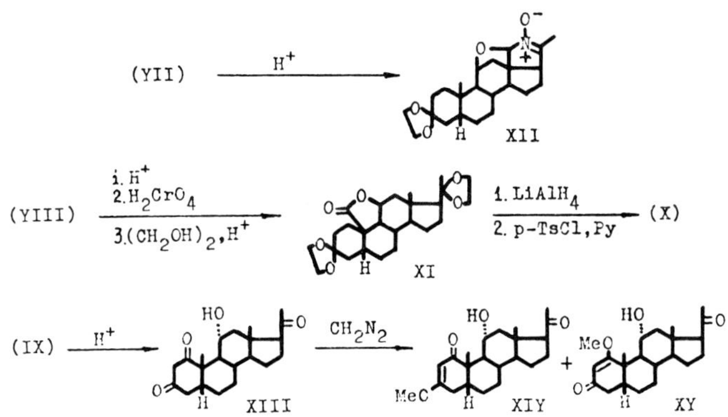

From the NMR-spectrum of oxime (YII) as well as from that of its isomer (YIII) one of the angular methyl groups is absent. However, there was a signal of an azomethine proton in the NMR-spectrum and an absorption band of the C=N double bond in the IR-spectrum. It shows that in the case of oxime (YII) the aldoxime exists independently and not as a hydroxylamine derivative. Acid treatment of oxime (YII) gives the nitrone (XII), that indicates the oxime group position.

The structure of the third isomeric product, oxime (IX), was elucidated after its full hydrolysis to the triketone (XIII). Triketone (XIII) contains the β-diketone moiety susceptible to enolization and after methylation with diazomethane it gives two isomeric enol ethers, (XIY) and (XY). These results hint at C-I as the position of the oxime group. The axial nature of the proton at C-II shows that in the course of photolysis inversion of IIβ-hydroxy group took place. This kind of hydroxy-group epimerization under radical reaction conditions is not unusual in steroid chemistry (8).

18-oxime (YII) isolated by chromatography (Alumina, $CHCl_3$) in 36 % yield from the mixture obtained after photolysis is a key intermediate product in the 7.8-dihydrobatrachotoxinin A(III) synthesis. Treatment of 18-oxime (YII) with nitrous acid in a biphasic system to avoid formation of the nitrone by-product (XII) followed by additional hydrolysis causes formation of the hemiacetal (XYI) which is oxidized to diketolactone (XVIII). Ketalization of (XVIII) using ethylene glycol and p-toluenesulfonic acid in a boiling toluene solution gave, unexpectedly, after subsequent lithium aluminium hydride reduction a mixture of two isomeric diols (XXII) and (XXIII) in a ratio close to 1:1. Susceptibility to isomerization of a side chain of this kind of lactones was noted, although to a smaller extent, by Wettstein (9). We found that the product of lactone (XVIII) ketalization which seemed to be chromatographycally homogeneous actually turned out to be a mixture of two substances: after removing the ethyleneketal groups two isomeric lactones (XVIII) and (XIX) were obtained again in a 1:1 ratio. Side chain isomerization which occurred during ketal formation under acidic condition and at high temperature was caused by mutual repulsion of the carbonyl group of a rigid lactonic ring and a side chain, forcing the latter to take a less hindered α-configuration. The interaction of this kind was visualized by circular

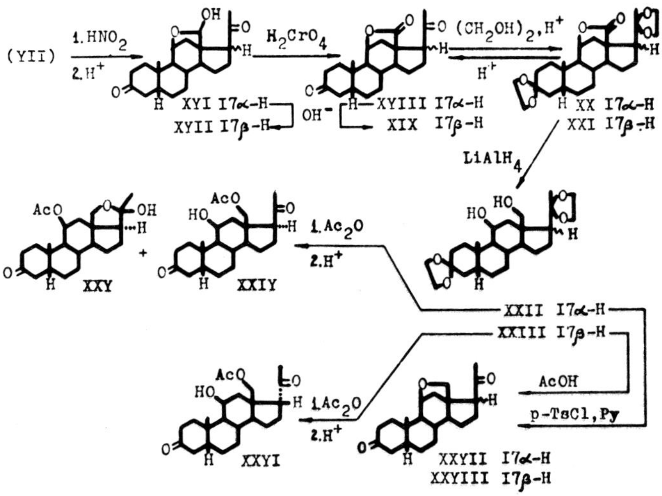

dichroism (CD). Fig.1 shows the CD curves for C-17 epimeric pairs of lactones (XVIII) and (XIX), monoacetates (XXIV) and (XXVI) and IIβ, 18-oxides had ordinary CD-curves, i.e. positive in case of 17β-side chain and negative in case of 17α-side chain (1o), the curves of both lactones were negative. This was due to the alteration of a preferable conformation of 17β-side chain of lactone (XVIII) the new conformation providing the negative CD-effect. The IIβ, 18-hemiacetal grouping apparently influenced the side chain much more. Thus, after treatment of the hemiacetal (XYI) with alkaline methanol and a number of subsequent transformations a mixture of the alcohols (XXII) and (XXIII) was obtained with the 17α-isomer prevailing in a ratio of4:1. Similar treatment of lactone (XVIII) gives isomeric alcohols in a ratio close to 1:1.

It was possible to avoid side chain isomerization during the process of ketal formation under more mild conditions (substitutingtoluene for benzene) and, by that, transforming smoothly the oxime (YII) to the diol (XXII). The next step in our synthesis was determined by the necessity to carry our selective protection of the 18-hydroxy group. Direct acetylation of diol (XXII) led to formation of a mixture of the monoacetates (XXIV) and (XXY) in spite of the well-known resistance of the

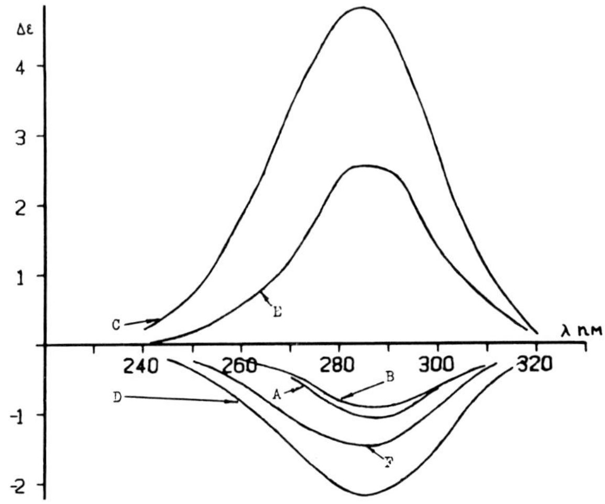

Fig.1 CD-curves, A - lactone (XVIII), B - lactone (XIX), C - monoacetate (XXIV), D - monoacetate (XXVI), E - 11β, 18-oxide (XXVII), F - 11β, 18-oxide (XXVIII).

axial IIβ, 18-diol, observed earlier by Wettstein(11), probably could be explained by transfer of the acyl residue from the 18- to the IIβ-hydroxy group. This process was promoted by the proximity of the bulky 17β-side chain and the 18-acetoxy group. In order to overcome this difficulty we decided to convert the side chain to the α-position. We believed that thus it would be possible to protect the 18-hydroxy group selectively by acetylation. We also assumed that the configuration at C-17 was not essential for biological activity because in the further synthesis this asymmetric centre was to be eliminated by introduction of the Δ^{16}double bond. For this purpose hemiacetal (XYI) was isomerized at C-17 and the mixture of isomers was subjected to successive transformation: oxidation, ketalization and reduction of the lactone ring. Then pure diol (XXIII) was isolated by chromatography and its acetylation under usual conditions gave without any difficulties a desired 18-monoacetate (XXYI). In this reaction no traces of IIβ-hydroxy group acetylation were observed. However the 17-isosystem showed resistance to the usual procedure of introducting the Δ^{16}-double bond, i.e. radical bromination with N-bromosuccinimide and dehydrobromination. Unfortunately, this way of differentiating IIβ- and 18-hydroxy groups was not successful. Since we believed that the mixture of monoacetates after acetylation of diol (XXII) was a result of simultaneous transacetylation it was decided to try another protecting group. The diol (XXII) was silylated under very mild conditions using trimethylsilyldiethylamine (12) and the 18-monosilyl ether (XXIX) was obtained with high selectivity. Dehydration of the IIβ-hydroxy group in (XXIX) using thionyl chloride and pyridine gave the corresponding $\Delta^{9(II)}$-derivative (XXX). 18-Silyl ether proved to be very stable and was cleaved only by boiling of (XXX) in 1% solution of potassium hydroxide in aqueous methanol. Following acetylation of alcohol (XXXI) and acid hydrolysis of the ketal groups resulted in endione acetate (XXXIII), which was further used for the preparation of the 7.8-dihydrobatrachotoxinin A (III) according to the Wehrli procedure.

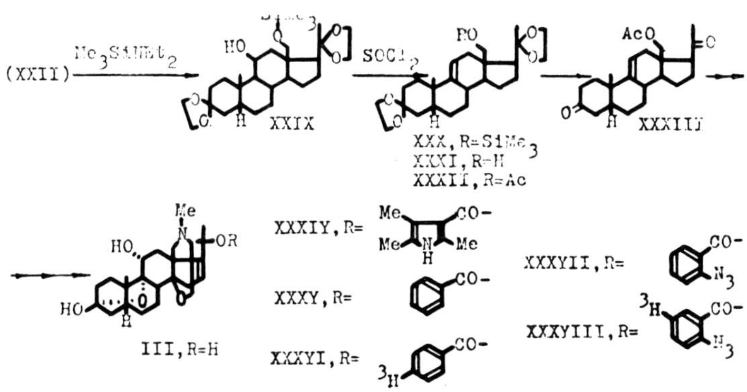

Esterification of 7.8-dihydrobatrachotoxinin A (III) was performed using the mixed anhydrides, which were obtained from the corresponding acids and ethyl chloroformate. The advantage of this method (4,13) is the exclusive esterification of the allylic hydroxyl group at C-2o in (III). This way 2,4,5-trimethylpyrrol carboxylate (XXXIY, benzoate (XXXY), p-(^{3}H)-benzoate (XXXYI), o-azidobenzoate (XXXYII) and (5-^{3}H)-2-azidobenzoate (XXXYIII)$^{+}$of 7.8-dihydrobatrachotoxinin A(III) were prepared in almost quantitative yield and purified by high-pressure liquid chromatography using the Du Pont 83o instrument. Radiochemical purity of esters (XXXYI) and (XXXYIII) was 95-97 % with specific activities of about 27 Ci/mmol and 29 Ci/mmol, respectively. All the esters (XXXIY) - (XXXYIII) showed high biological activity when interacting with excitable membranes and thus may be used as effective molecular instruments for the investigation of sodium channels (14).

$^{+}$(5-^{3}H)-2-azidobenzoic acid was obtained from (5-^{3}H)-2-aminobenzoic acid

Acknowledgement

We wish to thank Dr. N.F. Myasoyedov and Dr.V.P. Shevchenko for the p-(^{3}H)-benzoic acid and (5-^{3}H)-2-aminobenzoic acid prepared for us.

References

1) Märki,F., Witkop,B. (1963): Experientia 19, 239.

2) Albuquerque,E.X., Daly,J.W., Witkop.B. (1971): Science 172, 955.

3) Tokuyama,T., Daly,J.W., Witkop,B., Karle,I.L., Karle,J. (1968): J.Am.Chem.Soc.9o, 1917.

4) Tokuyama,T., Daly,J., Witkop,B. (1969): J.Am.Chem.Soc. 91, 3931.

5) Imhof,R., Gössinger,E., Graf,W., Berner-Fenz,L., Berner,H., Schaufelberger,R., Wehrli,H. (1973): Helv.Chim.Acta. 56, 139.

6) Graf,W., Gössinger,E., Imhof,R., Wehrli,H. (1972): Helv. Chim.Acta 55, 1545.

7) Barton,D.H.R., Beaton,J.M., Geller,L.E., Petchet,M.M. (1961): J.Am.Chem.Soc. 83, 4o76

8) Heusler,K., Kalvoda,J., Anner,G., Wettstein,A. (1963): Helv.Chim.Acta 46, 352

9) Schmidlin,J., Wettstein,A., (1962): Helv.Chim.Acta 45, 331

1o) Velluz,L., Legrand,M., Grosjean,M. (1965): Optical Circular Dichroism. Principles, Measurements, and Applications. Verlag Chemie, Weinheim/Bergstr., Academic Press Inc., New York and London.

11) Schmidlin,J., Wettstein,A., (1959): Helv.Chim.Acta 42, 2636

12) Weisz,J., Felföld,K., Kovacs,K., (1968): Acta Chim.Acad. Sci. Hung. 58, 189

13) Brown,G.B., Tieszen,S.C., Daly,J.W., Warnick,J.E., Albuquerque,E.X. (1981): Cell.Mol.Neurobiol.I, 19

14) Soldatov,N., Prosolova,T., Kovalenko,V., Petrenko,A., Grishin,E. : this issue

BATRACHOTOXIN - A TOOL FOR ELUCIDATING THE SODIUM CHANNEL FUNCTIONAL ORGANIZATION.

Boris Khodorov
Vishnevsky Surgery Institute, Moscow 113093, USSR

Introduction

Stereodical alkaloid batrachotoxin (BTX) belongs to the group of lipid-soluble toxins inducing a steady depolarization of excitable membranes due to a practically irreversible increase in its Na-conductance (see (1,2) for reviews).
Voltage-clamp experiments on frog myelinated fibres showed that this depolarization results from two principal effects of BTX: (a) a shift (by about 60 mV) in the voltage dependence of Na channels activation to more negative potentials; and (b) elimination of normal Na channels inactivation (3).
To reveal the origin of these changes in Na-activation and inactivation further investigations of the effects of BTX on Na channels gating machinery were carried out (4-7). The aim of the present paper is to summarize briefly the main results of these studies.

Effects of BTX on the rate constants of Na channels activation and deactivation.

Binding to open Na channels, BTX decreases dramatically the probability of their closing upon membrane repolarization.
This conclusion was derived (5) from the analysis of the effects of BTX on 'opening' (α_m) and 'closing' (β_m) rate constants of the activation process in frog node of Ranvier. Slow and approximately exponential kinetics of activation of BTX-modified

Toxins as Tools in Neurochemistry

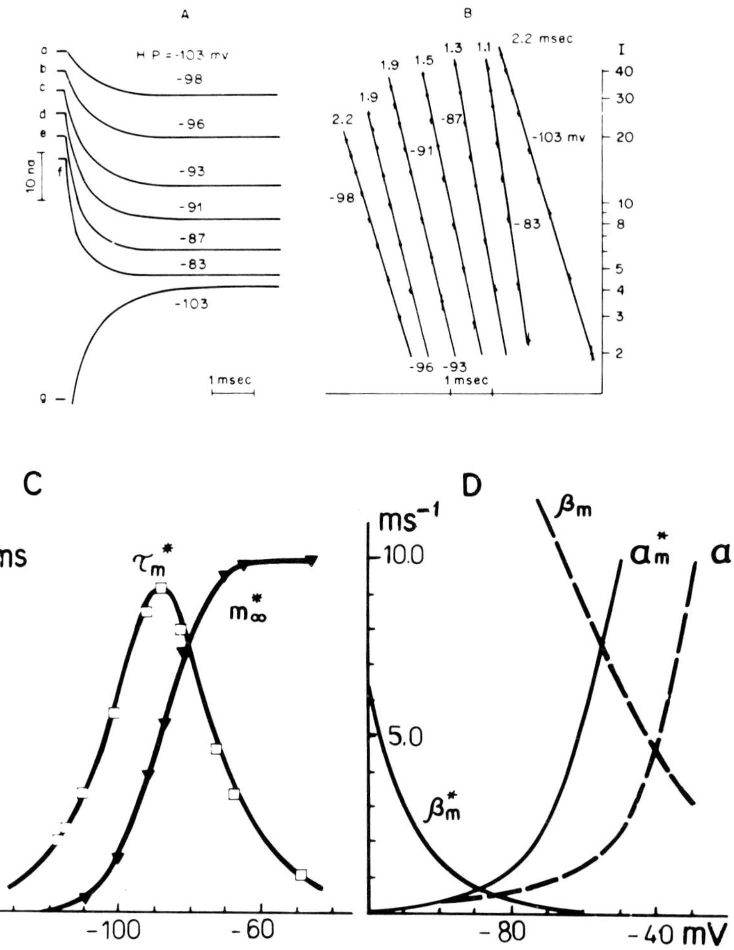

Fig. 1. Effects of BTX on the kinetic properties of Na channels A: Na currents in BTX-modified Na channels associated with step potentials to various E. The lower record is of-current after 5-ms voltage step to-83 mV. Holding potential (H.P) - 103 mV B: the same currents in logarithmic scale. The upper numbers are the time constants and the lower ones are E in mV. C: voltage dependence of the steady-state activation, m_{∞} and its time constant. The mid-point E=88 mV, z =-4, τ_{max}=0.9 ms. Temperature 20 °C (4). D: Voltage dependence of calculated rate constants of activation for normal (α_m and β_m) and BTX-modified (α_m^* and β_m^*) Na channels. The same experiment as in Fig. IA-C (5)

Na channels (Fig. 1 A,B) facilitated this analysis.
Fig. I C,D shows that BTX produces a drastical reduction of β_m shifting its voltage dependence by about 60 mV more negative potentials (E). α_m -E relationship is also shifted to more negative E, but only by about 20 mV. This drastical decrease in the closing rate constant is the main reason for the negative shift in the steady-state sodium activation (m_∞-E relationship). Indeed the fact that already at E=-90 mV about half of BTX-modified Na channels are activated (m_∞=0.5) is caused not by a negative shift of α_m-E relationship (according to calculation, α_m at -90 mV in BTX-modified Na channels is about the same as in normal ones (9)) but results from a reduction of β_m(E): here β_m/α_m= I while in normal nodal membrane β_m/α_m at -90 mV is about 40. Thus due to a decrease in the closing rate constant Na channels altered by BTX become capable of opening even at very negative E ($<$-100mV) where the probability of Na channel transition to the activated state (α_m) is still very small.
All these conclusion derived from the analysis of macroscopic ionic currents were confirmed by direct recording of single Na channel currents from isolated membrane patches of the neuroblastoma cell (10): BTX decreased drastically the probability of Na channels closing both during membrane depolarization and upon its termination.

Changes in the intramembrane charge movement

A study of the effects of BTX on Na channels intramembrane gating charge movement (Q) was carried out in voltage-clamped frog node of Ranvier under conditions where all Na channels were modified.(6,7).
First of all it has been found that BTX removed the depolarization-induced charge immobilization and slows down significantly the OFF charge movement. Fig. 2 A,B compares Q in control conditions (a) and after application of BTX(B). It can be seen that after BTX-treatment the depolarization-induced par-

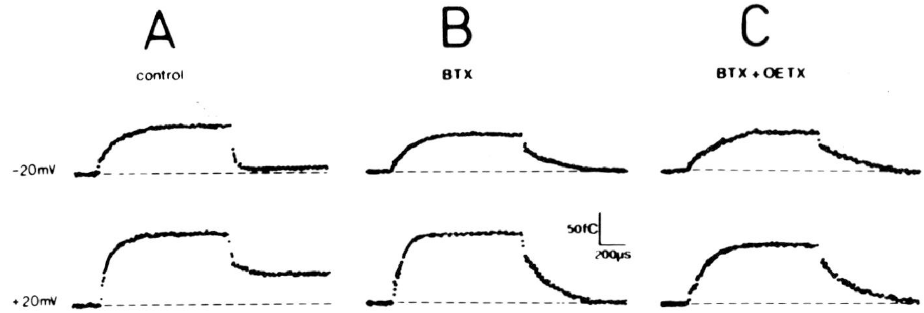

Fig. 2. Intramembrane charge movement (integral of asymmetrical capacity current) elicited by depolarizations to -20 and +20 mV in control conditions (A), after treatment with BTX (10^{-5}M) and after 10-min application of $4x10^{-5}$M oenanthotoxin (C). The averaged integral of the gating current was obtained in response to 12 depolarizing and 24 half-amplitude hyperpolarizing pulses from a holding potential -120 mV. Several determination of Q_{max} for E between 120 and +80 mV gave the following mean $\mp$ SEM values: control: 123$\pm$11 fC (n=5); BTX: 122$\mp$ 7 fC (n=9), OETX :111 $\mp$9 fC (n=3). Temperature 16°C (6).

Fig. 3. Elimination of charge immobilization by BTX. Superimozed traces of ON and OFF responses obtained with pulses of various duration to E =-50 mV. Charge movements were monitored directly as the transient non-linear component of integrated currents (see (13)). Tempareture 14°C (7).

tial immobilization of the gating change ($Q_{off} < Q_{on}$) desappears. It cannot be revealed even if the depolarizing pulse duration is increased up to 20 ms (Fig. 3). This gives new strong evidence in favour of the notion that the gating charge immobilization underlies Na current inactivation (11-15).

Fig. 2 and 3 also illustrate a dramatic deceleration of the backward ('closing') charge movement after BTX treatment of the nodal membrane. There is no doubt that it is just this change in the gating currents kinetics that underlies a decrease in the Na channels closing rate constant β_m (see Fig. I).

The second principal finding of this study is that after BTX treatment the parameters of the steady state Na conductance (g_{Na}) vs.E and of Q vs.E relationships become very similar (Fig. 4) in contrast to that observed in the normal membrane (13, 16-18). At first approximation both Q-E and g_{Na}-E relationships can be described by the equation

$$M = \frac{M_{max}}{1 + \exp(E - E)/k} \qquad (I)$$

for the two-state Boltzman model. Here M denotes the steady-state Q or g_{Na} values; M_{max} is their maximum; E is the midpoint potential and k is the steepness factor of the curve.

The original records of Q used to plot Q-E curve like that shown in Fig. 4, are presented in Fig. 5. Note that only Q_{off} records for pulses to -100 and -80 mV are shown. The reason for this is that for pulses to E negative to-60 mV the time course of ON charge movement is so slow that Q_{on} measurements by the integral method used are considered to be unreliable. At large E the kinetics of Q_{on} is control conditions and in the presence of BTX is about the same (see Fig. 2 A,B). This is consistent with the observations (4) that BTX increases maximum τ_m but does not affect appreciably τ_m at large E.

The similarity in the parameters of Q -E and g_{Na}-E curves in the presence of BTX (see Fig. 4) results from the following changes in these voltage-dependencies.

(I) The midpoint potential, E, of g_{Na} curve is shifted under

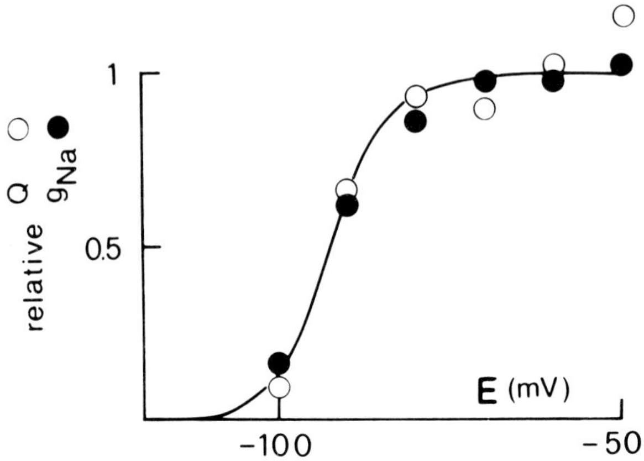

Fig. 4. Q-E and g_{Na}-E relationships in a BTX-treated node of Ranviner. g_{Na} was calculated from quasi instaneous current recorded at repolarization after pulses to various E whose duration was adjusted to give steady-state current during each voltage pulse. Q values are mean values of two Q_{off} determinations after pulses of the same duration as used for the sodium current recordings. The curve presents the non-linear least-squares fit of Eqn. I to the combined Q and g_{Na} values range -100 to 50 mV. Parameters values obtained from the fit were: relative Q_{max}, $g_{Na\ max}$=1 E= 92.1 mV, k =4.47 mV. Temperature: 13°C (7).

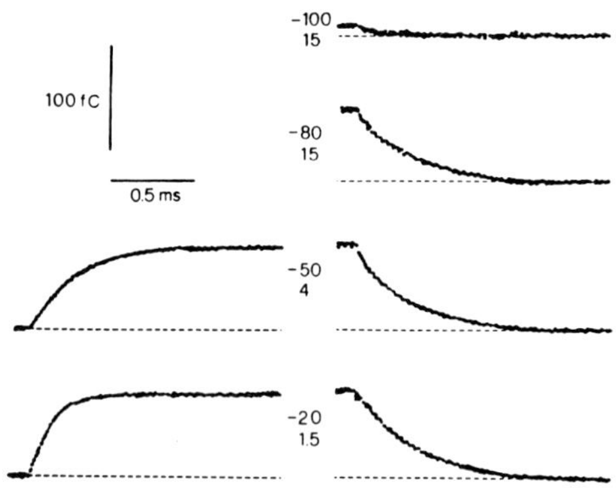

Fig. 5. Gating charge movement traces obtained from a fiber after BTX treatment. ON (left) and OFF (right) gating charge movements were recorded during and after pulses to various voltages for various durations (upper and lower nimbers, respectively, next to each trace) Temperature 13°C (7).

the action of BTX by about 60 mV to more negative E with-out noticeable changes in the steepness (I/k) of this curve.

(2) The negative (by about 50 mV) shift in E of Q -E curve is accompained by about three-fold increase of its I/k (7). The latter effect indicates an increase in the valence of the gating charged particles. the value k is given by

$$k=RT/Faz$$

with z being the valence of the charged gating particle and a the fraction of the membrane field that the particle traver-ses as it moves between its two possible location. Thus if a were same in control and BTX-treated nodes, the relative changes in k indicate an approximately three-fold increase in valence of the charged gating particle under the action of BTX. In spite of a such a marked increase in z, Q_{max} measured before and after BTX-treatment proved to be about the same (6,7). This result is consistent with the hypothesis that BTX induces an aggrega-tion of the charged gating particles associated with each chan-nel and causes them to move as a unit having approximately three times the average valence of the individual particle. Movement of this single unit would open the BTX-modified Na channel (4,7). An alternative explanation of BTX-induced changes in z and the kinetics of Na activation may be that the toxin decreases in some way the probability of occupancy of intermediate states in sequence of states leading from closed to an open channel. In either case the relults obtained suggest that the hypothe-tical 'voltage sensor' and the 'activation gate' itself are linked rigidly or even constitute a common structure (activa-tion subunit). This may hold true not only for BTX-modified but also for normal Na channels. However in normal conditions the independent movement of individual gating particles and complicated relations between 'inactivation -sensitive' (Q_1) and 'inactivation -insensitive' (Q_n) components of gating cur-rents obscur (14,18,19) true relationships between voltage sensor and the activation gate.

Effect of BTX on the drug-induced charge immobilization

BTX removes not only the inactivation of Q, but it also hinder the charge immobilization caused by some blocking drugs. So pretreatment of the nodal membrane with BTX protects Na channels from the blocking action of $4x10^{-5}$ oenanthotoxin (OETX) (6) which in normal conditions causes parallel suppression of both Na ionic and gating currents (20, 21). Fig. 2 C shows that superfusion of the BTX-treated node with OETX for a period 10-12 min did not significanly modify the charge movement.
The antagonizm between BTX and OETX proved to be mutual (6, see also (29)).
Phenomenologically the interactions between BTX and OETX is very similar to that between BTX and amine local anesthetics and related amine compounds (22-25). Since the inhibition of Na current caused by these drugs is accompanied by a partial charge immobilization, it seems very probable that BTX prevents not only the inhibition of I_{Na} but also the immobilization of gating charge.
It is necessary, however, to stress that in the nodal membrane BTX does not affect apparently the blocking action of benzocaine or butanol (6,22-25). There is no doubt that inhibition of the ionic currents by these drugs in BTX-modified Na channels, as well as in normal channels, is accompanied by a partial charge immobilization. Thus BTX only increases the resistance of Q to the blocking drugs but does not eliminate the charge immobilization at all.
Judging by available data, which cannot be discussed here in detail , in the Na channel of nerve membranes there exists allosteric interaction between BTX-receptor (R_{BTX}) and the receptors: for amine blocking drugs (R_a) and for the 'physiological blocking particle'- h-gate (R_h). Binding of BTX to its receptor decreases the affinity of both R_a and R_h ('negative heterotopic cooperativity'). This manifests itself in a decrease in the Na channels sensitivity to the blocking drugs and the removal (total (3,4,7) or partial (26)) of Na inactivation.

In turn the affinity of BTX-receptor decreases both under the action Na channels inactivation (4) and their drug-induced blockage. The latter conclusion is based both on electrophysiological studies of BTX-local anesthetics interactions (6,22, 23,27) and the study of the effects of various drugs on the binding of labelled BTX-A benzonate (28) or 7,8 dehydro-A benzoate (29).
By contrast, the relationship between R_h (II) and R_a could be characterized as 'positive heterotopic cooperative' ones : binding of the blocker to R_a enhances the inactivation; in turn, the inactivation enhances the drug-induced blockage (30,31).

"Key position" of the BTX-receptor in the Na channel

From an analysis of dose-response relations (4,28) it follows that simulataneous alteration of all the major properties of the Na channel produced by BTX results from its binding to a single receptor on the one-to-one stochiometric basis. This evidently indicated that BTX-receptor belongs to that subunit which plays a key role in structural interactions of various functionally important channel entities. Among them are activation and inactivation subunits, the selectivity filter (4,32, 33) receptors for polypeptide sea anemone and α-scorpion toxins (see (34-36) for reviews). The only structural entity of the Na channel which remains unaffected by BTX nerve fibre membrane is the tetrodotoxin receptor located in the external channel mouth (37).
According to Catterall (35) BTX shares a common receptor with three other lipid-soluble neurotoxins-veratridine, aconitine and grayanotoxin. These toxins, like BTX, produce simultaneous modification of Na channel activation, inactivation and ion selectivity (35). However these changes are not identical to those produced by BTX. One can assume therefore that the individual receptors for these four toxins overlap only partially.

Na channel activation promotes or is a prerequisite for BTX-receptor interaction (4). This conclusion is inferred from the fact that repetitive stimulation of the membrane but not a maintained depolarization accelerates greatly the process of Na channel modification by BTX. Judging by the analogous effects of repetitive pulsing on the rate of Na channel modification by aconitine (38), veratridine (39,40) or grayanotoxin (41) these toxins also prefer to interact with activated channels. Further experiments are required to clear up whether channel activation makes the receptor more accessible for these chemicals, or during the activation process BTX-receptor changes its conformation resulting in its conversion from 'precursor' to a'mature' form.

One can expect that further electrophysiologic analysis of the effects of BTX and radioligand binding studies of toxin-receptor interactions will provide new insight into the molecular organization of the voltage-sensitive Na channel.

References

1. Albuquerque, E.X.: Feder.Proc., 31, 1133-1138 (1972).
2. Albuquerque, E.X., Daly, J.W.: in The specificity and action of animal, bacterial and plant toxins (P.Caustercases, ed) 1, 297-338 (1976).
3. Khodorov, B.I., Peganov, E.M., Revenko, S.V., Shishkova, L.D.: Brain Res. 84, 541-546 (1975).
4. Khodorov, B.I.Revenko, S.V.: Neurascience, 4, 1315-1340 (1979)
5. Khodorov, B.I.: in: Structure and Function in excitable cells (D.Chang et al. eds.) (1983) in press.
6. Dubois, J.M., Khodorov, B.I.: Pflugers Arch. 395, 55-58 (1982)
7. Dubois, J.M., Schneider, M., Khodorov, B.I.: J.Gen.Physiol. (1983) (in press).
8. Revenko S.V.: Neirofisiologia (USSR) 9, 544-547 (1977).
9. Hille, B.: Thesis of the Rockefeller University (1967)
10. Quant, F., Narahashi, T.: Proc.Natl.Acad.Sci. USA, 79, 6732-6736 (1982).

11. Armstrong, C., Bezanilla, F.: J.Gen.Physiol. 50, 279-283 (1977).

12. Nonner, W.: J.Physiol. 299, 573-604 (1980).

13. Dubois, J.M., Schneider, M.: J.Gen.Physiol. 79, 571-602 (1982).

14. Khodorov, B.I.: Neuroscience, 4, 865-876 (1979).

15. Armstrong, C.:Physiol. Rev. 61, 644-683 (1981).

16. Meves, H.: Annals NY Acad. Sci. 303, 332-338 (1977).

17. Keynes, R.: adv.Physiol.Sci. (Salanski J. ed. 4 , 37-39, (1980) Budapest.

18. Peganov, E.M., Khodorov, B.I. Bull. exp. biol. med. (USSR) 11, 515-518 (1977).

19. Keynes, R., Greef, N., Helden, D. Proc.Roy.Soc.B. 215, 391-404 (1982).

20. Dubois, J.M., Schneider, M.: Nature, 289, 685-688 (1982).

21. Dubois, J.M., Schneider, M.: Toxicon, 20, 49-55 (1982).

22. Khodorov, B.I. in: Membrane transport prosesses (D.Tosteson, Yu.Ovchinnikov, R.Latorre, eds) 2, Raven Press New York,

23. Zaborovskaya L.D., Khodorov, B.I.: Gen Physiol.Biophys. 1, 283-285 (1982).

24. Zaborovskaya, L.D., Khodorov, B.I.: Neurofisiologia (USSR) 14, 636-643 (1982).

25. Khodorov, B.I. Prog.Biophys.Mol.Biol., 37, 49-89 (1982).

26. Mazhayeva, G.N., Naumov, A.P., Khodorov,B.I. in USSR-Sweden Symposium Physico-Chemical Biology, pp.221-222 (1981).

27. Albuquerque, E.X., Seyama, I., Narahashi, T.: J.Pharmacol. Exp.Therapeutics, 184, 308-314 (1973).

28. Creveling, C.R., McNeal, E.T., Daly, J.W., Brown, G.B.: Molecul.Pharmacol. (1983) in press.

29. Soldatov, N.M., Prosolova, T.K., Kovalenko, V.A., Petrenko, A.G., Grishin, E.V., Ovchinnikov, Yu.A. This Symposium.

30. Khodorov, B.I., Shishkova, L.D., Peganov, E.M., Revenko, S.V.: Bioch. Biophys. Acta. 433. 409-435 (1976).

31. Hille, B.: J.Gen.Physiol., 69; 497-515 (1977).

32. Mozhayeva, G.N., Naumov, A.P., Khodorov, B.I.: Gen. Physiol. 1. 453-455 (1982).

33. Mozhaeva, G.N., Naumov, A.P., Khodorov, B.I.: Gen.Physiol. 1, 281-282 (1982).

34. Catteral, W.A.; J.Biol.Chem. 252, 8669-8676 (1977)

35. Catteral, W.A.: Ann.Rev.Pharmacol.Toxicol. 20, 15-43 (1980).

36. Jaques, Y., Fosset, M., Lazdunski, M.: J.Biol.Chem. 253, 7383-7392 (1978)

37. Mozhaeva, G.N., Naumov, A.P., Khodorov, B.I.:Gen.Physiol. Biophys., 1, 221-223 (1982).

38. Mozhaeva, G.N., Naumov, A.P., Negulayev, Yu., A., Nosyreva, E.D.: Bioch.Biophys. Acta 466, 461-473 (1977)

39. Hille, B.: J.Gen.Physiol., 51, 199-219 (1968).

40. Peganov, E.M., Timin, E.N., Khodorov, B.I.: in Biophysica Membrane (USSR) 502-507 (1973) Kaunas

41. Seyama, I., Narahashi, T.: J.Pharmacol.Exp. Therapeutics 219, 614-624 (1981).

IDENTIFICATION OF SODIUM CHANNEL COMPONENTS INTERACTING WITH NEUROTOXINS

Nickolay Soldatov, Tat'yana Prosolova, Victor Kovalenko, Alexander Petrenko, Eugenii Grishin, Yuri Ovchinnikov

Shemyakin Institute of Bioorganic Chemistry, USSR Academy of Sciences, 117988 Moscow, USSR

Introduction

Extensive biochemical studies showed that the fast sodium channel of excitable membrane is a high molecular weight glycoprotein (1,2). It's major functional parts are i) the ionic pore including the spatially separated anionic (presumably carboxyl) groups both in the selectivity filter and external mouth of the channel (3), and ii) the potential sensitive gating machinery regulating the ionic transport (4). In fact the sodium channel molecular characteristics are restrained by these data.
The study of the sodium channel is based mainly on the usage of a number of neurotoxins whose specific binding and a concomitant change of the channel function properties make these drugs very selective molecular probes of the channel. In this paper the use of neurotoxins for identification of sodium channel receptors in rat brain synaptosomes is described.

Results

The binding of tetrodotoxin (TTX) in a one-to-one molar ratio to the external mouth of the channel results in reversible blocking of ionic transport (6) thereby making TTX one of the most commonly used probes of the sodium channel. The synthesis of new radioactively labeled TTX derivatives suitable for receptor iden-

Toxins as Tools in Neurochemistry

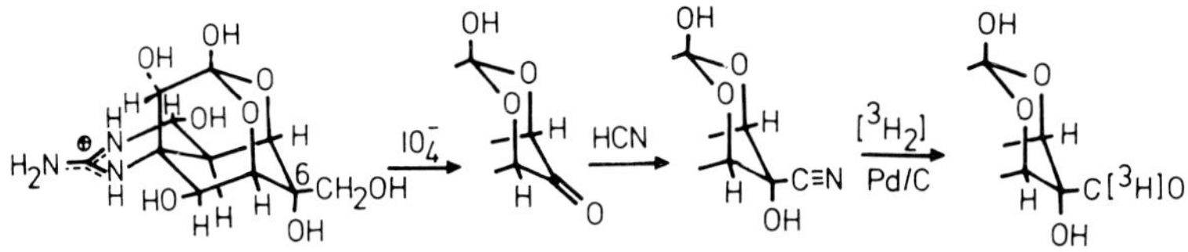
I II III IV

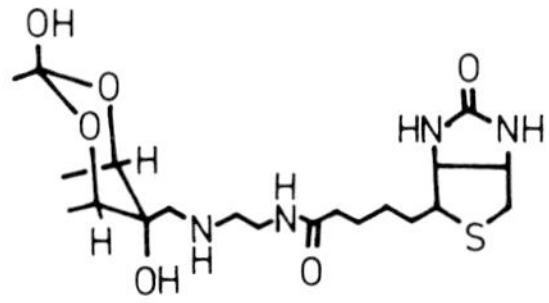
V

tification was started with mild periodic oxydation of TTX molecule (I) (7). The treatment of 6-ketoderivative (II) by HCN led to respective cyanohydrine (III) transformed into a number of radioactive TTX analogs by catalytic hydrogenation with tritium gas (8). The following Bio-Rex 70 chromatography allowed to isolate the major labeled aldehyde (IV) (/^{3}H/A-TTX) with total yield of 4-9% and specific radioactivity up to 10 Ci/mmol.

The indispensible step in TTX receptor enrichment is its solubilization, for example, by means of 1% Lubrol PX (LPX) (9). The binding of /^{3}H/A-TTX to solubilized synaptosomal membranes was characterized by affinity constant 3,6 nM (4°C) and maximal binding capacity up to 1,5 pmol/mg of protein (8). So, /^{3}H/A-TTX is an effective tool for a radioligand assay of the sodium channel receptor.

The presence of aldehyde group in /^{3}H/A-TTX allows to use it for the synthesis of other labeled TTX derivatives. Thus /^{3}H/A-TTX was coupled to biotin by reductive amination without a loss of affinity both to receptor and avidine. Now we try to identify the TTX receptor using prepared derivative (V) within the biospecific biotin-avidine system.

/^{3}H/A-TTX was applied with a success to the receptor assay during its purification and reconstitution according to the scheme in fig. 1. The pretreatment of synaptosomal membranes with 0,075% LPX allowed to separate up to 10% of total protein with near quantitative retention of the TTX receptor in pellets (P_1). The

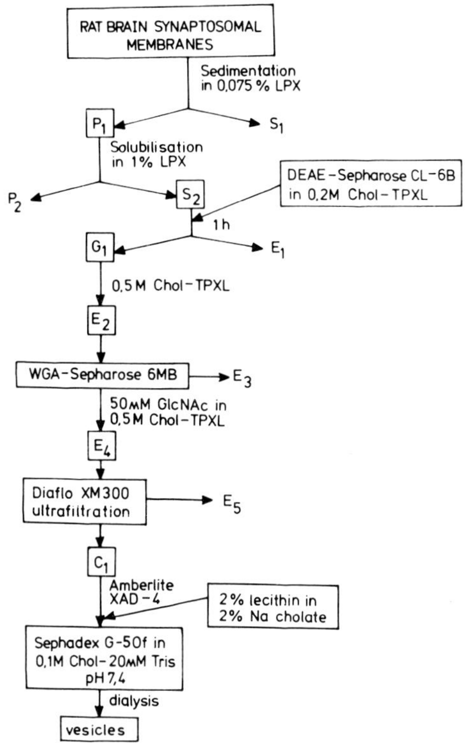

Fig. 1. Flow diagram for enrichment and reconstitution of TTX receptor. TPXL - 20 mM Tris-HCl, pH 7,4, 0,1% LPX, 0,0265% egg yolk lecithin; Chol - choline-HCl; GlcNAc - N-acetyl-D-glucosamine. For detailes see text.

subsequent solubilization of P_1 with 1% LPX led to extraction of near 50% of the membrane protein including 60% of TTX binding sites (S_2). The receptor was stabilized both by addition of exogenic lecithin and reduction of LPX concentration up to 0,1%. It resulted in enhancement of the half-time of spontaneous receptor inactivation from 2,3 h to about 1 day. Receptor fraction G_1 was adsorbed on DEAE-Sepharose followed by high ionic strength elution buffer washing (E_2) and subsequent chromatography on WGA-Sepharose. Because of the high affinity of the glycopeptide part of the receptor to the wheat germ lectin this step completed in 45-fold enrichment of TTX receptor (E_4) with total yield about 60%. The ultrafiltration of the fraction E_4 through Diaflo XM-300 membrane allowed additionally to enrich TTX receptor (C_2) near 2,3-fold. Chromatography of C_2 on Amberlite XAD-4 was used to remove LPX before receptor reconstitution in phosphatidylcholine liposomes (10), which was performed from dispersion of egg yolk lecithin in sodium cholate (11). After cholate separation by gel-filtration the binding of /^{3}H/A-TTX by vesicles reached 37-45% of the initial level and was attended by decrease of temperature sensitivity of the receptor. Its binding activity completely retained even after 4 h of incubation at 25°C. An additional 46% of TTX binding sites were recovered upon bilayer destruction with 0,5% LPX. So the reconstitution of the partially purified TTX receptor resulted in its random orientation in the artificial membrane. The above method of reconstitution into lipid bilayer vesicles homogeneous in size permitted to stabilize the TTX receptor at any step of purification for subsequent investigation of other sodium channel receptors.

Perhaps among such receptors a binding site of alkaloid toxins (aconitine, veratridine and batrachotoxin) is a most intriguing component of the fast sodium channel. The interaction of these lipophilic drugs with the sodium channel is accompanied with the persistent change of dimensions of the selectivity filter centered in the membrane (12). Simultaneously the change of potential sensitivity of the channel gating machinery is induced (13). So, the alkaloid toxin receptor is supposed to be an intramembrane fragment of the ionic pore of the channel that is spatially separated from the TTX binding site.

Previously we showed that veratridine-Sepharose specifically binds a fraction of the solubilized and partially purified TTX receptor (C_1 or E_4 in fig. 1) which could be eluted by an excess of free veratridine or aconitine at constant ionic strength (14). That result testified to the presence of both receptors in the same membrane macromolecule which does not dissociate in non-ionic detergents.

The reconstituted TTX receptor also posessed the affinity to veratridine-Sepharose. The elution by an excess of aconitine (fig. 2) resulted in 10 to 15-fold enhancement of specific TTX

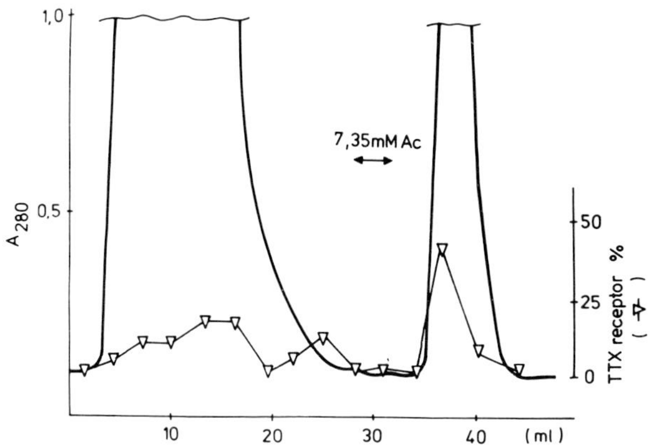

Fig. 2. Veratridine-Sepharose chromatography of enriched TTX receptor reconstituted in lecithin vesicles. Arrow shows an impulse of aconitine applied to the column at constant ionic strength and pH of elution buffer.

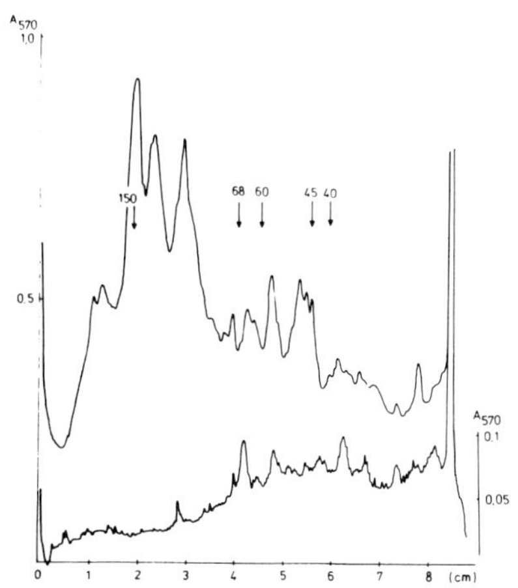

Fig. 3. SDS-electrophoretic profiles of partially purified TTX receptor fraction C_1 of fig. 1 (above curve) and its incorporated in vesicles fraction eluted by aconitine from veratridine-Sepharose (lower curve). Arrows show molecular weights of reference proteins.

binding activity accompanied with an effective separation according to the polypeptide composition (fig. 3).

Because of nearly equal efficiency of alkaloid toxins from both external and axoplasmic sides, the interaction of veratridine with the sodium channel is usually thought to be performed from the lipid phase after its penetration into the membrane. The possible depth of immobilized toxin inclusion into the channel macromolecule is restricted by the length of a spacer arm between a Sepharose particle and a veratridine molecule. In consists of the 13-atom chain, so that the toxin binding site can be located no further than the second alfa-helix segment of the channel protein from the lipid-protein boundary.

The direct radioligand assay of the alkaloid toxin receptor in the membrane as well as solubilized forms is hampered by very high lipid solubility of aconitine and veratridine (14,15) and by relatively low affinity of these toxins to the channel (about 10^{-5} M). These difficulties in studying the membrane-bound alkaloid toxin receptor were overcome by use of 7,8-dihydrobatrachotoxinin A 20α -/^{3}H/benzoate (/^{3}H/DBTX) synthesized by Dr. E.A. Yelin et al. (16).

The specific binding of /^{3}H/DBTX to rat brain synaptosomes lineary increased with protein concentration up to 1,2 mg/ml and reached an equilibrium at 25-30 min (25°C). Fig. 4 shows the effects of a nonlabeled dimethylpyrrol analog of DBTX and aconi-

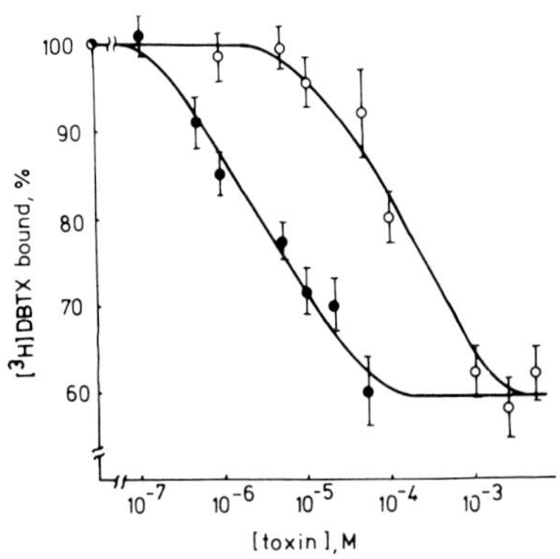

Fig. 4. Competitive displacement curves of /^{3}H/DBTX binding by dimethylpyrrol analog of DBTX (left) and aconitine (right) Binding of 80 nM /^{3}H/DBTX was measured at 22°C in the presence of 1 μM of TTX in 130 mM Chol-HCl, 50 mM HEPES-Tris, pH 7,4, 5,4 mM KCl, 0,8 mM $MgSO_4$ and 5,5 mM glucose.

tine on $/^3H/DBTX$ binding. Both toxins completely inhibited specific $/^3H/DBTX$ binding, values of $K_{0.5}$ being $2.5 \cdot 10^{-6}$ M for DBTX and $1.3 \cdot 10^{-4}$ M for aconitine. Results of Hill plots analysis of these data were consistent with the presence of a single class of noninteracting binding sites.

Polypeptide neurotoxins from scorpion and sea anemone venoms can allosterically enhance the specific binding of alkaloid toxins (5). Fig. 5 shows the dependence of $/^3H/DBTX$-receptor interaction on the concentration of Anemonia sulcata toxin II (ATX II).

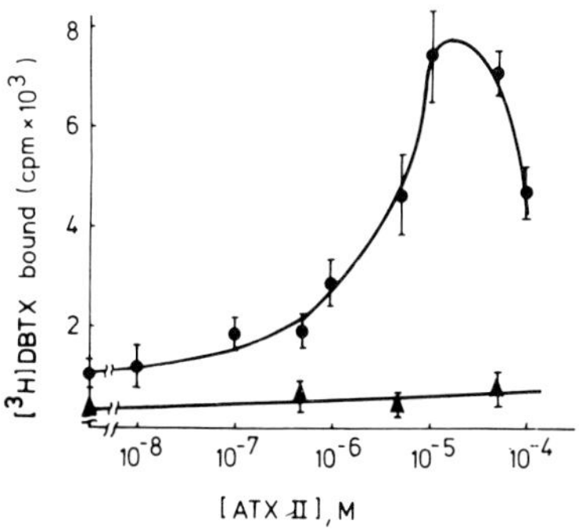

Fig. 5. Enhancement of specific $/^3H/DBTX$ binding by ATX II, measured in the presence of 1 μM TTX and 80 nM $/^3H/DBTX$.

$K_{0.5}$ value of ATX II effect was about $3 \cdot 10^{-6}$ M that considerably exeeded the affinity of ATX II to its receptor (17). In the presence of 10^{-5} M ATX II the affinity of $/^3H/DBTX$ to the receptor increased 3.5-fold, and a portion of specific binding was about 70-80% as a result of 7 to 7.5 fold enhancement of DBTX interaction with its receptor.

Previously we described the isolation of a number of neurotoxins from the scorpion Buthus eupeus venom (18). Those miniproteins like ATX II slowed down inactivation of sodium channels (19). Table 1 illustrates the concentration dependence of maximal enhancement of specific $/^3H/DBTX$ binding by some of these toxins (B_{max}), as well as the effects of the Leiurus quinquestriatus venom ("Sigma", USA), toxic fraction of the scorpion Orthochirus scrabiculosus venom and A. sulcata toxins I-III. The comparative

Table 1. Polypeptide toxins induced enhancement of specific /^{3}H/-DBTX binding by rat brain synaptosomes (details see in text).

Toxin	$B_{10^{-6}}$	B_{max}	$K_{0.5}$ (M)	K_d (M)	ref.
-	1	1	-	-	
B.eupeus, M_3	3.4±0.5	4.5±0.5	$5\cdot10^{-6}$	$\gg 10^{-6}$	(19)
M_6	2.3±0.5	-	-	$\geqslant 5\cdot10^{-7}$	
M_7	1.7±0.8	1.8	10^{-7}	$1.7\cdot10^{-7}$	(19)
M_8	1.4±0.5	2.0±0.1	10^{-7}	$0.9\cdot10^{-9}$	(19)
M_9	1.1±0.2	2.5	10^{-5}	$5\cdot10^{-8}$	
M_{10}	1.6±0.6	-	-	$6\cdot10^{-8}$	(20)
L.quinquestriatus	-	8.0	3*	-	
O.scrabiculosus	-	12,8	3*	-	
A.sulcata, ATX I	1.3±0.5	-	-	$7\cdot10^{-6}$	(17)
ATX II	3.3±0.1	7.5	$3\cdot10^{-6}$	$2\cdot10^{-7}$	(17)
ATX III	1.7±0.5	2	$2.5\cdot10^{-6}$	$\gg 10^{-5}$	(17)

*µg/ml

values of /^{3}H/DBTX binding enhancement by 10^{-6} M of these toxins ($B_{10^{-6}}$) and literature data on their own affinities to the sodium channel receptor (K_d) are also included.

All studied toxins induced a saturable increase of /^{3}H/DBTX binding, but no correlation between $K_{0.5}$ values for the effect and K_d ones for peptide toxins binding was observed. At saturable concentrations of peptide toxins the increase of equilibrium concentrations of /^{3}H/DBTX-channel complexes did not reach a common level. This difference is supposed to be an indication of different steady-state inactivation parameters of modified channels. From this point of view polypeptide toxins could be subdivided into "effective" ones (ATX II, Leiurus and Orthochirus toxins) and a "weak" ones, independent on the toxin affinities to the sodium channel.

According to available data the concentration of "effective" toxin binding sites is about 10 times higher than that of "weak" ones (17,21-23) despite of the common mechanism of action. Moreover, a photosensitive derivative of toxin M_{10} was shown (22) to

be bound to M_r 77 and 51 kD synaptosomal proteins, the former being capable of interacting with veratridine-Sepharose (14). Another "weak" toxin ATX I formed a covalent complex with M_r 86 and 53 kD proteins of the crayfish nerve membrane (23). In spite, the "effective" toxin of Leiurus venom was bound to synaptosomal proteins of M_r 240 and 32 kD (24). Such discrepancies of molecular characteristics of polypeptide toxin receptors might be accounted for the intrinsic complexity of molecular organization of the channel gating machinery which could include an overlapping but discrete polypeptide toxin binding sites. Anyhow this problem seemed to be related to elucidation of the chemical nature of intramembrane mobile charge of the sodium channel.
The dependence of alkaloid toxin binding on the channel gating regulation is manifested in the effects of local anesthetics and their functional analogs. The interaction of these drugs with the sodium channel leads to its blocking promoted by inactivation process (13). Independently on the order of action and on the presence of polypeptide toxins those drugs (table 2) inhibited competitively the specific /^{3}H/DBTX binding to synaptosomal

Table 2. Apparent dissociation constants (K_i) of inhibitors of specific /^{3}H/DBTX binding to synaptosomes.

Inhibitor	K_i (M)
Benzocaine	$3.1 \cdot 10^{-4}$
Trimecaine	$3.0 \cdot 10^{-5}$
Tetracaine	$5.7 \cdot 10^{-7}$
N-Propylaimalin	$1.5 \cdot 10^{-7}$
Yohimbine	$3.1 \cdot 10^{-6}$
Strychnine	$1.0 \cdot 10^{-6}$
Oenanthotoxin	$2.7 \cdot 10^{-6}$
Cyanpermethrin	$3.3 \cdot 10^{-7}$
Lubrol PX	$2.5 \cdot 10^{-5}$

vesicles. Since an increase in benzocaine, strychnine, trimecaine, tetracaine or N-propylaimalin level in the presence of one of them at constant concentration resulted in an increase of apparent K_i values, the conclusion was made that all the drugs interacted competitively within the single receptor site of the sodium channel (according to a joint work with B.Khodorov). The inhibition of /^{3}H/DBTX binding by a number of membrane active drugs (table 2) as well as by cytotoxin from the Naja naja oxiana venom correlated well with a simple bimolecular mechanism of interaction with the sodium channel.

The elimination of polar headgroups of glycerophospholipids by phospholipase C or an acylphosphate bond splitting by phospholipase D did not significantly change the specific binding of /^{3}H/DBTX but effectively decreased its non-specific interaction. On the other hand, bee venom phospholipase A_2 was an effective inhibitor of the toxin receptor. This result suggested that glycerophospholipid acyl chains should be present in the /^{3}H/DBTX binding site. The involvement of membrane lipids was also supported by the analysis of photoinduced modification of synaptosomal vesicles by 20α-(o-azido)benzoate derivative of DBTX (/^{3}H/DBTX-N_3). Upon UV-light irradiation at least 70% of specifically bound /^{3}H/DBTX-N_3 was found to be in a covalent complex with membrane lipids.

Nevertheless, the specific /^{3}H/DBTX binding was characterized by high sensitivity to proteolytic degradation which augmented with the decrease of proteases specificity (cf. pronase vs. trypsin). 1-10 mM of N-ethylmaleimide, dithiotreitol or reduced glutathione also inhibited the specific /^{3}H/DBTX binding to about 50-70% of its normal level. The involvement of S-S or SH-groups in toxin binding, however, is still under the question.

The results of this work clearly show that the alkaloid toxin binding site is composed both of protein and lipid components, and hence the ionic pore is located in contact with membrane lipids. The high sensitivity of the TTX receptor in the external end of the pore to lipid environment (25) argues in favour of this conclusion. Such a location of the ionic pore undouptedly has a fundamental meaning and allows to connect the ionic transport regulation with the heightened conformational lability of polypeptide chains of the ionic passway exposed to the lipid environment.

We hope that the study of the polypeptide part of /^{3}H/DBTX-N_3

receptor will contribute to understanding of the molecular organization of the fast sodium channel.

Acknowledgements

We are very grateful to Dr. L.Beress for his gift of anemone toxins, Dr. G.Mozhayeva and Dr. A.Naumov for the voltage clamp studies of scorpion toxins, and Prof. B.Khodorov for providing inhibitors of DBTX binding, and discussions, and suggestions.

References

1. Cohen, S.A., Barchi, R.L.: Biochim. Biophys. Acta 645, 253-261 (1981).
2. Levinson, S.R.: Dev. Biochem. 19, 315-331 (1981).
3. Sigworth, F.J., Spalding, B.C.: Nature (London) 283, 293-295 (1980).
4. Armstrong, C.M.: Quart. Rev. Biophys. 7, 179-210 (1975).
5. Catterall, W.A.: Ann. Rev. Pharmacol. Toxicol. 20, 15-43 (1980).
6. Cuervo, L.A., Adelman, W.J.,jr.: J. Gen. Physiol. 55, 309-335 (1970).
7. Chicheportiche, R., Balerna, M., Lombet, A., Romey, G., Lazdunski, M.: J. Biol. Chem. 254, 1552-1557 (1979).
8. Kovalenko, V.A., Pashkov, V.N., Grishin, E.V., Ovchinnikov, Yu.A., Shevchenko, V.P., Myasoedov, N.F.: Bioorg. Khim. 8, 710-712 (1982).
9. Agnew, W.S., Levinson, S.R., Brabson, J.S., Raftery, M.A.: Proc. Natl. Acad. Sci. USA 75, 2606-2611 (1978).
10. Kovalenko, V.A., Pashkov, V.N., Grishin, E.V.: Bioorg. Khim. 7, 1828-1837 (1981).
11. Brunner, J., Skrabal, P., Hauser, H.: Biochim. Biophys. Acta 455, 322-331 (1976).
12. Mozhayeva, G.N., Naumov, A.P., Negulyaev, Yu.A.: Biochim. Biophys. Acta 643, 251-255 (1981).
13. Khodorov, B.: Progr. Biophys. Mol. Biol. 37, 49-89 (1981).
14. Soldatov, N.M., Prosolova, T.K., Kiyatkin, N.I., Grishin, E.V.: IV FRG-USSR Symposium on Chemistry of Peptides and Proteins, Tubingen (1982). Submitted for publication in "Chemistry of Peptides and Proteins". de Gruyter, Berlin·NY.

15. Balerna, M., Fosset, M., Chicheportiche, R., Romey, G., Lazdunski, M.: Biochemistry 14, 5500-5511 (1975).

16. Yelin, E.A., Leonov, V.N., Tikhomirova, O.B., Torgov, I.V.: This issue.

17. Schweitz, H., Vincent, J.-P., Barhanin, J., Frelin, C., Linden, G., Hugues, M., Lazdunski, M.: Biochemistry 20, 5245-5259 (1982).

18. Grishin, E.V., Soldatov, N.M., Tashmukhamedov, B.A., Atakuziev, B.U.: Bioorg. Khim. 4, 450-461 (1978).

19. Mozhayeva, G.N., Naumov, A.P., Soldatov, N.M., Grishin, E.V.: Biofizika (USSR) 24, 235-241 (1979).

20. Grishin, E.V., Soldatov, N.M., Ovchinnikov, Yu.A., Mozhayeva, G.N., Naumov, A.P., Zubov, A.N., Nisman, B.Ch.: Bioorg. Khim. 6, 724-730 (1980).

21. Ray, R., Morrow, C.S., Catterall, W.A.: J. Biol. Chem. 253, 7307-7313 (1978).

22. Soldatov, N.M., Kovalenko, V.A., Grishin, E.V.: Chemistry of Peptides and Proteins, vol. 1 (eds. W.Voelter, E.Wunsch, Yu.A.Ovchinnikov, V.T.Ivanov), 399-405. de Gruyter, Berlin· New York, 1982.

23. Hucho, F., Stengelin, S., Bandini, G.; Recent Advances in Receptor Chemistry (eds. F.Gualtieri, M.Giannella, C.Melchiorre), 37-58. Elsevier, Amsterdam, 1979.

24. Beneski, D.A., Catterall, W.A.: Proc. Natl. Acad. Sci. USA 77, 639-643 (1980).

25. Agnew, W.S., Raftery, M.A.: Biochemistry 18, 1912-1919 (1979).

VOLTAGE-DEPENDENT Na-CHANNELS: COMPARATIVE STUDIES IN ELECTROPHORUS ELECTRICUS PREPARATIONS

Hans-Heinrich Grünhagen

Institut für Physiologische Chemie
Med. Fak. der Universität des Saarlandes
D-6650 Homburg/Saar

(Present address: BASF AG, Hauptlabor B9, D-6700 Ludwigshafen)

Introduction

Modulation of membrane permeability is widely used by nature to control cell function. Specific ion channels are major elements in this concept. Most cells establish Na-gradients across their plasma membranes. Accordingly, Na-channels have evolved in many biological systems. Pharmacologically and most likely molecularly quite different Na-channels are found e.g. in epithelial cells, in rods of the retina, in acetylcholine-receptor associated synaptic proteins, and on membranes conducting an action potential. Although other Na-channels may reveal some voltage-dependent properties, too, the latter ones are characterized by an extremely steep voltage dependence of their gating.

Voltage-dependent channels are present in excitable membranes of e.g. nerve cells, striated muscle cells, heart cells, Electrophorus electroplaques and Sternarchus electrocytes. Toxins - especially tetrodotoxin (TTX), saxitoxin (STX), peptide toxins from scorpion or sea anemone venoms, and batrachotoxin - have been useful tools to demonstrate the close relationship between the members of this voltage-dependent group of channels. However, they have also revealed quantitative or even qualitative pharmacological differences: e.g. much higher concentra-

Toxins as Tools in Neurochemistry

tions of TTX are necessary to block channels in heart cells than in axons, and voltage-dependent Na-channels on striated muscles appear to lack TTX-sensitivity during early embryonic development. In recent years, interesting molecular features of voltage dependent Na-channels have been worked out, for a review cf refs (1,2). *In vitro* approaches will help to understand whether different pharmacological properties result from different protein structures or represent alternative states of related or even identical protein sequences. For this purpose the comparative investigation of a given Na-channel in different preparations appears useful. This paper presents corresponding approaches with material from *Electrophorus electricus* electroplaques, and discusses intrinsic advantages and limitations of the chosen experimental techniques.

Methods and Results

1) Electrophysiological studies

At first sight, the morphology of *Electrophorus electricus* electroplaques appears favourable for electrophysiological experiments: an almost macroscopic thickness of the cross section facilitates electrode impalement, and a low impedance of the non-excitable face causes the voltage to drop primarily across the excitable membrane when passing current through the cell. However, electroplaques have a large excitable area and deep invaginations on their plasma membrane. Therefore only a limited number of voltage-clamp studies has been carried out (3,4). In collaboration with E. Bartels-Bernal and D. Hof improved electrophysiological methods have been applied to analyze neurotoxin effects in voltage-clamped electroplaques (5). Having possibly a limited kinetic resolution compared to other electrophysiological preparations, this approach is nevertheless extremely important to determine dose-response curves for neurotoxin effects *in vivo*.

When electroplaques from a series of Electrophorus were analyzed, significant variations between different animals, but only minor differences for cells from a given eel were observed. In all cases considerably more TTX was necessary to block Na-channels than in axons. The apparent K_D was 7 nM in the most sensitive and typically between 10 and 40 nM in the other animals (5).

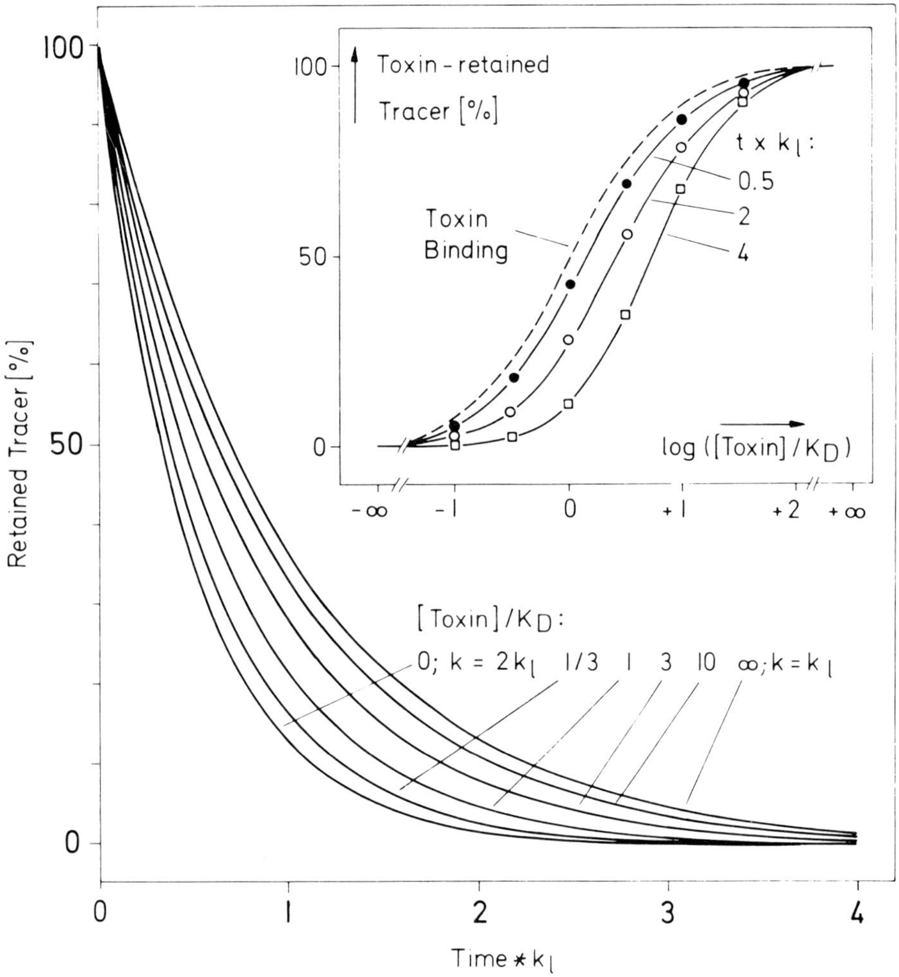

Fig. 1 : Calculated interference between leak and specific flux in vesicle efflux experiments (cf text).

2) Binding studies with membrane fragments

With centrifugation techniques, membrane fragments with 5 pmol TTX-binding / mg protein can be obtained. Additional electrophoresis or lectin affinity chromatography yields specific binding up to 30 pmol/mg (6). Using chemically tritiated TTX with high specific activity (7), a $K_D=20\pm10$ nM is obtained for binding <u>in vitro</u>. This value is in agreement with the blocking concentration of TTX <u>in vivo</u> and is significantly higher than K_D-values for binding to axons (cf (1)).

3) ^{22}Na-efflux from membrane vesicles

Purified Na-channel-rich membrane fragments form vesicles (6). ^{22}Na-efflux from these vesicles can be studied to analyze functional properties <u>in vitro</u> (8). This technique depends on a physical separation of intra- from extravesicular tracer, usually achieved by filtration/washing or gel filtration chromatography. Hence time resolution is limited. In general only the amplitudes in the lower second and minute time range rather than the initial slopes of the efflux curves can be evaluated.

Taking into account leak efflux besides specific efflux with kinetic constants k_l and k_s, respectively, the concentration of retained tracer is (c_i = initial internal concentration):

$$c_t = c_i * e^{-(k_l+k_s)*t}$$

Neglecting intravesicular diffusion (10), channel inactivation and channel distribution on vesicles, k_s can be assumed proportional to the number of open channels. Fig. 1 shows the resulting efflux curves calculated for the presence of a blocking toxin Tx with a dissociation constant K_D , where

$$k_s([Tx]) = k_s * \frac{K_D}{K_D+[Tx]} \quad \text{and } k_s=k_l \text{ for } [Tx]=0 .$$

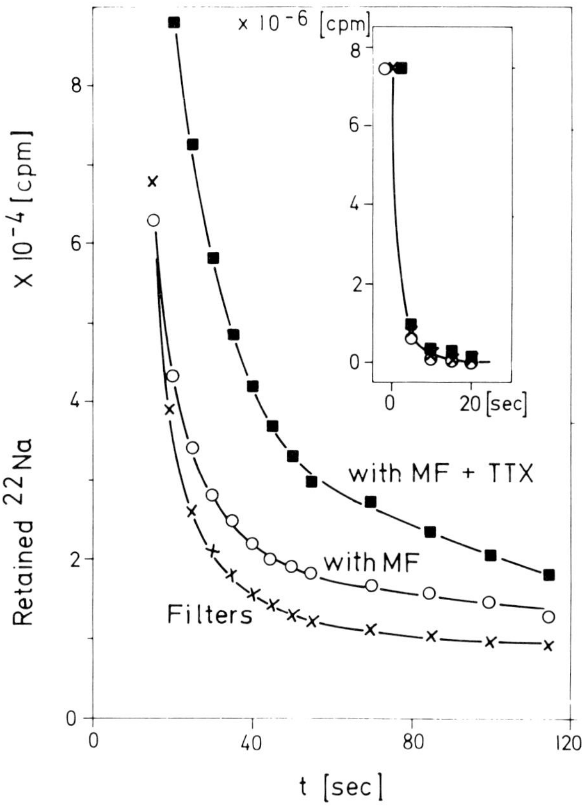

Fig. 2 : ^{22}Na-Efflux from vesicles as measured by controlled elution (cf text). MF: membrane fragment vesicles.

Evaluating amplitudes at instants of the order of the reciprocal leak constant or later, yields a systematic deviation of the apparent half-saturation toxin concentration towards higher values (cf inset of Fig. 1).

The technique of controlled elution (9) allows to measure efflux in the second time range. Fig. 2 shows experiments where Na-channel-inactivation had been inhibited by a transient treatment (1 min, 1 mM) of vesicles with N-bromoacetamide. To establish bulk ionic gradients, vesicles were incubated with 160 mM NaCl, 5 mM KCl and transferred at the beginning of the efflux experiment to 160 mM KCl, 5 mM NaCl. In the sample with TTX, the vesicle preparation had been incubated with the toxin overnight at a concentration of 10^{-6} M. Quantitative analyses of the blocking effect of TTX in this system yield values of $K_{app} \leqslant 5*10^{-8}$M (10).

4) Spectroscopic analysis of alkaline ion flux

In contrast to filtration and chromatographic separation, spectroscopic techniques offer the intrinsic advantage of high kinetic resolution. Their applicability depends on specific optical signals. In aqueous phase the interaction of ligands with Na-ions is very unfavourable, and no useful Na-indicator is known today. However, Na- and K-fluxes across vesicle membranes can be visualized indirectly: in the presence of bulk ionic gradients different Na- and K-permeabilities lead to the establishment of membrane potentials. The magnitude of these effects is

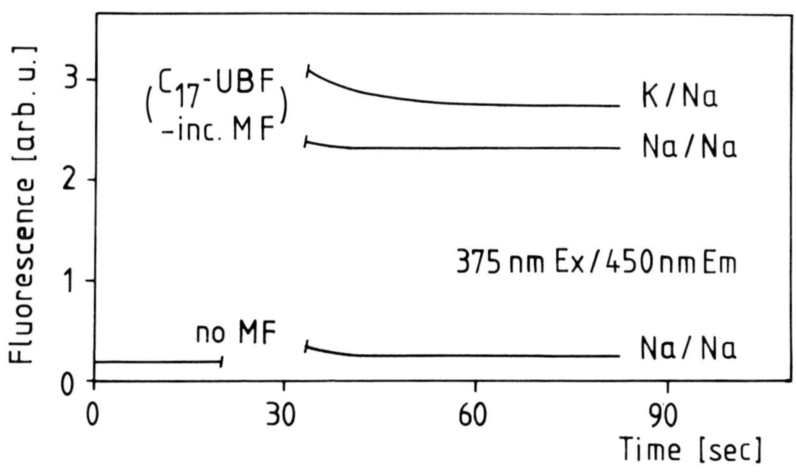

Fig. 3 : Fluorescence effects observed in the presence or absence of ionic gradients (cf text). 0.1 ml membrane fragments (MF), incub. with 1 nmol C_{17}-UBF, after G-50 chromatography diluted into 1 ml. Ionic milieu (160 mM) as indicated.

calculated to be of physiological significance (10). In recent years a whole series of potential-sensitive dyes has been characterized(11). The specific contributions from different ion species to the membrane potential can thus be measured optically, if permeabilities can be controlled pharmacologically or by specific toxins.

Many different physical effects may account for potential-sensitive optical phenomena, and the signals may show up as e.g. absorption or fluorescence changes. Fig. 3 shows the fluorescence of a long chain derivative of umbelliferone (C_{17}-UBF, Molecular Probes) incorporated in Na-channel-rich membrane fragment (MF) vesicles. Establishing a Na/K-gradient across these vesicle membranes yields a fluorescence change on dilution different from the control experiment without ion exchange (Na/Na) or in the absence of membrane fragments.

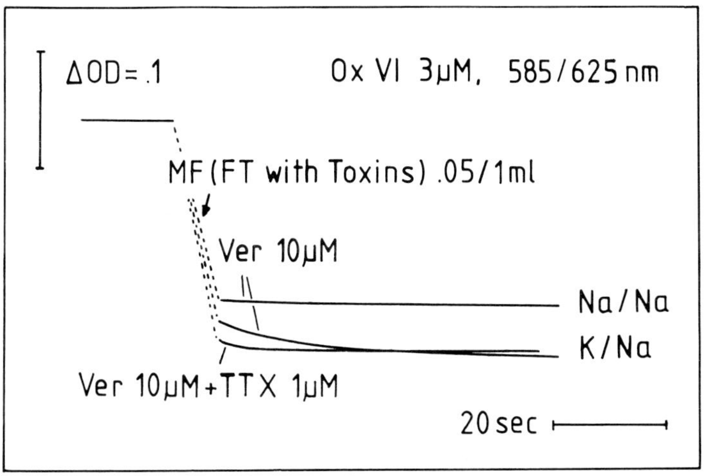

Fig. 4 : Differential absorption changes of Oxonol VI (Ox VI) in Na-channel-rich membrane fragment (MF) vesicles freeze-thawed (FT) with the toxins TTX and veratridine (Ver) as indicated. Exchange of ionic media (160 mM) by dilution of 0.05 ml into 1 ml.

Fig. 4 shows the differential absorption changes of Oxonol VI (12) for one experiment without ionic gradient and two experiments with Na/K-gradient. In the latter ones, Na-channel-inactivation was slowed down by incubation of the vesicles with 10 μM veratridine. The additional presence of 1μM TTX has a significant influence on the time course of the potential changes.

Figs. 3 and 4 may serve to demonstrate the occurence of membrane potentials after exchange of ionic milieus. However, the time resolution of the chosen setup (manual cuvette mixing in an Aminco double wavelength spectrometer or in a SLM spectrofluorimeter) is not sufficient for quantitative analyses. To overcome this problem, a fast preparative, multistep mixing system (13) has been combined with a kinetic fiberoptics spectrometer for absorption and fluorescence measurements. Fig. 5 shows a stopped flow experiment with this apparatus.

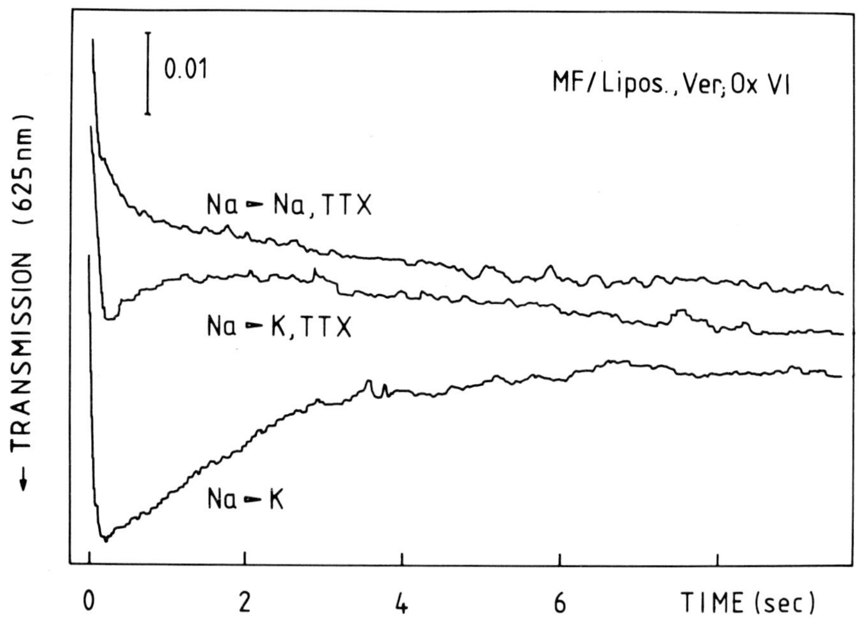

Fig. 5 : Fast absorption changes of Oxonol VI (Ox VI), 3 μM. Single shot experiments, 10 X dilution of Na-channel-rich vesicles (MF) fused with soybean liposomes (Lipos) and incub. with 10 μM veratridine (Ver). 1 μM TTX and ionic milieu (160 mM) as indicated.

After slowing down of Na-channel inactivation by veratridine, establishment of ionic gradients (Na → K) causes a fast transient of the membrane potential (lower trace). TTX blocks this transient almost completely (middle trace) to yield a signal comparable to conditions without ionic gradients (Na → Na).

5) Analysis of Na-channels in planar lipid membranes

For the experiments of Fig. 5, Na-channel-rich membrane fragments were fused with soybean liposomes to reduce leak permeability. In general, not only the bulk membrane phase, but also the channel properties themselves could be influenced by the lipid environment. To study such effects directly, transfer of channels into planar host membranes is very attractive. It is

Fig. 6 : Na-channels from purified membrane fragments transferred to a planar bilayer membrane (H. Schindler, H. H. Grünhagen: in preparation). The membrane was separating two compartments a and b ; a: lipid vesicles (of soybean lipid / cholesterol 10:1) / membrane fragment vesicles 1000:1, cevadine 5×10^{-5} M; b: lipid vesicles; both 500 mM NaCl, 20 mM HEPES, pH 7.0; recorded after potential change from -100 mV to + 100 mV. Channels can be blocked by TTX in compartment a.

possible now (14) to achieve this without solubilization of proteins and without dissolution of lipids in solvents. In collaboration with H. Schindler Na-channels from purified membrane fragments have been incorporated into planar bilayers. Fig. 6 shows single channel events measured with this method.

Discussion

Electrophysiological techniques have long been the only means to study excitable membranes. Most functional knowledge about these systems is still based on such analyses. On the other hand biochemical studies of ion channels have progressed rapidly in recent years and many molecular properties will be known in the near future. To relate phenomenological and molecular data, investigation of membrane function _in vitro_ appears useful. Among _in vitro_ methods spectroscopic and planar bilayer techniques are most promising. Both approaches have a kinetic resolution overlapping with the electrophysiological one. The strength of spectroscopy may be its potential to record structural transitions from large numbers of purified channels, whereas planar bilayers allow to control and to analyze single channel properties.

References

1. Ritchie,J.M., Rogart,R.B.: Rev. Physiol. Biochem. Pharmacol. 79 (1977) 1-50
2. Agnew, W.S., Moore, A.C., Levinson, S.R., Raftery, M.A., in G. Matsumoto and M. Kotani (Eds.), Nerve Membrane , University of Tokyo Press, Tokyo, 1981, pp 25-44
3. Nakamura, Y., Nakajima, S., Grundfest, H.: J. Gen. Physiol. 49(1965) 321
4. Lester, H.A.: J. Gen. Physiol. 72(1978) 847
5. Bartels-Bernal, E., Hof, D., Grünhagen, H.H.: in preparation
6. Grünhagen, H.H., Dahl, G., Reiter, P.: Biochim. Biophys. Acta 642(1981) 267-285
7. Grünhagen, H.H., Rack, M., Stämpfli, R., Fasold, H. Reiter, P.: Arch. Biochem. Biophys. 206(1981) 198-204
8. Kasai, M., Changeux, J.P.: J. Membr. Biol. 6(1971) 1-80
9. Grünhagen, H.H.: Anal. Biochem. 109(1980) 18-26
10. Grünhagen, H.H.: Proceedings of the 36 th Int. Meeting of the Société de Chimie Physique (Sept. 27 - Oct. 1st, 1982) in Paris; "Biochemistry and Bioenergetics", Elsevier, in pr.
11. cf review in Nature 292(1981) 801-802
12. Smith, J.C., Russ, P., Cooperman, B.S., Chance, B.: Biochemistry 15(1976) 5094-5105
13. Cash, D.J., Hess, G.P.: Anal. Biochem. 112(1981) 39-51
14. Schindler, H.G.: FEBS Letters 122(1980) 77-79

THE INTERACTION OF NEUROTOXINS WITH THE SOLUBLE PROTEINS OF THE EXCITABLE TISSUES

Valeryi Lishko
Institute of Biochemistry of the Ukrainian Academy of Sciences, Kiev, USSR

Introduction

The modern hypothesis on membrane biogenesis describes the mechanism of incorporation of protein molecules into phospholipid matrix as a cotranslational process (1). It starts with the synthesis of a hydrophobic N-terminal signal sequence of the protein destined to span the bilayer. The signal peptide binds together with the ribosome to a specific receptor to the endoplasmic reticulum. It is supposed that the force of polypeptide elongation drives the chain through the bilayer. The signal peptide is cleaved off when the protein enters the membrane.

A second model of the assembly of protein into membranes supposes an ability of hydrophilic proteins to cooperate with the phospholipid bilayer. As a result of such cooperation the polypeptide changes its conformation adequately to the hydrophobic environment and is integrated with the bilayer (2).

This second model supposes the existence of water soluble precursors of membrane proteins localized in the cell cytoplasma. There are many confirmations of this idea. For example, some mitochondrial proteins are coded by nuclear genes, synthesized outside the mitochondria on cytoplasmic ribosomes, released in the cytoplasma and then associated with the mitochondrial membrane (2).

Catalase and uricase are two peroxisomal enzymes which are synthesized in liver cell cytoplasma and subsequently transferred into the organelles (3).

Toxins as Tools in Neurochemistry

Besides the endogenous proteins being able to penetrate into the bilayer there are other polypeptides that have been shown tó associate with a membrane. These include mainly toxins: diphtheria toxin (4), latrotoxin from black widow venom (5), cholera toxin (6) and many others.

We report here data which show that in the cytoplasma of the excitable cells there is a soluble TTX-sensitive protein. In view of the high specificity of TTX this protein could be a precursor of membrane voltage-dependent sodium channels.

Results

At the beginning of these experiments we worked out a method for incorporating TTX-sensitive protein from nerve and muscle membranes into liposomes. This method is based on preparing proteoliposomes by adding solubilized brain microsomal fractions to a cholate-phospholipid suspension followed by removal of the detergent by gel chromatography (7). We had observed a stimulation of the efflux of sodium from these vesicles in presence of veratridine. TTX inhibited the veratridine-induced increment. We supposed that this method could be used for the identification of sodium channels or at least its TTX-receptor protein.

Unexpectedly for us we had found TTX-sensitive structures even in soluble fractions of brain, heart and skeletal muscle homogenates. Initially we prepared proteoliposomes by sonication of a phospholipid suspension with the supernatants of tissue homogenates. But subsequently it was shown that TTX-sensitive liposomes could be obtained even after simple preincubation of the vesicles with the extracts.

A typical experiment was performed as follows: Tissues were homogenized in equal volume of a medium containing 25o mM sucrose/1o mM Tris-HCl (pH 7.4)/o.2 mM EDTA. The homogenate was centrifuged at 1oo ooo g for 6o min. The supernatants were used in experiments. All steps were carried out at $4^{o}C$.

Total lipids from brain as well as lecithine and phosphatidylserine were obtained by standard procedure. To obtain liposomes, 25 mg lipids were dried under vacuum, suspended in 1 ml of a solution containing 1oo mM NaCl/1o mM Tris-HCl (pH 7.4)/o.2 mM EDTA. The suspension was sonicated at room temperature for 5 min at 44 kHz. 1 ml of tissue was added to sonicated liposomes usually to a final concentration of 2-3 mg protein per ml. The suspension was incubated either with or without TTX and veratridine at a concentration of 2oo nM and 2oo μg/ml respectively. Sodium influx was started by addition of 1o μCi of ^{22}NaCl and at a fixed time an aliquot of the suspension was percolated through a column of CM-Sephadex C-25 to remove the extravesicular radioactivity. The liposomes with associated ^{22}Na were

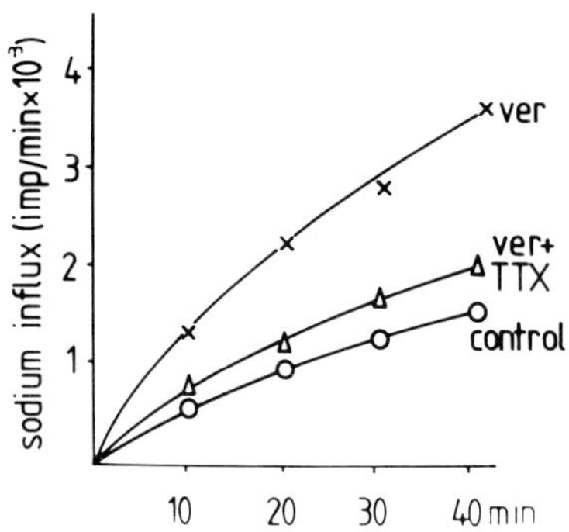

Fig.1 Effect of TTX and veratridine on sodium influx into liposomes treated with cattle brain soluble fraction.

eluted with the void volume. Fig.1 shows the effect of TTX and veratridine on the sodium influx into liposomes preincubated with the soluble fraction of brain homogenate. One can see that the ^{22}Na influx was increased by veratridine and this increment was reduced to about 3o % by TTX. Fig.2 shows that similar results were obtained with liposomes pretreated with an extract of heart muscle. To find out whether this effect was specific for extracts from excitable tissues, a soluble fraction of

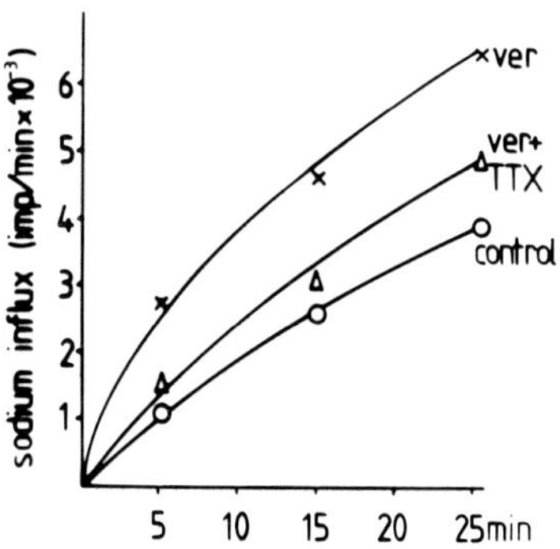

Fig.2 Effect of TTX and veratridine on sodium inlux into the liposomes treated with soluble fraction of heart muscle

added to liposomes and ^{22}Na influx was studied in the presence of TTX and veratridine. Fig.3 shows that the drugs had no effect on the sodium permeability of these vesicles. The possibility cannot be excluded that in supernatants there are small fragments of plasma membranes which can fuse with the phospholipid vesicles. However, we were not able to detect any Na-K-ATPase activity in the supernatants. Besides, prolonged centrifugation of diluted extracts with reduced density did not change their ability to induce TTX-sensitivity of the liposomes.

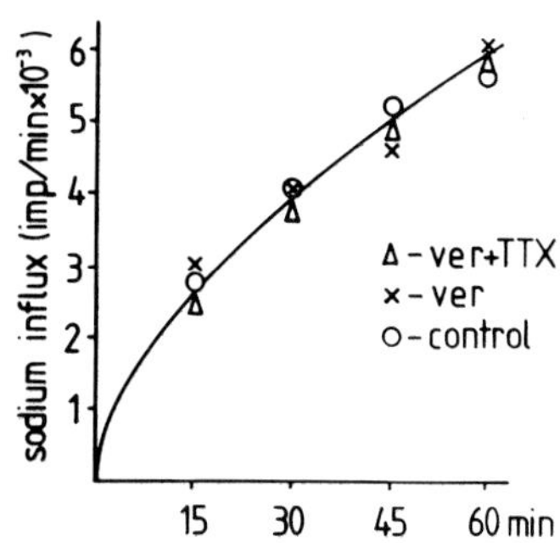

Fig.3 Sodium influx into liposomes treated with liver soluble fraction

On the basis of these experiments we supposed that in extractable fractions of brain and muscle there are hydrophilic proteins (probably related to voltage dependent sodium channels) which are able to react spontaneously with phospholipid bilayers and to penetrate them forming a TTX-sensitive channel. It was interesting to study how many proteins were bound to the liposomes from the soluble fraction. To clarify this, the liposomic suspension after preincubation with the supernatants was chromatographed on Sephadex G-2oo to separate the vesicles from unbound proteins. After such chromatography the liposomes still contained 4o - 6o mg of protein per mg phospholipid and still responded to veratridine and TTX. The proteins attached to the liposomes were precipitated by TCA. The precipitate was washed with a methanol-ether mixture to remove the phospholipids. The proteins were solubilized in 1 % SDS and separated by electrophoresis on 7.5 % polyacrylamide gel.

Comparing electropherograms of brain total extract and proteins associated with the liposomes one can see that only some proteins were bound to the vesicles (Fig.4).

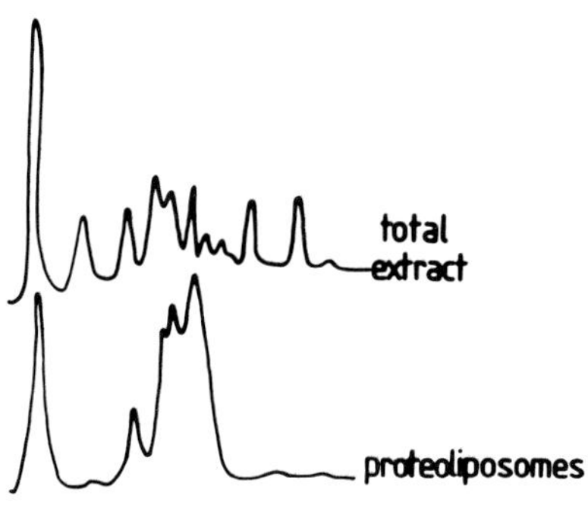

Fig.4 Electrophoresis of brain soluble proteins

Three of them are found to be prevailing but there are also several minor fractions. And Fig.5 shows the electrophoresis of heart muscle supernatants in comparison with the fraction attached to the liposomes. In this case it is also possible to see that only some proteins of the muscle supernatant interacted with the phospholipid vesicles. Probably some of them are simply adsorbed on the surface of the liposomes. Of course studying the channel activity, first of all we were interested in those proteins which span the bilayer.

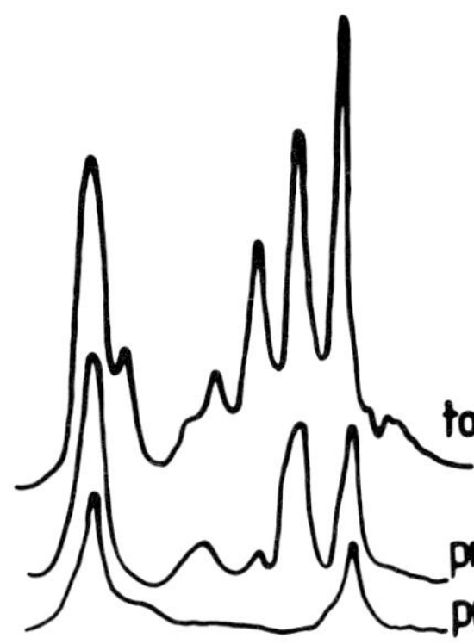

Fig.5 Electrophoresis of heart soluble proteins

Is there any transmembrane protein among those demonstrated by electrophoresis? After chromatography on Sephadex G-2oo proteoliposomes were treated with pronase at a concentration of 1oo μg/ml at 37° C for one hour. All electrophoretic bands but one were practically eliminated from the liposomes (Fig.5). Probably the resistant protein is buried in the phospholipid bilayer and protected against pronase digestion.

Another proof of integral incorporation of soluble proteins into the liposomic membrane was obtained from experiments with small angle X-ray scattering. Fig.6 shows the electronic density profile of liposomes and proteoliposomes, prepared with

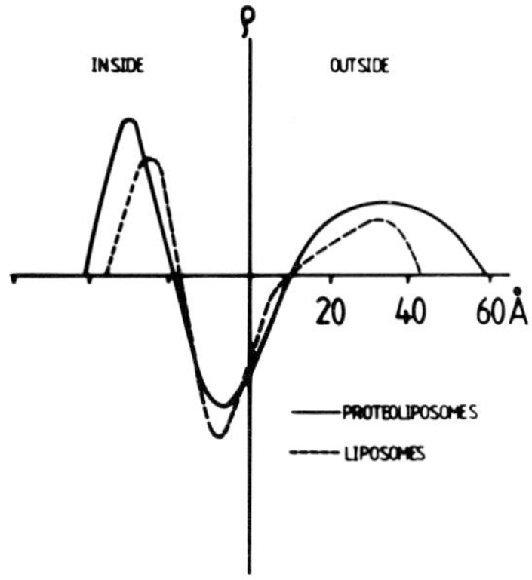

Fig.6 X-ray small angle scattering of the liposomes and the proteoliposomes

total brain soluble fraction. Comparing the two curves one can see that the electron density of the inner region of the phospholipid bilayer was not changed significantly from the control. But the elcetron density was changed at the outside as well as at the inside surface of the proteoliposomes. This means that some components of the supernatants are not only adsorbed on the liposomes but also penetrate through the bilayer .

It is known that incorporation of proteins into liposomic membrane changes its permeability. This effect can be explained by a disorder of the phospholipids when the polypeptide spans the bilayer. We have studied the permeability of the vesicles during preincubation with supernatants. Fig.7 shows the results of these experiments. One can see that the influx of sodium into the liposomes was not changed after one hour incubation without or with tetrodotoxin. But in the presence of the brain soluble fraction the liposomes become more permeable to sodium. The permeability reached a constant value after 6o min of preincubation.

This experiment was performed with liposomes consisting of

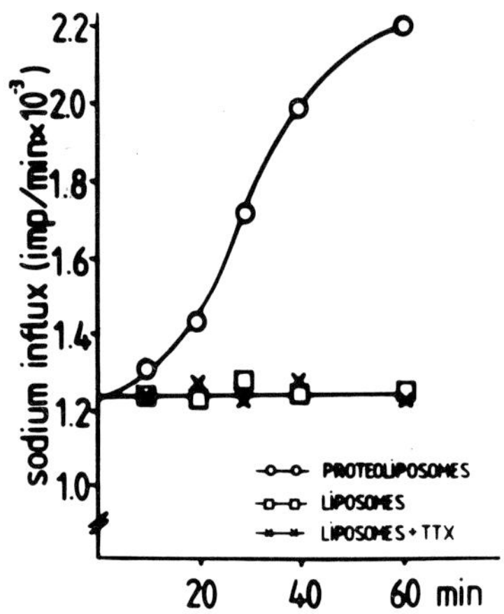

Fig.7 ^{22}Na influx into proteoliposomes as a function of preincubation time

phosphatidylcholine, phosphatidylserine and cholesterol (2:1:1). The possibility could not be excluded that the permeability change was caused by phospholipases which were present in the tissue extracts. But on thin layer chromatograms we could not detect any difference in the phospholipid composition of the liposomes before and after preincubation with supernatants.

More probably the increase in permeability was due to protein-liposome interaction. If this explanation is correct this process must be slow.

In the next set of experiments we have studied the influence of phospholipid composition of the liposomes on the induction of the TTX-sensitivity.

All experiments described before were performed with liposomes prepared from total brain lipids. But when we proceeded to commercial phospholipids we could not obtain any effect of TTX or veratridine. Fig.8 shows that the TTX-sensitivity appears only after addition of cholesterol to the phospholipid mixture.

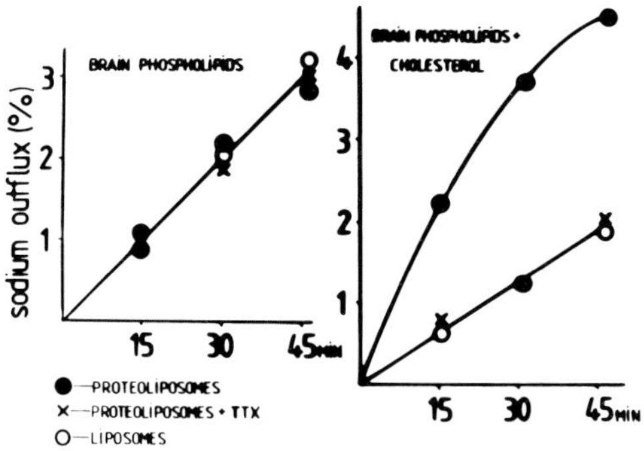

Fig.8 Induction of TTX sensitivity on brain phospolipid (A), and brain phospholipids + cholesterol 3o%, liposomes, preincubated with brain extracts (B)

The same results were obtained with lecithin-phosphatidylserine liposomes. The striking result of these experiments was the strictly asymetrical action of TTX. Fig.9 shows the efflux of sodium from lecithin-phosphatidylserine-cholesterol liposomes containing TTX inside the vesicles or when TTX was present in the medium . One can see that only internal TTX inhibited sodium efflux. Now it is difficult to speculate about the reason of such vectorial action of TTX. Perhaps in crude lipid proteoliposomes the TTX-sensitive protein has a chance orientation. The asymetrical incorporation of TTX-sensitive protein into lecithine-phosphatidylserine liposomes corresponds to what has to happen in nerve cells. It is well known that the TTX-binding site of the sodium channel is localized on the outside membrane surface.

If the mechanism of incorporation of TTX-sensitive protein into cell membranes and into liposomes is the same, it means that in the latter case the TTX-binding site should be localized inside the vesicle.

The existence of a cytoplasmic precursor of voltage dependent sodium channels seems to be reasonable in view of some data

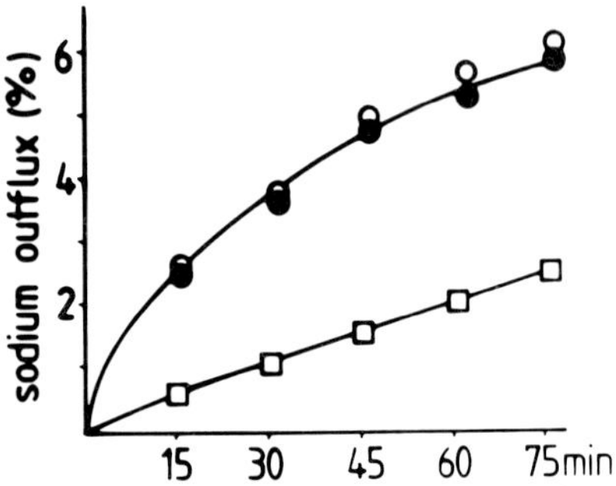

Fig.9 The effect of the internal and external TTX on the ^{22}Na outflux from the proteoliposomes.
O - Na outflux from the control liposomes
O - liposomes preincubated with brain extracts.TTX was added in the buffer
- the same as O but TTX was introduced inside the liposomes

from the literature. The most direct evidence comes from work of Inoue (8) and Krawczyk (9). In these experiments a protoplasmic drop isolated from a Nitella cell was found to become excitable in response to an electrical stimulus. It was also shown that Nitella protoplasma forms Ca-Na-channels when added to black lipid films.

An interesting result has been obtained recently with a behavioral mutant of Paramecium which looses the ability to move backward (1o). It was shown that after tight contact between the membranes of wild type and mutant cells, the latter became electrically excitable and they regained this mobility. The same effect could be achieved by the microinjection of wild type cell cytoplasm in the mutant cells.

All these data support the idea of the existence of a hydrophilic sodium channel precursor. In this case neurotoxins could be a useful instrument for the identification and purification of this protein.

References

1) Blobel,G. (198o): Proc.Natl.Acad.Sci.USA 77, 1496-15oo.

2) Waksman,A., Hubert,P., Crémel,G., Rendon,A., Burgun,C. (198o): Biochim. Biophys.Acta 6o4, 249-296.

3) Goldman,B.M., Blobel,G. (1978): Proc.Natl.Acad.Sci.USA 75, 5o66-5o7o.

4) Boquet,P., Silverman,M.S., Pappenheimer,A.M., Vernon,W. B. (1976): Proc.Natl.Acad.Sci.USA 73, 4449-4453.

5) Finkelstein,A., Rubin,L.L., Tzeng,M.C. (1976): Science 193, 1oo9-1o11.

6) Fishman,P.M., Brady,R.D. (1976): Science 194, 9o6-915.

7) Malysheva,M.K., Lishko,V.K., Chagovetz,A.M. (198o): Biochim. Biophys. Acta 6o2, 7o-77.

8) Inoue,I., Ueda,T., Kobatake,Y. (1973): Biochim. Biophys. Acta 298, 653-661.

9) Krawczyk,S. (1978): Nature 273, 5657-5658.

1o) Hiwatashi,K., Haga,N., Takahashi,M. (198o): J.Cell Biol. 84, 476-48o.

II. Palytoxin

CARIBBEAN PALYTOXIN - A NEW TOOL IN MEMBRANE RESEARCH

Minireview

Lâszlo Bêress
Institute of Toxicology, University of Kiel (Germany)

To Dr. István Bêress dedicated on his 80 th birthday

Introduction

Toxins isolated from several Palythoa species, e.g. P. toxica (28), P. tuberculosa (17,23), P. caribaeorum and P. mammillosa (4) and a tahitian Palythoa specimen (27,30) are the most potent known biologically active substances of marine origin (20,22,34,35). All share the common name "palytoxin" (PTX) though this term was introduced primarily by Scheuer for the toxin of the zooanthid P. toxica (31).
All isolated palytoxins exhibit the same U.V. spectra (λ max. 233,263 nm), all are nonprotein compounds with molecular weights ranging from 2500 to 3000 Dalton (4,17,23,28,30). The structure of three palytoxins of different origin (P. toxica, tahitian Palythoa sp. and P. tuberculosa) have been elucidated recently (25,33). The absolute stereochemistry of the palytoxin from P. toxica has also been established (26). Palytoxin depolarizes every exitable organ investigated like cardiac (3,9,19) skeletal (1,9) and smooth muscle (18,29) as well as nerve fibres (11) initiating contracture. It is the most potent coronary vasoconstrictor substance known (20,21,24). A dose dependent decrease in contractile force on innervated and denervated rat diaphragms has also been observed after treatment with palytoxin (10,32).
An improved method for the isolation of the caribbean palytoxin has been elaborated recently (5). Toxin isolated by this procedure enabled the initiation of new studies concerning the mode of action of palytoxin on the cell membrane.

Toxins as Tools in Neurochemistry

Chemistry

According to recent investigations by Moore the structure of the caribbean toxin turned out to be identical with the structures of the toxins from P. toxica and P. tuberculosa (R.E. Moore, personal communication). The palytoxins isolated from P. toxica, tahitian Palythoa sp. and P. tuberculosa are all chemically quasi identical compounds consisting of a long partially unsaturated (eight carbon-carbon double bonds) aliphatic chain, interspaced cyclic ether, 40-42 HO- and two amide groups. The chain starts with an HO-group and ends with an NH_2-group. According to Moore and Bartolini (25) the structure of the tahitian Palytoxin is given in Fig.1.

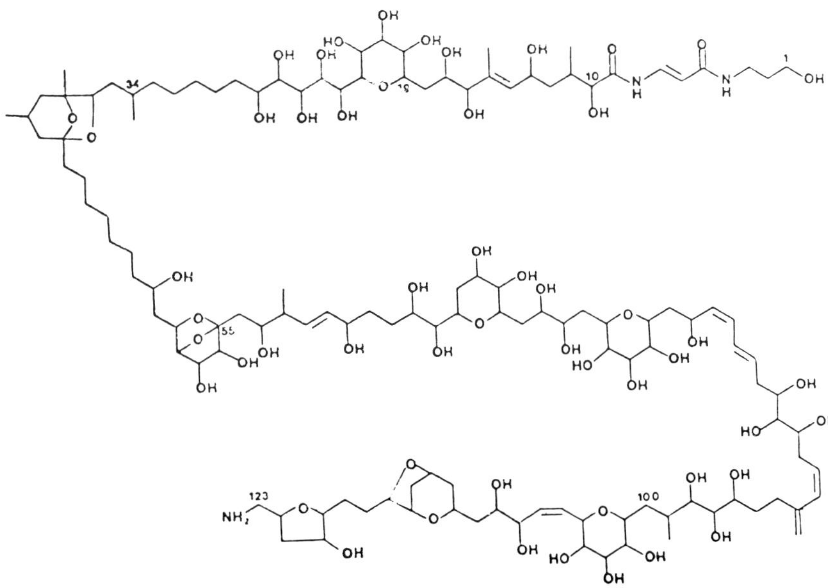

Fig. 1: The structure of the tahitian Palytoxin (25)

The molecular formulas are $C_{129}H_{221}N_3O_{54}$ (Mw 2659) for palytoxin isolated from the tahitian Palythoa sp. and $C_{129}H_{223}N_3O_{54}$ (Mw 2677) for the two palytoxins from Palythoa toxica (25). The palytoxin isolated from the okinawan Palythoa tuberculosa has the same gross structure as the major component in the palytoxin from the hawaiian Palythoa toxica (25,33). The subtle differences in the palytoxins from the different Palythoa species are exclusively due to structural differences in the hemiketal ring at the position C_{55} in the molecule (s.Fig.1). The palytoxin from the P. toxica has 64 chiral centers; sixty of them have been determined (26).

Mode of action of the caribbean palytoxin on the cell membrane

Because of its very high toxicity (LD_{50} in shore crab Carcinus maenas 62,5 ng/kg (5), LD_{50} in mice 250 ng/kg (14)) and of its action on contractility of cardiac (3) and skeletal (1) muscle we started our work with the hypothesis that the caribbean palytoxin is a neuro- and cardiotoxin. Recent investigations, however, indicated, that palytoxin is in fact a potent haemolysin which has a slow course of action and causes a large prelytic potassium loss (14). It haemolizes the erythrocytes from several animal species. The haemolysis is triggered by a primary membrane lesion induced by the toxin. The early damage manifests itself by the immediate start of the potassium loss which becomes manifest long before the haemolysis starts. Depending on the haemolysing conditions used the ED_{50} of palytoxin ranges between 10 picogramm and 2 nanogramm per milliliter (14). Palytoxin decreases the osmotic resistance of the erythrocytes in a concentration-dependent manner, so that osmotic influences are negligible for K^+ release but considerable in haemolysis. External K^+ inhibits the haemoglobin release and Rb^+ inhibits the release of K^+ and haemoglobin. Ca^{++} (over 20 μM) and borate (over 5 μM) enhance the loss of K^+ and haemoglobin. With both Ca^{++} and borate present the efficacy of palytoxin is raised about 10 000 fold. Under these conditions, about 15 palytoxin molecules per human red cell trigger 50% K^+ loss over a wide range of cell concentration (2).
Palytoxin formes small pores in the erythrocyte membranes, which allow

the entrance of Na^+ and choline, whereas inositol is largely, Ca^{++} as well as sucrose and inulin are completely excluded. Amphotericin B resembles palytoxin in its ability to cause a considerable prelytic K^+ loss and to form small pores. However, it is about 1000 times weaker than palytoxin, it is not inhibited by K^+, Rb^+ and not activated by Ca^{++} and borate. Thus palytoxin represents a novel typ of cytolysin; its effect is reversible (2). Ouabain in concentrations as low as 5×10^{-8}M completely prevents the palytoxin effect (16). The glycosid effect is specific because of its small concentration range, rat erythrocytes are less sensitive than human cells to ouabain, potassium release due to amphotericin B or the Ca^{++} ionophore A 23178 is not influenced by ouabain, and dog erythrocytes which lack the Na^+-K^+ ATPase are resistent to palytoxin as well as to ouabain (15). Palytoxin and ouabain share a common receptor. Its occupation by ouabain leads to the inhibition of the Na^+-K^+ ATPase, whereas its interaction with palytoxin results in formation of small pores in the neighbouring membrane areas probably forced by the surface activity of the palytoxin. Palytoxin itself has no direct influence on the enzyme (15). Recent studies on the action of palytoxin on erythrocytes and also on resealed ghosts revealed, that palytoxin once bound to the erythrocyte membrane transforms their sodium pump into a pore allowing the transport of small ions through the membrane. This process is assisted by ATP from inside, whereas Ca^{++} promotes from outside the efficacy of palytoxin (7,8). At our present state of knowledge the mode of action of palytoxin on the red cell membrane can be described in the following way (Fig. 2):

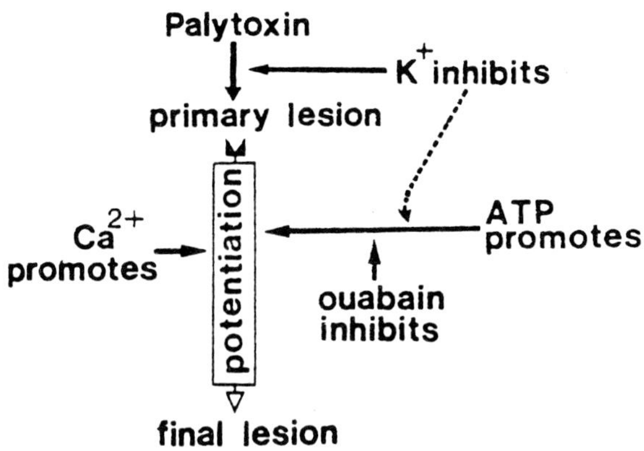

Fig.2: The mode of action of palytoxin on the red cell membrane (7).

Palytoxin is a potent releaser of histamine from the mast cells. The concentration-response curve is bell-shaped with its range between 0.05 ng/ml and 5 μg/ml and its maximum at 50 ng/ml. The release of histamine is specific because heat inactivated mast cells are insensitive to palytoxin and extracellular calcium is essential. Concerning the effect of ions on the palytoxin induced histamine release mast cells behave like erythrocytes; extracellular Ca^{++} and borate promotes and K^{+} inhibits the histamine release (6). Palytoxin concentrations as low as 10^{-11} to 10^{-12}M promote the shift of lipophilic (^{3}H)-tetraphenylphosphonium ion from particulate brain corte of guinea pigs and rats and from crude synaptosomes of rats which indicates depolarization. The shift is neither influenced by tetrodotoxin nor by the calcium channel blocker nimodipin, nor by substitution of choline for Na ions. It is increased by Ca^{++} and by borate, the latter interacting with the toxin itself (12).

The caribbean palytoxin is the most potent membrane active agent known so far. Because of the specificity of its action on the cell membrane it might represent a new tool in membrane research (13).

Acknowledgement:

The author is very grateful to Prof. Dr. E. Habermann, Giessen, for offering all his scientific reprints and preprints for the preparation of the manuscript.

Supported by the Deutsche Forschungsgemeinschaft (Be 554/16-5).

References

1. Agena, G., Alsen, C. and Draijer, F.: Naunyn-Schmiedeberg's Arch. Pharmacol. 313, R 44 (1980)
2. Ahnert-Hilger, G., Chhatwal, G.S., Hessler, H.J. and Habermann, E.: Biochim. Biophys. Acta 688, 486-494 (1982)
3. Alsen, C., Agena, G. and Béress, L.: Toxicon 20, 57 (1982)
4. Attaway, D.H. and Ciereszko, L.S.: Proceedings of the Second International Coral Reef Symposium, Vol. 1. Great Barriar Reef Committee Brisbane pp. 497-504 (1974)

5. Béress, L., Zwick, J., Kolkenbrock, H.J., Kaul, P.N. and Wassermann, O.: Toxicon in press

6. Chhatwal, G.S., Ahnert-Hilger, G., Béress, L. and Habermann, E.: Int.Arch. Allergy and appl. Immunol. 68, 97-100 (1982)

7. Chhatwal, G.S., Hessler, H.J., Béress, L. and Habermann, E.: Toxicon in press

8. Chhatwal, G.S., Hessler, H.J. and Habermann, E.: Naunyn-Schmiedeberg's Arch. Pharmacol. submitted

9. Deguchi, T., Urakawa, N. and Takamatsu, S.: In Animal Plant and Microbial Toxins (Ohsaka, A., Hayashi, K. and Sawai, Y. eds.) Vol. 2 pp. 379-394 Plenum Press, New York (1976)

10. Draijer, F., Suvanto, P., Tesseraux, I. and Béress, L.: Toxicon 20, 65 (1982)

11. Dubois, J.M. and Cohen, J.B.: J. Pharmacol. Exp. Ther. 201, 148-155 (1977)

12. Habermann, E.: Naunyn-Schmiedeberg's Arch.Pharmacol. Submitted

13. Habermann, E.: Personal communication

14. Habermann, E., Ahnert-Hilger, G., Chhatwal, G.S. and Béress, L.: Biochim. Biophys. Acta 649, 481-486 (1981)

15. Habermann, E. and Chhatwal, G.S.: Naunyn-Schmiedeberg's Arch. Pharmacol. 319, 101-107 (1982)

16. Habermann, E., Chhatwal, G.S. and Hessler, H.J.: Naunyn-Schmiedeberg's Arch. Pharmacol. 317, P 374

17. Hirata, Y., Uemura, D., Ueda, K. and Takano, S.: Pure Appl. Chem. 51, 1875-1883 (1979)

18. Ito, K., Karaki, H. and Urakawa, N.: Eur. J. Pharmacol. 46, 9-14 (1977)

19. Ito, K., Karaki, H. and Urakawa, N.: Japan J. Pharmacol. 29, 467-476 (1979)

20. Kaul, P.N.: In Food-Drugs from the Sea Proceedings (Webber, H.H. and Ruggieri, D.S. eds.) pp. 311-323, Mar. Techn. Soc. (1974)

21. Kaul, P.N.: Fed. Proc. 40, 10-14 (1981)

22. Kaul, P.N., Farmer, M.R. and Ciereszko, L.S.: Proc. West.Pharm.Soc. 17, 294-301 (1974)

23. Kimura, S. and Hashimoto, Y.: Publs. Seto mar. biol. Lab. 20, 713-718 (1973)

24. Kulkarni, S.K.,Kirlin, W.G. and Kaul, P.N.: in Food and Drugs from the Sea; myth or reality? (Kaul, P.N. and Sindermann, C.J. eds.) pp. 73-80 University of Oklahoma Press (1978)

25. Moore, R.E. and Bartolini, G.: J. Am. chem. Soc. 103, 2491-2494 (1981)

26. Moore, R.E., Bartolini, G. and Barchi, J.: J. Am. chem. Soc. 104, 3776-3779 (1982)

27. Moore, R.E., Dietrich, R.F., Hatton, B., Higa, T. and Scheuer, P.J.: J. Org. Chem. 40, 540-542 (1975)

28. Moore, R.E. and Scheuer, P.J.: Science, 172, 495-497 (1971)

29. Ohizumi, Y. and Shibata, S.: J. Pharmacol. Exp. Ther. 214, 209-212 (1980)

30. Quinn, R.J., Kashiwagi, M., Moore, R.E. and Norton, T.R.: J. pharm. Sci. 63, 257-260 (1974)

31. Scheuer, P.J.: Fortschr. Chem. Org. Nat.stoffe 22, 265 (1964)

32. Suvanto, P. and Tesseraux, I.: Naunyn-Schmiedeberg's Arch. Pharmacol. 319, R 34 (1982)

33. Uemura, D., Ueda, K., Hirata, Y., Naoki, H. and Iwashita, T.: Tetrahedron Lett. 22, 2781-2784 (1981)

34. Vick, J.A. and Wiles, J.S.: Toxicology and Appl. Pharmacol. 34, 214-223 (1975)

35. Wiles, J.S., Vick, J.A. and Christensen, M.K.: Toxicon 12, 427-433 (1974)

PHYSIOLOGICAL AND MORPHOLOGICAL EFFECTS OF PALYTOXIN (PALYTHOA CARIBAEORUM) ON SKELETAL MUSCLE

I. Tesseraux[*], J.B. Harris[**] and S.C. Watkins[**]
[*] Dept. of Toxicology, University of Kiel, D-2300 Kiel, West-Germany
[**] Muscular Dystrophy Group Laboratories, Newcastle General Hospital, Newcastle upon Tyne NE 4 6 BE, England

Introduction

Palytoxin (PTX) has been isolated from the coral Palythoa caribaeorum by Beress and coworkers (1). PTX is a highly potent nonpeptide toxin with a molecular weight of 2659-2677 (2).
Exposure of isolated phrenic-nerve-diaphragm preparations of mice (3) and rats (4) to PTX results in an early increase and a later decrease in the twitch tension of the muscle. A similar direct action of PTX may be seen with rat denervated diaphragm (5). These effects have been correlated with the depolarizing action of PTX observed in frog sartorius muscle (3). In this study the electrophysiological effects of PTX have been investigated on two different mammalian muscles, Extensor digitorum longus, EDL (a fast-twitch muscle) and Soleus, SOL (a slow-twitch muscle). Morphological effects of the toxin were studied on PTX-treated SOL muscles.

Methods

Female rats weighing approximately 200 g were sacrificed. The EDL and SOL muscles were removed and mounted in an organ bath containing an oxygenated saline maintained at room temperature (20-21°C). Electrical properties of the muscle fibres were recorded using a standard intracellular microelectrode technique. Action potentials were generated after locally hyperpolarizing the muscle membrane to about -90 mV. Dantrolene sodium (2.5x10^{-5}M) was added to prevent mechanical artefacts (6). For morphological studies the muscles were incubated in a bathing

Toxins as Tools in Neurochemistry

fluid containing PTX for 60 min. Membrane potentials were recorded throughout this time. The muscle tissue was first fixed in 5 % glutaraldehyde in 0.1 M phosphate buffer and post-fixed in 1 % osmium tetroxide. Ultrathin sections were prepared, stained and examined with a Siemens 102 electron microscope.

Results

PTX caused a dose-dependent depolarization of both EDL and SOL muscle fibres. The threshold concentration for this effect was 10^{-12}M (Tab.1). The effect dose dependent and variable increased with the time after application of the toxin. EDL muscles were slightly more affected than SOL muscles. The ionic basis of the depolarizing activity of PTX was investigated. In the absence of Na^+ (Na^+ replaced by choline) there was little depolarization (Tab.2). By contrast, the sodium channel blocking agent tetrodotoxin (TTX 10^{-6}M) did not prevent the PTX-evoked depolarization (Tab.2). Similar experiments have been made by replacing Ca^{2+} in the bathing fluid either by Co^{2+} or Mg^{2+}. Using Co^{2+} as a substitute for Ca^{2+} the depolarization caused by PTX was reduced at low concentrations of PTX (10^{-10}M), but at higher concentrations of PTX (10^{-8}M) or when Mg^{2+} was used to replace Ca^{2+} the PTX-effect was not altered. No difference was noted between EDL and SOL muscles.
Action potentials were generated in the presence of PTX, whenever it was possible to pre-set the membrane potential to a level of -90 mV. Under such conditions action potential properties were not significantly changed. Only action potential overshoot appeared to be slightly decreased in some cases in EDL muscles (Tab. 3), but this effect was variable and it is unclear whether the decrease was due to the influence of the toxin. The rate of rise of the action potentials did not differ from the control values in either EDL or SOL muscles.
The electron micrographs taken from SOL muscle incubated with 10^{-9}M PTX for 60 min. revealed different stages of muscle damage. The degree of this damage varied within the same muscle. The first sign of PTX-induced damage was a dilatation of sarcoplasmatic reticulum (Fig.1). At this

Tab. 1: Effect of PTX on resting membrane potential (mV) on extensor digitorum longus (EDL) and soleus (SOL) muscles, measured at 10, 30 and 60 min after application of the toxin.

EDL	15 min	30 min	60 min
control	-71.9 ± 0.35(142)	-	-
10^{-12}M PTX	-51.7 ± 1.76(6)	-	-45.2 ± 2.50(5)
10^{-10}M PTX	-59.8 ± 6.93(45)	-58.2 ± 2.70(40)	-45.0 ± 3.02(6)
10^{-9} M PTX	-33.1 ± 1.34(15)	-35.7 ± 3.11(35)	-
10^{-8} M PTX	-26.8 ± 2.01(55)	-	-
10^{-6} M PTX	-	-18.6 ± 1.16(20)	-

SOL	15 min	30 min	60 min
control	-70.0 ± 0.35(139)	-	-
10^{-12}M PTX	-60.8 ± 1.20(5)	-	-58.7 ± 1.79(6)
10^{-10}M PTX	-61.3 ± 1.94(38)	-58.0 ± 0.69(40)	-44.0 ± 1.18(6)
10^{-9} M PTX	-56.8 ± 1.47(12)	-44.4 ± 2.50(34)	-
10^{-8} M PTX	-27.7 ± 2.04(67)	-	-
10^{-6} M PTX	-	-18.2 ± 1.49(20)	-

The results are quoted as mean ± s.e. The figures in parentheses represent the number of fibres studied.

Tab. 2: Effect of PTX on resting membrane potential (mV) on extensor digitorum longus (EDL) and soleus (SOL) muscle fibres in the presence of tetrodotoxin (10^{-6}M TTX) and in the absence of Na^{+} (replaced by choline).

EDL	15 min	30 min	60 min
control	-71.9 ± 0.35 (142)	-	-
10^{-12}M PTX + TTX	-33.6 ± 3.12 (25)		-29.9 ± 5.17(16)
10^{-10}M PTX + TTX	-	-46.4 ± 3.89(20)	-26.2 ± 1.75(40)
10^{-8} M PTX + TTX	-23.1 ± 2.10 (20)	-18.1 ± 1.52(20)	-
Sodium free fluid	-65.8 ± 0.980(10)	-	-
Sodium free fluid + 10^{-12}M PTX	-60.2 ± 0.983(20)	-54.9 ± 1.00(21)	-

SOL	15 min	30 min	60 min
control	-70.0 ± 0.35(139)	-	-
10^{-12}M PTX + TTX	-	-56.0 ± 1.69(20)	-50.5 ± 2.72(22)
10^{-10}M PTX + TTX	-44.0 ± 2.90(21)	-34.2 ± 2.25(20)	-20.9 ± 0.98(40)
10^{-8} M PTX + TTX	-16.0 ± 8.94(20)	-12.6 ± 0.55(20)	-
Sodium free fluid	-64.2 ± 0.85(10)	-	-
Sodium free fluid + 10^{-12}M PTX	-61.9 ± 0.55(20)	-60.1 ± 0.80(19)	-

The results are quoted as mean ± s.e. The figures in parentheses represent the number of fibres studied.

Tab. 3: Effect of PTX on some properties of action potentials elicited from a standard resting potential (-90 mV) using a double microelectrode technique in extensor digitorum longus (EDL) and soleus (SOL) muscle fibres.

OVERSHOOT (mV)

EDL	15 min	60 min	90 min
control	34.4 ± 1.7(38)	26.9 ± 0.6(16)	-
10^{-12}M PTX	13.0 ± 2.7(6)	10.0 ± 3.9(5)	11.6 ± 2.5 (5)
10^{-10}M PTX	16.5 ± 2.2(13)	22.7 ± 2.4(6)	19.0 ± 3.3 (6)
SOL			
control	27.2 ± 1.5(39)	23.6 ± 1.6(15)	-
10^{-12}M PTX	22.4 ± 3.5(5)	21.3 ± 2.3(6)	11.0 ± 3.3 (6)
10^{-10}M PTX	20.7 ± 2.6(6)	12.7 ± 1.6(6)	13.3 ± 2.3 (6)

V_{max} (Vs^{-1})

EDL	15 min	60 min	90 min
control	302 ± 14 (38)	234 ± 17 (16)	-
10^{-12}M PTX	217 ± 17 (6)	260 ± 23 (5)	236 ± 19 (5)
10^{-10}M PTX	244 ± 14 (13)	220 ± 28 (6)	233 ± 32 (6)
SOL			
control	232 ± 8 (39)	204 ± 14 (14)	-
10^{-12}M PTX	268 ± 32 (5)	-	220 ± 20 (5)
10^{-10}M PTX	247 ± 14 (6)	210 ± 7 (6)	217 ± 19 (6)

The results are quoted as mean ± s.e. The figures in parentheses represent the number of fibres studied.

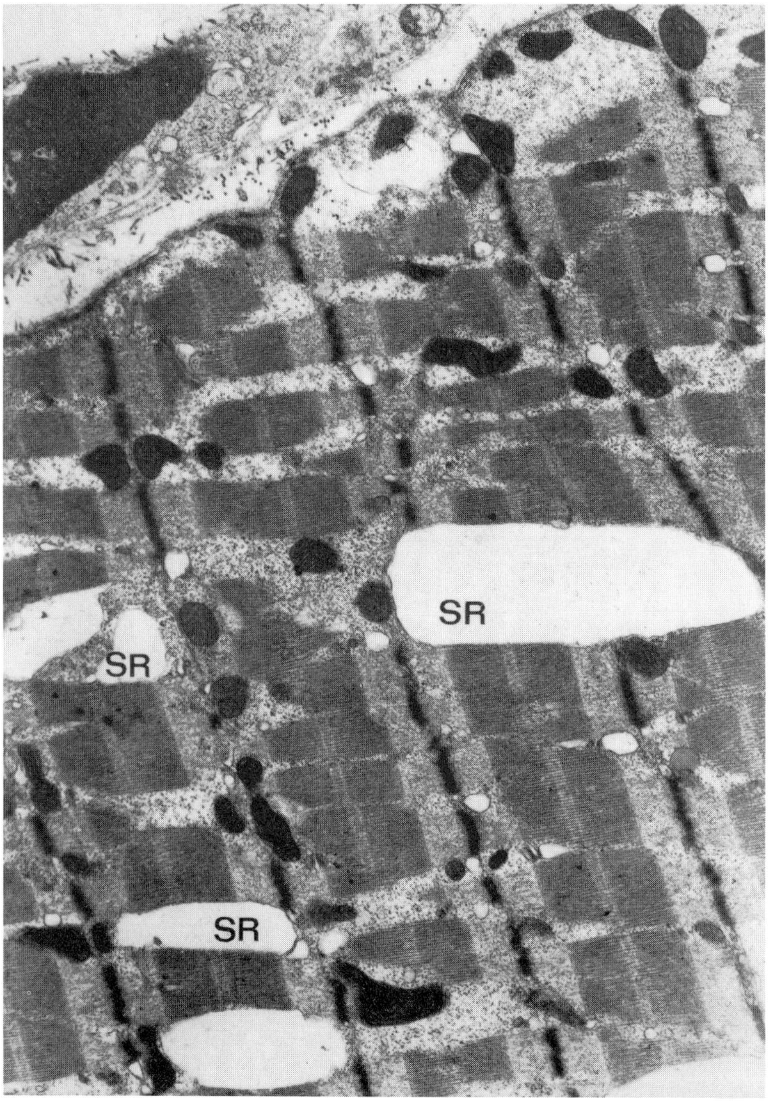

Fig. 1: Electron micrograph of soleus muscle incubated with 10^{-9}M PTX showing massive dilatation of sarcoplasmatic reticulum (SR). Note that fine structure of myofibrils is maintained. Magnification 15 000 x

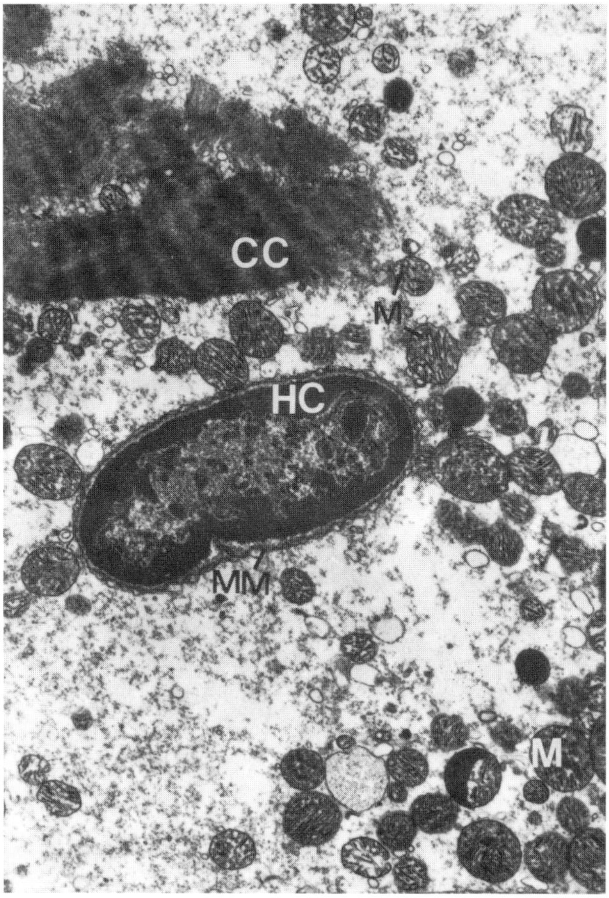

Fig. 2: Electron micrograph of soleus muscle incubated with 10^{-9}M PTX. Dense christae structure in mitochondria (M), contraction clots (CC) was seen. Note that the membrane of the myonucleus (MM) appears separated from the heterochromatin (HC). Magnification 15 000 x

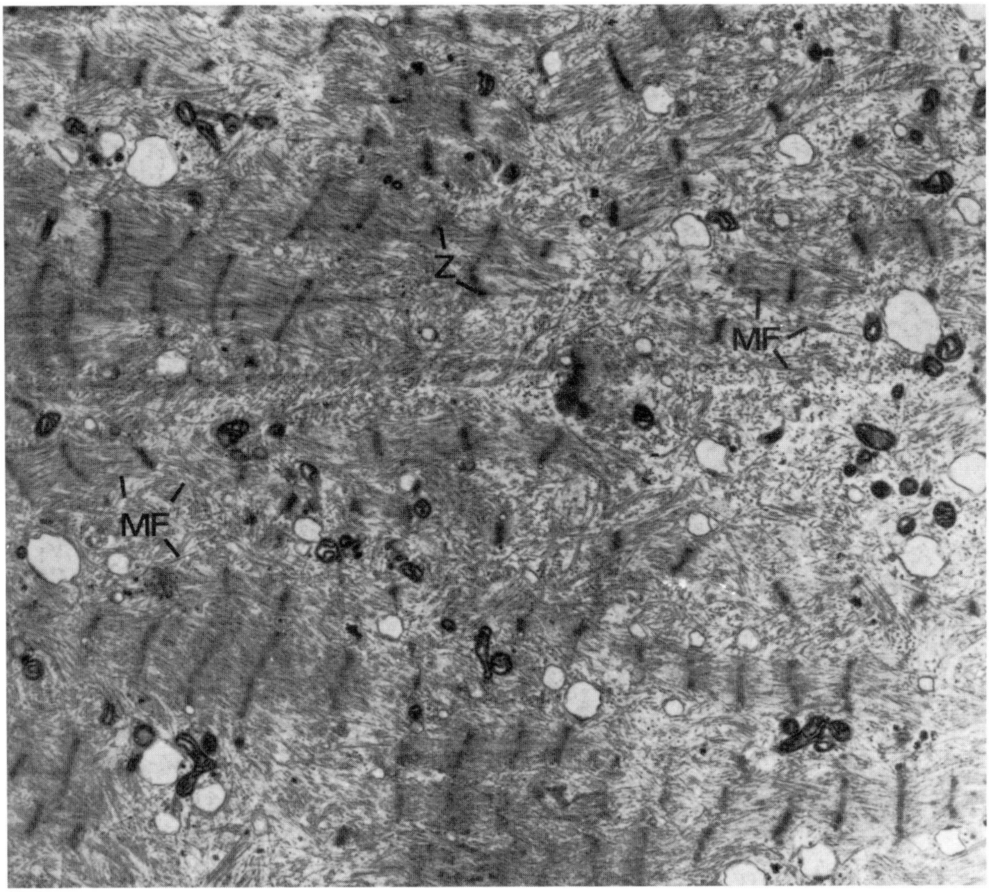

Fig. 3: Electron micrograph of soleus muscle incubated with 10^{-9}M PTX. Disruption of myofibrils (MF) and cell lysis occur. Z = Z-lines. Magnification 12 000 x

stage, the fine structure of myofibrils in the same region appeared to be normal. Later signs of toxin induced damage were the appearance of dense cristae in mitochondria, in some cases appearing as three layers (Fig.2). This may indicate an accumulation of Ca^{2+}-ions. The myonuclei appeared unchanged, though in some cases the membrane seemed to be abnormally separated from the electron dense heterochromatine (Fig. 2). In some fibres, particularly in the periphery of the muscle more severe damage was observed. The fine structure was destroyed and contraction clots occurred (Fig.3). It was also noted that in a given region of muscle, damaged and undamaged muscle fibres could co-exist.

Discussion

The depolarization of the muscle caused by exposure to PTX is Na^+ dependent, but the entry of Na^+ is not inhibited by TTX. This would indicate that PTX either initiates a general increase in Na^+ permeability or that it disrupts the normal fast Na^+ channel in such a way that the resting Na^+ conductance is increased and that the binding or orientation of TTX is disturbed. Since PTX has no early effect on the characteristics of directly elicited action potentials, it would seem unlikely that its "target" is the fast Na^+ channel. The depolarization also occurs in the absence of Ca^{2+}, and is clearly not dependent on an influx of Ca^{2+} activating endogenous peptide hydrolases or phospholipases.
We conclude that PTX probably acts by causing a non-specific increase in membrane permeability, and that this gives rise to a loss of membrane potential and the disruption of the muscle. The toxin acts at such a low concentration that it seems probable that it has a specific binding site. The nature of the binding site requires elucidation.

References

1. Béress, L.: Toxicon, in press, 1983
2. Moore, R.E., Bartolini, G.: J.Am.Chem.Soc. 103, 2491-2494 (1981)

3. Deguchi, T., Urakawa, N., Takamatsu, S.: Animal, Plant and Microbial toxins 2, eds. A. Ohasaka, K. Hayashi and Y. Sawai, Plenum Publishing Corp., New York, 379-394 (1976)

4. Draijer, F., Suvanto, P., Tesseraux, I., Bêress, L.: Toxicon 20, 65 (1982)

5. Suvanto, P., Tesseraux, I.: Naunyn-Schmiedeberg's Arch. Pharmacol. 319, R 34 (1982)

6. Alsen, C., Harris, J.B., Tesseraux, I.: Br.J.Pharmac. 74, 61-71 (1981)

PALYTOXIN - A CATION IONOPHORE?

S.Stengelin[1+], L.Béress[2], L.Lauffer[1], and F.Hucho[1]

1 Institut für Biochemie, Freie Universität, Fabeckstr.34, D-1ooo Berlin 33
2 Institut für Meereskunde der Universität Kiel, Düsternbrookerweg 2o, D-23oo Kiel

Introduction

Palytoxin (PTX) is an extremely poisonous agent produced by various coelenterates of the genus Palythoa. Recently its structure has been determined (1,2).

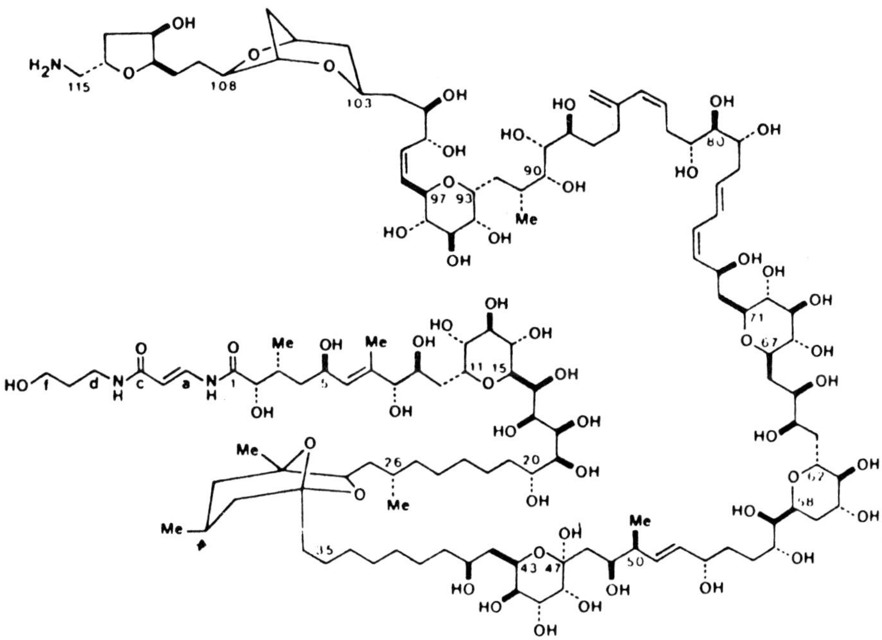

Fig.1 The structure of palytoxin (from ref.2).

With frog myelinated fibres (3) as well as with unmyelinated axons (4) PTX causes a depolarization by increasing the resting sodium conductance. Since this effect appears to be irreversible and can only be partly antagonized by the specific sodium channel blocker tetrodotoxin, a general membrane da-

+ Present address: The Salk Institute, San Diego, CA 92138, USA

Toxins as Tools in Neurochemistry

mage as mode of action of PTX seemed likely. To investigate this possibility, we determined the effects of PTX isolated from Palythoa caribaeorum (5) on $^{22}Na^+$-and Li^+-efflux from membrane vesicles prepared from (i) crayfish nerves (Astacus leptodactylus) and (ii) Torpedo californica electric organ, (i) being axonal and (ii) being postsynaptical material. Furthermore, we investigated if PTX increased the Li^+-permeability of artificial liposomes.

Materials and Methods

$^{22}NaCl$	2oo µCi/ml, NEN, Dreieich
$Na^{125}I$	1oo mCi/ml, Amersham Buchler Braunschweig

All other chemicals and reagents were of the highest purity commercially available.

Iodination of sea anemone toxin (ATX I) from Anemonia sulcata: ATX I was isolated as described in (6). Its radiolabeling was performed by the lactoperoxidase method and is described in detail in (7).

Binding of ^{125}I-ATX I to whole crayfish nerves: The procedure is described in (7).

Preparation of axonal membrane fragments from Astacus leptodactylus walking leg nerves and preparation of acetylcholine receptor-rich membrane fragments from Torpedo californica: The methods are described in (8) and (9), respectively.

Preparation of artificial liposomes: The procedure is described in detail in (9) and is an adaptation of the method developed by Brunner et al.(1o).
A mixture of 2.62 mg phosphatidylcholine, 1.76 mg phosphatidylethanolamine, o.8o mg phosphatidylserine, o.4o mg sphingomyeline, o.11 mg phosphatidylinositol and o.85 mg cholesterol is dissolved in 1.5 % sodium cholate. After stirring for

3o min at room temperature the detergent is removed by gel filtration on a column (1.7cm x 25 cm) of Sephadex G 5o. During filtration, liposomes are formed and eluted approximately with the void volume. When Li^+-efflux (see below) is to be measured the lipids are dissolved in a Tris-buffer (1o mM, pH 7.4), containing 1oo mM NaCl, 3oo mM LiCl and 1.5 % sodium cholate.

$^{22}Na^+$-efflux from membrane fragments: A detailed description can be found in (9). In general, with Astacus- and Torpedo-membrane vesicles, small aliquots (5o-1oo μl) of a concentrated membrane suspension preincubated with 2o-5o μCi/ml $^{22}Na^+$ are diluted 5ofold in a medium which is isosmotic but contains no radioactive Na^+. After dilution 1 - 2 ml are filtered at distinct time intervals through membrane filters (Schleicher and Schüll, type BA 85) and the radioactivity remaining on the filters is determined in a liquid scintillation counter.

Li^+-efflux from membrane vesicles

Torpedo membranes are incubated in Tris-buffered (1o mM, pH 7.4) Ringer's solution (16o mM NaCl, 5 mM KCl, 2 mM $CaCl_2$) containing additionally 2oo - 4oo mM LiCl and are diluted 5ofold isosmotically in a medium containing NaCl instead of LiCl. Aliquots are filtered at distinct time intervals through membrane filters (see above); after filtration the filters are washed overnight with 2 ml 1% Triton X-1oo in H_2O and the Li^+-content of the wash solution is determined by flame emission spectroscopy in an atomic absorption spectrophotometer (Varian type AA-275). With artifical liposomes larger aliquots (5oo-1ooo μl) are diluted and correspondingly larger aliquots are filtered in order to retain enough Li^+ on the filters to be measured by flame emission spectroscopy.

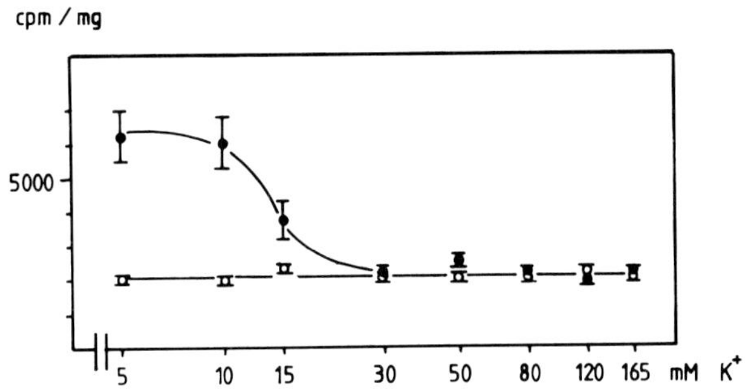

Fig.2 Binding of ^{125}I-ATX I to crayfish nerves in the presence of variing concentrations of external K^+. Pieces of main walking leg nerves (5-15 mg) were incubated 16 hours at 4°C with 1oo nM ^{125}I-ATX I (-•-) or 1oo nM ^{125}I-ATX I and 3 mM NaN_3 (-o-). NaN_3 blocks specific binding completely as has been shown in (7).

Results

PTX and the binding of ATX I to whole crayfish nerves: ATX I is a neurotoxin specifically acting on the voltage dependent sodium channel (11). Its binding to whole nerves depends on the potential existing at the axonal membrane. Substantial but indirect evidence for this fact has been published previously (7) but it can also been shown in a direct experiment: replacing external Na^+ by K^+ and thereby depolarizing the axonal membrane prevented specific binding of ^{125}I-ATX I completely (Fig.2).
A similar effect was observed when PTX instead of K^+ was present in the incubation medium (Fig.3).

PTX prevented ATX I binding. Although we cannot totally rule out a direct interaction of PTX with ATX I-binding sites, this possibility seems rather unlikely: half maximal inhibition of ^{125}I-ATX I-binding occured at concentrations of PTX as low as $1o^{-11}$ - $1o^{-1o}$ M, whereas ^{125}I-ATX I was present at a con-

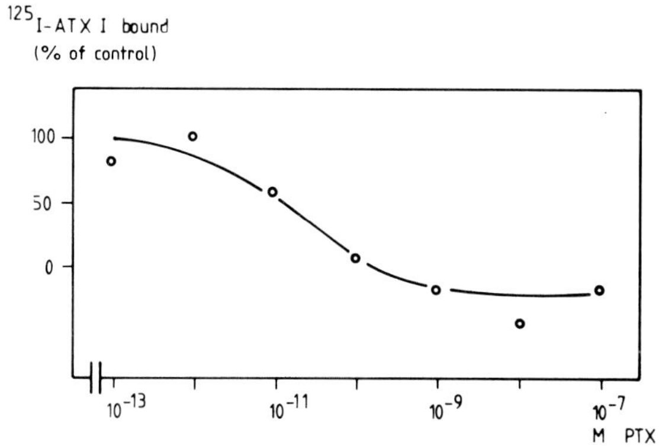

Fig.3 Binding of ^{125}I-ATX I to crayfish nerves in the presence of PTX. Concentration of ^{125}I-ATX I was 9o nM. As control (1oo%) specific binding in the absence of effector was determined by subtracting unspecific binding (obtained from samples containing 9o nM ^{125}I-ATX I and 3 mM NaN_3) from total binding (obtained from samples containing only 9o nM ^{125}I-ATX I).

centration of 9 x $1o^{-8}$ M. Specific binding of ^{125}I-ATX I in the experiment in Fig.3 was in the order of 1o fmol/mg; taking the density of the nerve as 1g/ml this amounts to a concentration of binding sites of about $1o^{-8}$ M; in other words, at a concentration of PTX ($1o^{-9}$M), where ^{125}I-ATX I-binding was completely inhibited there were fewer PTX molecules present than ATX I-binding sites. This reasoning provides strong evidence for PTX inhibiting ATX I-binding indirectly via depolarization of the axonal membrane, presumably by increasing the resting sodium conductance.

PTX and the cation permeability of Astacus nerve membrane vesicles and Torpedo acetylcholine receptor-rich membrane vesicles:

PTX markedly increased $^{22}Na^{+}$-efflux from Astacus nerve membrane vesicles. The PTX-induced $^{22}Na^{+}$-efflux was not inhibited by the sodium channel blocker tetrodotoxin (Fig.4).

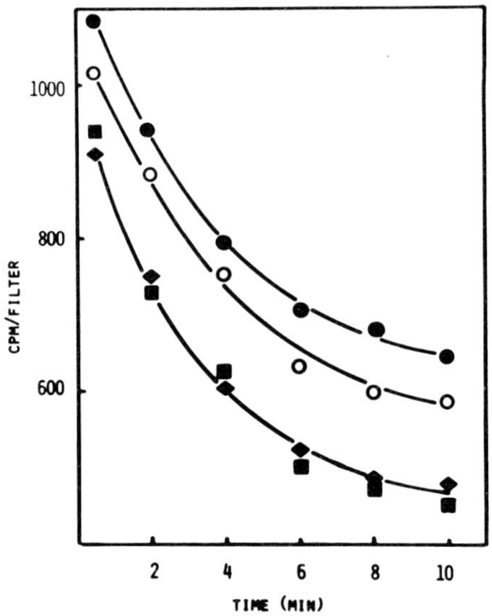

Fig.4 $^{22}Na^{+}$-efflux from membrane vesicles of Astacus leptodactylus walking leg nerves in the presence of PTX and tetrodotoxin; efflux was determined as described in Methods. Protein content of membranes was 4.2 mg/ml. Prior to dilution, aliquots were incubated 3o min at 2o °C with 14o nM tetrodotoxin (-●-), 14o nM PTX (-■-) or 14o nM both PTX and tetrodotoxin (-◆-). (-o-): control, no addition of toxins.

In a similar experiment PTX increased the $^{22}Na^{+}$-permeability of Torpedo acetylcholine receptor-rich membrane vesicles. The PTX-effect could not be blocked by the potent cholinergic antagonist α-bungarotoxin (Fig.5). PTX also enhanced Li^{+}-permeability of Torpedo membrane vesicles, though higher concentrations of PTX seemed to be required for maximal effect (Fig.6). The lipophilic cation triphenylmethylphosphonium (10^{-4} M), an ion channel blocker of the acetylcholine receptor complex (12), did not inhibit the Li^{+}-efflux induced by 10^{-6}M PTX, whereas the glykoside Ouabain inhibited PTX-induced Li^{+}-efflux from the membrane vesicles to a high degree (Fig.7).

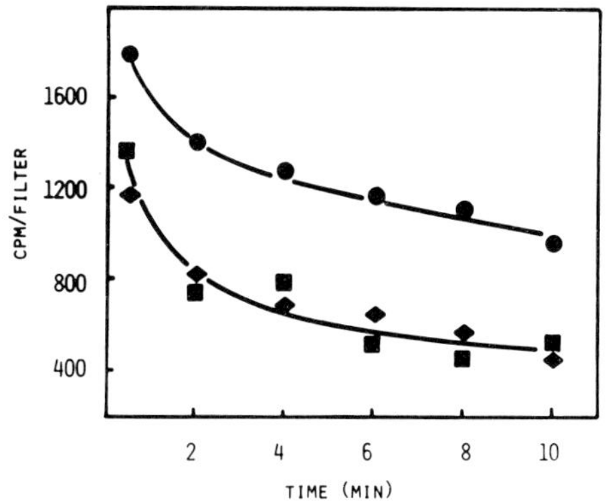

Fig.5 $^{22}Na^+$-efflux from acetylcholine receptor-rich membrane vesicles in the presence of PTX; protein content of membrane vesicles was 7.2 mg/ml. Prior to dilution, aliquots were incubated 3o min at 2o °C with 14o nM PTX (-■-) or 14o nM PTX together with 18o µM α-bungarotoxin (-◆-). (-●-): control, no addition of toxins.

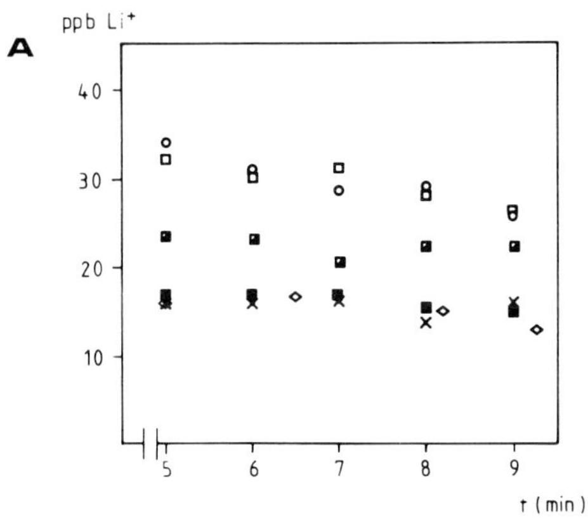

Fig.6 Li^+-efflux from acetylcholine receptor-rich membrane vesicles in the presence of PTX; A) Direct representation of the data. Li^+-efflux was measured as described in Methods. Protein content of membrane vesicle suspension was 4 mg/ml. Prior to dilution, aliquots were incubated 3o min at 2o °C with PTX at concentrations of $1o^{-8}$ M (-□-), $1o^{-7}$ M (-◪-), $1o^{-6}$ M (-■-), $1o^{-5}$ M (-x-) and $1o^{-4}$ M (-◇-). (-o-): control, no addition of PTX.

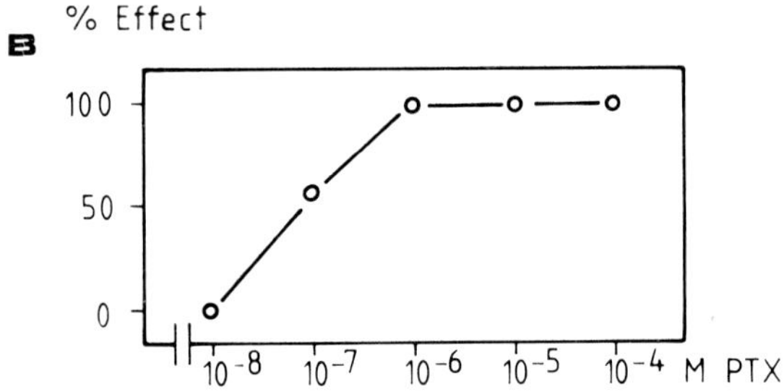

Fig.6 B) Dose response curve for the action of PTX as taken from the data shown in A). For this representation, the amplitudes of the observed Li^+-fluxes were taken (mean of the measured values, the values for 10^{-6} M PTX were taken as 1oo % response.

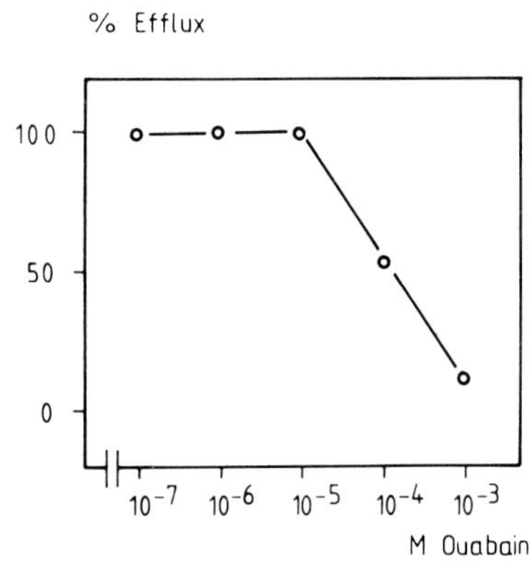

Fig.7 Dose response curve for the inhibition of the PTX-induced Li^+-efflux from acetylcholine receptor-rich membrane vesicles by Ouabain; the data were compiled as described in the legend to Fig.6 B). "O% efflux" was determined in the absence of effectors, "1oo% efflux" was taken as mean of the values in the presence of 10^{-6} M PTX. The concentration of Ouabain was varied in the presence of 10^{-6} M PTX. Ouabain is virtually not soluble in water. To avoid unspecific effects caused by addition of samples dissolved in ethanol, we gave Ouabain aliquots into test tubes and evaporated the solvent (ethanol). After addition of membranes

Fig.7,cont.: and PTX, the test tubes were vigorously shaken for 3o min at 2o °C. For this reason we cannot be sure about the actual concentrations of Ouabain in the assays.

PTX and Li^+-permeability of artificial liposomes

PTX did not affect the Li^+-permeability of artificial liposomes , whereas a large amount of trapped Li^+ was set free by the antibiotic Gramicidin or by osmotic shock. A typical experiment is shown in Fig.8.

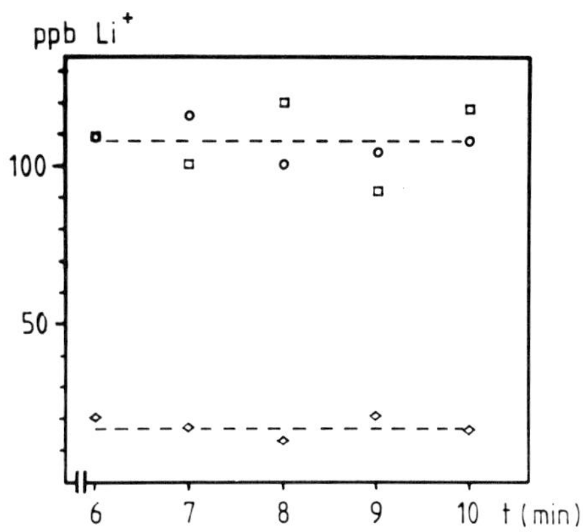

Fig.8 Li^+-efflux from artificial liposomes; liposomes were prepared and efflux was measured as described in Methods. (-o-): control, no further additions; (-□-): prior to dilution, liposomes were incubated 3o min at 2o °C with 2oo nM PTX or (-◇-): with 2oo µg/ml Gramicidin. Broken lines indicate the arithmetic mean of the Li^+-determinations (amounts to the same value in the control- and the PTX-curve).

Discussion

In the present work we have shown that PTX at very low concentrations (10^{-11} M) completely blocks specific voltage dependent ATX I-binding to whole crayfish nerves. Most likely this is due to the depolarization of the axonal membrane caused by increased sodium permeability rather than to a direct competition for binding sites. Furthermore, we found that PTX made membrane vesicles of both Torpedo electric organ and Astacus walking leg nerves permeable for $^{22}Na^+$.

At higher concentrations a PTX-stimulated Li^+-efflux from Torpedo electric organ vesicles could be observed as well (so far nerve membrane vesicles have not been investigated for PTX effects on Li^+-permeability). Habermann and coworkers reported a K^+-loss from red blood cells caused by PTX (13). The pores formed or mediated by PTX were permeable even for inositol but higher PTX-concentrations were required for maximal inositol-permeability. Since the Li^+-ion in its hydrate shell has a larger diameter than the hydrated Na^+-ion, these two observations may hint at a dose-dependent change of the PTX pore dimensions.

No effect of PTX on the permeability of artificial liposomes for both $^{22}Na^+$ and Li^+ could be demonstrated. This negative result shows that PTX per se does not form an ion-pore. Rather, membrane proteins seem to be required for the PTX effect. This requirement may be explained in two possible ways: firstly, PTX could perturb the lipid-protein interface thereby rendering it permeable for cations; secondly, PTX "opens" ionophors present in certain membrane protein complexes. Since the sodium channel blocker tetrodotoxin and the acetylcholine receptor regulated ion channel blocker triphenylmethylphosphonium do not inhibit PTX-induced cation flux, the ion channels of these two membrane protein complexes are unlikely candidates to be affected by PTX. Furthermore, PTX is not a cholinergic agonist itself since its effect on the cation permeability of Torpedo membrane vesicles cannot be blocked by the cholinergic antagonist α-bungarotoxin.

A third candidate for PTX-action is a protein complex present in all excitable cells: the Na/K-dependent ATPase. Habermann et al. showed that K^+-loss from erithrocytes is inhibited by Ouabain just like Li^+-efflux from acetylcholine receptor-rich vesicles in our experiments; therefore it was concluded that PTX renders the Na/K-dependent ATPase into an ionophor (14). We cannot exclude that a very small amount of ATPase is present in the acetylcholine receptor-rich membrane vesicles, though our preparation is rather pure and free of ATPase containing membranes. Nevertheless even a small number of ATPase-molecules could account for the observed effects, if only the permeability of the pore induced by PTX was high enough.

In conclusion, we can state that PTX needs membrane proteins to be effective, it does not act on pure liposomal membranes. Whether the membrane protein required is in fact the Na/K-dependent ATPase as hinted at by the inhibitory effect of Ouabain or whether this effect can be explained by a direct interaction of Ouabain with the PTX-induced pore remains to be investigated. In future we will perform experiments with vesicles containing membrane proteins but no Na/K-dependent ATPase, e. g. reconstituted membranes with affinity purified acetylcholine receptor, to exclude even traces of ATPase from the membrane system. Reconstitution of artificial liposomes containing other completely unrelated membrane proteins should indicate whether our hypothesis is correct that PTX causes leakiness in membranes by perturbing protein-lipid interfaces which have nothing to do with physiologically important ion channels.

Acknowledgement: This work was supported by the Deutsche Forschungsgemeinschaft, the Fonds der Chemischen Industrie and the Land Berlin.

References:

1) Moore,R.E., Bartolini,G.J. (1981): J.Am.Chem.Soc. 1o3, 2491-2494.

2) Cha,J.K., Christ,W.J., Finan,J.M., Fujioka,H., Kishi,Y., Klein,L.L., Ko,S.S., Leder,J., McWhorter,W.W.Jr., Pfaff, K.-P., and Yonaga,M. (1982): J.Am.Chem.Soc. 1o4, 7369-7371.

3) Dubois,J.M., Cohen,J.B. (1977): J.Pharm.Exp.Ther.2o1, 148-155.

4) Pichon,Y. (1982): Toxicon 2o, 41-47.

5) Béress,L., Zwick,J., Kolkenbrock,H.J., Kaul,P.N., and Wassermann,O. (1983): Toxicon in press .

6) Béress,L., Béress,R., Wunderer,G. (1975): FEBS Lett. 5o, 311-314 .

7) Stengelin,S., Hucho,F. (198o): Hoppe Seyler's Z.Physiol. Chem. 361, 577-585.

8) Denburg,J.L. (1972): Biochim. Biophys. Acta 282, 453-458.

9) Schiebler,W., Hucho,F. (1978): Eur.J.Biochem. 85, 55-63.

1o) Brunner,J., Skrabal,P., Hauser,H. (1976): Biochim.Biophys. Acta 688, 486-494.

11) Catterall,W.A. (198o): Ann.Rev.Pharmacol. Toxicol.2o, 15-43.

12) Lauffer,L., Hucho,F. (1982): Proc.Natl.Acad. Sci.USA 79, 24o6-24o9.

13) Ahnert-Hilger,G., Chhatwal,G.S., Hessler,H.-J., Habermann, E. (1982): Biochim. Biophys. Acta 688, 486-494.

14) Habermann,E. Chhatwal,G.S. (1982):Naunyn-Schmiedeberg's Arch. Pharmacol.319, 1o1-1o7.

III. Acetylcholine Receptors

VENOMS AND TOXINS IN NEUROCHEMICAL RESEARCH OF INSECTS

Heinz Breer

Universität Osnabrück
Abt. Zoophysiologie
D-4500 Osnabrück

Introduction

Insects are of enormous interest and importance as agricultural pest, as vectors in infectious and parasitic diseases, however also as beneficial pollinators and predators. In order to control harmful insects while sparing their beneficial relatives and other creatures, a more detailed knowledge of molecular processes in the nervous system of insects is urgently needed (1); as a sound basis for understanding the targets and mechanisms of action of insecticides and insect behaviour-modifying substances; and also to better understand the phenomena involved in selectivity of and resistance to these substances. Because many of these agents probably act on synapses it is especially important to unravel biochemical and pharmacological details of synaptic transmission in insects.

However, in contrast to the enormous progress that has been made in exploring the electrophysiological correlates of behaviour, insects have largely been ignored by neurochemists in the past and our knowledge in biochemistry of insect nervous system and especially of synapses and neuronal membranes has unfortunately lagged behind other animal groups (2). One main reason was probably the failure of subcellular fractionation techniques, so successfully used in research of the vertebrate nervous system, when applied to insects. The availablity of isolated nerve terminals from insects however would allow to analyse functionally intact synaptic components and compare their basic properties with that of vertebrate; another major advantage of synaptosomes would be the possibility to employ toxins and drugs with well established effects on vertebrate, with-

Toxins as Tools in Neurochemistry

out the problems of access and the influence of surrounding structures.

Subcellular fractionation of insect nervous tissue

We have developed a microscale procedure for subcellular fractionation of nervous tissue, designed specifically for use with insects (3). Briefly, ganglia were chopped and gently homogenized in isotonic solution with a microhomogenizer. The P_2-fraction, produced by differential centrifugation, was resuspended in isotonic sucrose solution containing Ficoll; during centrifugation of this mixture the particulate material accumulated on the bottom as well as on top of the gradient. The resulting floting pellicle was highly enriched in various synaptosomal markers and consisted of numerous synaptosomal profiles (Fig. 1). Detailed analysis of the synaptosomal fraction revealed that isolated nerve endings showed a considerable viability in vitro and retained most of the original morphology, i. e. they are obviously well sealed structures containing spherical vesicles, mostly of the electron-lucent type. Furthermore these preparations were greatly enriched in cholinergic elements, indicating that the insect synaptosomal fraction contained high proportion of cholinergic nerve endings, which are provided with the enzymatic and structural equipment for the synthesis and storage of ACh. Thus this fraction was obviously a good preparation to study various aspects of cholinergic transmission in insects. Since the use of venoms and toxins as investigate probes of neuronal function has been most profitable in vertebrate neurochemistry, we have used several neurotoxins in various approaches in order to evaluate properties of synaptosomes and neuronal membranes from insects.

Modulation of choline uptake by neurotoxins

Isolated locust nerve endings accumulated exogenous choline in a linear, temperature-sensitive process. Concentration dependent experiments showed saturation, and analysis of the transport kinetics revealed the existence of a high affinity system (4); high affinity choline transport is supposed to be unique to cholinergic nerve endings and to be the rate limiting, regulatory step in the synthesis of acetylcholine.

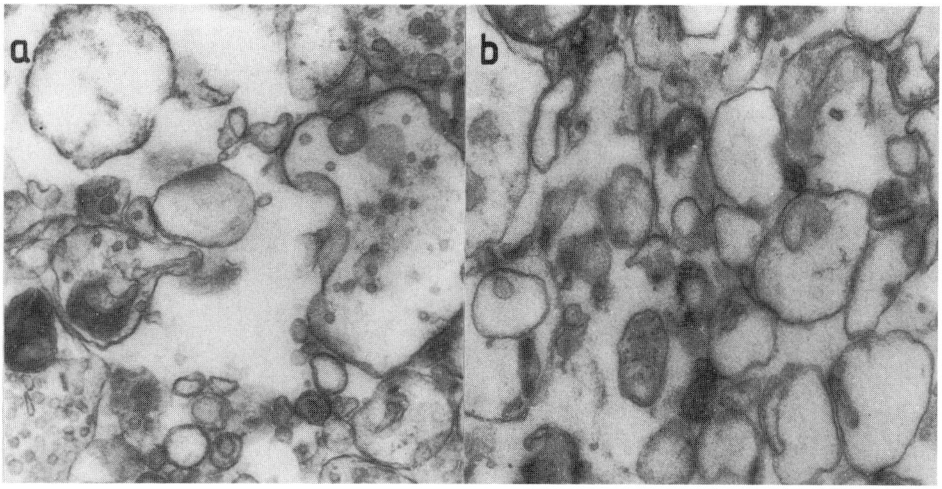

Fig. 1. Electronmicrographs of a synaptosomal fraction from insect ganglia (a) and membrane vesicles (b) derived from insect synaptosomes.

Various observations have led to the proposal that ion gradients, generated primarily by the Na^+, K^+-ATPase, are the immediate driving force for choline accumulation. However, the interpretation of transport studies on intact nerve terminals have always been complicated by intracellular metabolic activities and the sequestration of choline in subcellular compartments; therefore empty sacs which retain only the limiting membrane and which are provided with a different ion composition inside and outside would be much more appropriate. To get such a preparation, synaptosomes were lysed in buffered solution, the membranes were resealed in isotonic potassium buffer, and the resulting vesicle were obtained with artificially imposed ion gradients: $>[Na^+]_{out}$, $>[K^+]_{in}$. Those vesicles retained the capability to accumulate choline via a high affinity process, which is mainly driven by the Na^+-gradient (5).

In view of these observations it was of considerable interest, whether the choline transport in insect membrane vesicles was modulated by neurotoxins which are known to affect sodium ion channels in vertebrates.

The uptake of choline into membrane vesicle was significantly inhibited by veratridine, a plant alkaloid, which activate Na^+-channels; the inhibition effect increases progressively with time. In addition, it is shown that

tetrodotoxin, a specific blocker of Na-channels, which by itself has not much of an effect, almost completely prevented the inhibition by veratridine. The effects of vertridine and TTX are not restricted to choline only, but also the transport of GABA was effected. Thus a dissipation of the sodium gradient rather than effects on the membrane potential is obviously responsible for the reduced choline accumulation, caused by veratridine; this conclusion is further supported by the finding that the transport was inhibited to a similar extent, wether or not the proton ionophore CCCP, which would collapse any preexisting membrane potential, was present.
To further extent this aspect, the influence of venom from the scorpion, Tityus serrulatus was studied; this venom is known to contain components, which cause a delayed inactivation of Na^+-channels. If these toxins are active also on insect preparations, they should intensify the effect of veratridine: because a major consequence of a prolonged activation of the Na-ionophore would be a rapid increase in the intravesicular Na^+-concentration, and this of course would reduce the Na^+-gradient, which is supposed to energize the choline uptake.

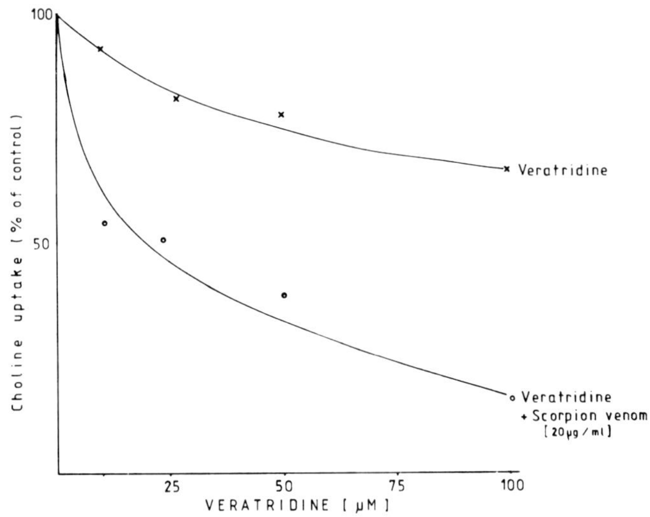

Fig. 2. Inhibition of choline uptake by different concentrations of veratridine in the absence or presence of scorpion venom.

As shown in Fig. 2 the combined actions of veratridine and scorpion venom caused a drastically reduced uptake rate for choline; the potentiating effect of scorpion venom was most obvious at low veratridine concentration.

The specificity of the venom effect was supported by the observation that TTX almost completely prevented the inhibitory influence. Analysis of the transport data in double-reciprocal plotes indicates, that the toxin-induced inhibition was of the non-competitive type; in contrast to HC-3 which competitively inhibits the high affinity choline uptake. These results emphasize the different inhibitory action of HC-3 and the natural toxins and provide additional support for the notion that choline uptake is a sodium gradient coupled process. As a more direct approach to analyse for the presence and functionability of sodium ion channels in insect membranes, the sodium flux into membrane vesicles was studied using tracer ^{22}Na in the medium; in preliminary experiments it could be shown that the Na^+-flux was increased approximately 5-fold over basal uptake when 100 µM veratridine was present; this effect strongly suggests that voltage-dependent Na^+-channels are operating in membrane vesicles, and can be activated _in vitro_.
The application of various toxins thus has dicisively contributed to our understanding that the high affinity choline transport in insects is mainly driven by a sodium gradient and that membranes of insect nerve endings are obviously endowed with channels, which show some pharmacological properties similar to their counterparts in vertebrates.

Toxin induced transmitter release

Isolated nerve endings from insect are provided with the enzymatic machinery for the synthesis of ACh and with organelles for its storage. Release of neurotransmitters from preloaded insect synaptosomes was studied using a very small and simple perfusion technique, which prevents re-uptake and allows a continuous measurment of the released material from small samples of synaptosomes. The addition of 100 µM veratridine to isotonic perfusion medium induced a considerable release of labelled material. The released labelled compounds were identified as mostly acetylcholine. The extent of release was shown to be concentration dependent; maximal release was achieved at about 100 uM veratridine. In order to specify the veratridine-action (Fig. 3), synaptosomes were rapidly perfused with sodium-free Ringer-solution containing depolarizing levels of veratridine. Under these

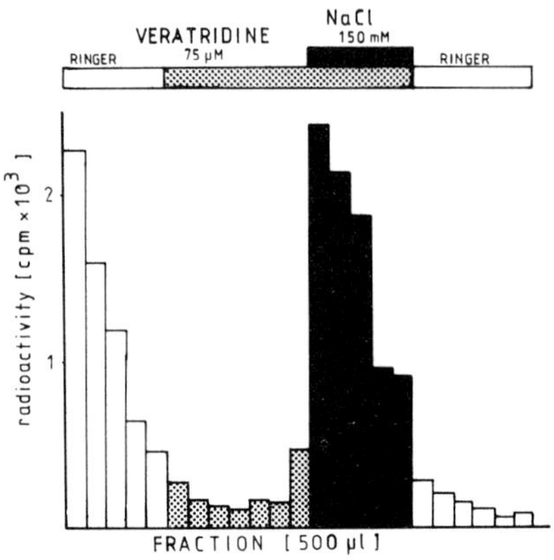

Fig. 3. Transmitter-release from preloaded insect synaptosomes by veratridine in the absence and presence of sodium chloride.

conditions no release was induced. Changing to NaCl containing Ringer at this point increased the efflux rate drastically. After the initial peak, the release rate decreased to the baseline. These results clearly demonstrate the sodium-dependence of veratridine action in insects.
There are a few toxins which have recently been used as releasing agents in vertebrates. The best characterized is a component of the venom from the black widow spider Latrodectus mactans: a protein named latrotoxin (LTX); it has been shown to cause an increase in the frequency of miniature endplate potentials, a depletion of synaptic vesicles from the terminal and ultimately block neuromuscular transmission (6). It therefore was of considerable interest to apply latrotoxin, which originate from an insect predator to insect synaptosomes. Preloaded synaptosomes were immobilized in the perfusion device and washed. A typical pattern of efflux of radiolabelled ACh from synaptosomes is shown in Fig. 4. After application of 5 µg LTX in isotonic solution, the release rate increased and reached a maximum of about 10 fold: it then gradually returned to the base line. In experiments using Ca^{2+}-free perfusion media, LTX induced a similar increase in release. Very recently we have found that another venom-component, different from LTX, the 'fraction E' which is obviously ineffective in vertebrates (7), has an very pronounced effect on transmitter release

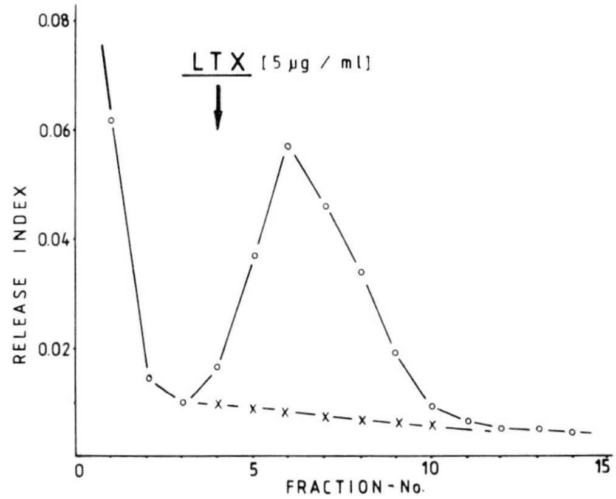

Fig. 4. Release of acetylcholine from insect synaptosomes induced by latrotoxin.

from insect synaptosomes. This result may be considered as a clue to look for insect-specific components in the spider venom; since insect belong to the natural prey of these spiders.

In the hemolymph of certain insects, beetles of the genus Leptinotarsa, Dr. Hsiao found a rather toxic protein, called Leptinotarsin (8), which also caused on increase in the frequency of miniature endplate potentials. We have applied this insect toxin to insect synaptosomes, and the releasing activity of this toxin was indicated by a significantly increased levels of radioactivity found in the eluate fractions after toxin treatment; similar to LTX, the action of Leptinotarsin seems also very little Ca^{2+}-dependent.

All these experiments have shown that insect synaptosomes can be used to study transmitter release in insects, that the release process is basicly similar to what is known from vertebrates and that toxins exist to induce transmitter release in insects.

Toxins as probes for putative cholinergic receptors

It is meanwhile well established that ACh is a major excitatory neurotransmitter in the CNS of insects; however, only little is known about the properties of cholinergic receptors in insects. Such information however would be of great interest not only because of its toxicological implications, but also because it may shed some light on comparative and phylogenetic aspects, concerning the evolutionary stability of the acetylcholine receptor.

The neurotoxin α-bungarotoxin, a polypeptid, purified from the venom of the snakes Bungarus multicinctus which produces a postsynaptic neuromuscular blockade by acting on the acetylcholine receptors has been extensively employed as very specific label to study the nicotinic acetylcholine receptors of mammalian skeletal muscle and fish electroplax. However, there is a controversy as to wether the α-toxin binding component in neuronal preparations is really a functional receptor.

Nevertheless we have employed this toxin as well as the powerful muscarinic antagonist quinuclidinyl bezilate (QNB) to study the putative nicotinic and muscarinic receptors in the ganglia of insects. When comparing the number of binding sites in locust and mouse, it is obvious that the insect central nervous tissue is endowed with a much higher concentration of putative receptors; furthermore in the locust ganglia the BGT-binding components are predominant, the concentration being more than one order to magnitude higher than QNB binding sites. In contrast, muscarinic receptors are the predominant types in the mouse brain. The observation that both insects and mammals, which are far apart on the evolutionary scale, contain in their nervous tissue nicotinic and muscarinic binding components, suggests that they share a common ancestor and that the 'classical' dichotomy of the cholinergic systems evolved early during evolution. The binding properties of ganglionic preparations from insects for α-bungarotoxin were determinded in filter binding assays. The binding isotherm for the toxin shows saturation in the nanomolar range and analysis of binding data in a Scatchard plot revealed an apparent dissociation constant of about 10^{-9} M and a maximal number of binding sites of more than 2000 fmol/mg protein (Fig. 5). A Hill coefficient of 0.98 characterized the receptor-toxin-interaction as an essentially non-cooperative binding. The bin-

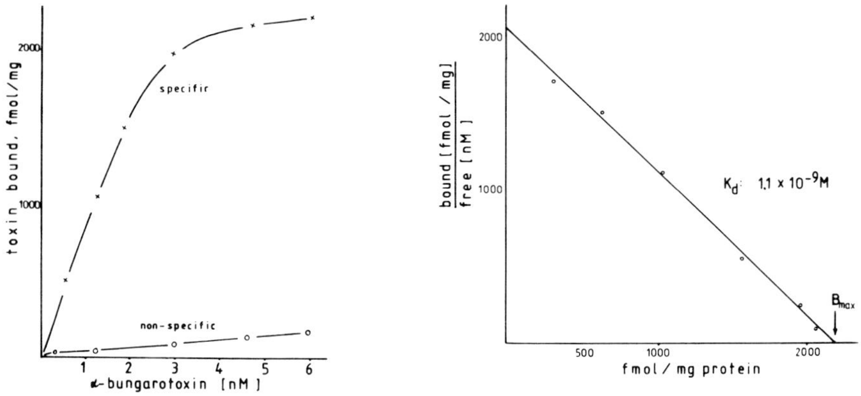

Fig. 5. Binding isotherm and scatchard plot of α-bungarotoxin binding to ganglionic preparations from locusts.

ding of α-BGT to its specific sites in insect neuronal membranes, differs from binding to electroplax-membranes in be-ing reversible; however the halftime for dissociation of the toxin-receptor complex is sufficiently long for accurate assays.

In order to establish the cholinergic nature of binding the cholinergic nature of binding entities, the specificity and pharmacological properties of α-bungarotoxin binding sites were studied: In competitive equilibrium studies,where aliquotes of membrane suspension were incubated with radio-labelled toxin and cholinergic agonists and antagonists at multiple concentrations, it was demonstrated that cold native toxin is the most potent competitor for labelled toxin; nicotinic and α-tubocurarine are the next potent agonists and antagonists (9). The most potent cholinergic agonist in vivo acetylcholine, itself was relatively ineffective in blocking toxin sites on coincubation, its competition effectiveness however, was markedly increased when preincubated with binding sites prior to addition to labelled toxin. These results confirm the cholinergic nature of α-toxin binding sites.

We have found that treatment of membrane fractions from locust ganglia with deoxycholate was relatively efficient in solubilizing the toxin-binding components; the binding parameters for the solubilized components were similar to those obtained for membrane bound. Affinity-chromato-graphy has been a powerful tool for the purification of ACh-receptors using snake toxins as affinity ligands. We have incubated deoxycholate extracts from locust CNS with Sepharose 4 B covalently linked with α-toxin. The bulk of the applied protein was eluted in the washingsteps. The adsorbed material was recovered by elution with tubocurarine; these fractions contained the specific toxin-binding sites.

Autoradiographic studies with radiolabelled toxin were performed by various groups (10) in order to demonstrate in situ the presence of putative receptor in the CNS of insects. Autoradiography of sections treated with labelled toxin revealed that the toxin-binding sites are heavily concentrated in the neuropile. Region of perikarya, which are supposed to be devoid of synapses in insects, as well as other synapses free areas, e. g. the chiasmata, show considerably fewer silver grains than the neuropile. The selective labelling pattern may indicate that the toxin binds to synaptic receptors.

Recently Sattelle's group have clearly demonstrated that nanomolar concentrations of bungarotoxin are able to block transmission at synapses between cercal afferent neurons and giant interneurons in the 6th abdominal ganglion of insects (11), indication that the α-toxin-binding component is a constituent of an ACh receptor that is functional in this monosynaptic pathway. Thus the results of experiments using α-toxin in various approaches supports the concept that the central nervous system of insects do contain acetylcholine receptors, and suggests the equivalence of a toxin binding site and a true nicotinc AChR.

Conclusion

It has been shown that synaptosomes and neuronal membranes can be isolated from the central nervous tissue, which provide a very useful in vitro system to study neurochemical aspect of synaptic transmission in insects. Essential prerequisites for progress in this research are the availability and application of specific toxins, as indispensable tools to establish the properties of functional entities in insect neuronal membranes.

Such approaches with toxins as specific probes not only will help to under-

stand molecular processes in insect nervous tissue, but also may contribute to develop and standardize simple assay systems for screening the targets and mode of action for modified or newly synthetized chemicals, putative insectizides. Thus more rational approaches for pesticide discovery might be possible than has been typical in the past; the evaluation of critical sites for chemicals in insects and especially comparative aspects, can finally lead to more selective and safer compounds.
In this context it is of particular interest that toxins have recently been discovered in the venom from scorpions and spiders, which are obviously highly selective for arthropods and have no detectable effects on vertebrates. Furthermore in the venom of solitary wasps, toxins have been detected that are paralytic only to a certain group of insects (the larve of butterflies) but not to other insect orders. These findings clearly show that nature has already realized the concept of toxin-selectivity and a comprehensive analysis of target and mode of action of those compounds may evaluate their potential as chemical models for novel control agents, as has already witnessed by the development of cartap, a derivatized nereistoxin, from marine annelids.

References

1. Corbett, J. R.: Trends in Neurosci. 3, I-IV (1980).
2. Lafon-Cazal, M.: Annêe Biol. 17, 489-528 (1978).
3. Breer, H.: Neurochemistry International 3, 155-163 (1981).
4. Breer, H.: J. Neurobiology 13, 107-117 (1982).
5. Breer, H.: FEBS-Letters 153, 278-283 (1983).
6. Meldolesi, J.: J. Neurochemistry 38, 1559-1569 (1982).
7. Frontali, N., Ceccarelli, B., Gorio, A., Mauro, A., Siekwitz, P., Tzeng, H. Ch., Hurlbut, W. P.: J. Cell Biol. 68, 462-479 (1976).
8. McClure, W. O., Abbott, B. C., Baxter, D. E., Hsiao, T. H., Satin, L. S., Siger, A., Yoshino, J. E.: Proc. Natl. Acad. Sci. 77, 1219-1223 (1980).
9. Breer, H.: Neurochemistry International 3, 43-52 (1981).
10. Schmidt-Nielsen, B. K., Gepner, J. I., Teng, N. N. H., Hall, L. M.: J. Neurochem. 29, 1013-1029 (1977).
11. Harrow, I. D., Hue, B., Pelhate, M., Sattelle, D. B.: J. Physiol. 295, 63-64 (1979).

ARE THERE NICOTINIC ACETYLCHOLINE RECEPTORS IN INVERTEBRATE GANGLIONIC TISSUE?

Gregor Fels[1], Heinz Breer[2] and Alfred Maelicke[1]

[1]Max-Planck-Institut für Ernährungsphysiologie, Rheinlanddamm 201, D-4600 Dortmund, FRG

[2]Universität Osnabrück, Zoophysiologie, Seminarstr. 20, D-4500 Osnabrück, FRG

Abstract

We have employed monoclonal antibodies raised against the acetylcholine receptor from Torpedo marmorata to test whether there exist structural similarities between this nicotinic receptor and the α-bungarotoxin binding component of head ganglionic tissue of the locust Locusta migratoria. As shown by the positive ELISA tests with several antibodies and the typically cholinergic competition pattern of antibodies directed against the ligand binding region of the receptor, Locusta ganglionic tissue indeed contains sites exhibiting specific nicotinic binding characteristics.

Introduction

Ever since binding sites for α-bungarotoxin (αBTX) have been discovered in the brain (1,2), it has been debated whether these represent true nicotinic acetylcholine receptors such as found in muscle endplates or not. This question still remains controversial: (i) Although nicotinic cholinergic synapses have been demonstrated in various areas of the brain by electrophysiological methods (3,4), αBTX does not gener-

Toxins as Tools in Neurochemistry

ally block transmission at these sites (4-6). (ii) Direct biochemical studies are hampered by the lack of sufficiently defined brain preparations that ideally should contain only a single type of (nicotinic) synapse. (iii) At present the best characterized preparations for biochemical and immunological studies are derived from sympathetic ganglia or related cells (7,8) and therefore do not directly address the problem under discussion. As a further weekness of many previous studies, these were focussed on either ligand binding or immuneserum cross-reactivities but did not employ both methods in combination.

Prompted by the availability of monoclonal antibodies against well defined regions (including the two nicotinic ligand binding regions) of the acetylcholine receptor from Torpedo marmorata (9), we have begun to tackle this problem using a combination of assays and several different cell preparations with more or less similarities to brain tissue. Here we describe the basic assays and also provide an initial characterization of the αBTX-sites found in the head ganglionic tissue of the orthopterian locust Locusta migratoria (10). Based on these data, these αBTX-sites indeed belong to true nicotinic receptors.

Materials and Methods

Torpedo membrane fragments and the head ganglionic tissue of Locusta migratoria were prepared as described previously (11, 12). The stable clones of antibody producing cells were produced and characterized by Dr. D. Watters of our laboratory (9). ^{125}I-α-bungarotoxin (17,1 μCi/μg) was obtained from NEN, all other ligands and buffer materials were from standard sources (9,13).

For the ELISA assays (14) an assay kit (NEN) was used employing peroxidase-labelled anti-mouse IgG and o-phenylenediamine as the major reagents. Gilford cuvettes (cuvette paks) were coated with 200 μl suspensions of *Torpedo* membranes (dilution 1:200) or *Locusta* membranes (dilution 1:100) in 0.1 M sodium carbonate buffer pH 9.0 overnight at 4°C. If fixation of the membranes was required, the cuvettes were filled with 0.25% glutaraldehyde in H_2O and left for 30 min at room temperature. After removing the excess of antigen by washing twice with washing buffer, 200 μl of cell culture supernatants were applied into the cuvettes and incubated for 3 h at 37°C. After two additional washes with washing buffer, peroxidase labelled anti-IgG (1:1000 in dilution buffer) was added and incubated for 1 h at 37°. The plates were than washed three times with washing buffer followed by two washes with H_2O and then were incubated with the substrate solution prepared according to the NEN assay kit procedure. After further incubation for 30 min at 37°C, the reaction was stopped by addition of 100 μl 4.5 M H_2SO_4 and the absorbance at 490 nm was read in a Gilford EIA manual reader.

Serial dilution (2^1-2^{10}) of culture supernatants were achieved as follows: First, the 10 cuvettes of a pak were loaded with 200 μl of PBS (phosphate buffer saline: 8.0 g NaCl, 0.2 g KH_2PO_4, 2.9 g Na_2HPO_4 x 12 H_2O, 0.2 g KCl, 0.2 g NaN_3 in 1 l H_2O, pH 7.4) or the appropriate ligand in PBS. Then 200 μl of cell culture supernatant was added to the first well and mixed. 200 μl of this mixture were then transfered to the second well. This procedure was repeated for each well, the 200 μl of the last well being discarded.

For ^{125}I-αBTX binding assays flexible microtiter plates (Dynatech) were coated with 150 μl of membrane suspension as described for the ELISA-assay. Excess antigen was removed by washing with PBS and, after fixation, the wells were preincubated for 3 h at room temperature with PBS (control curve) or antibody solution, both in the presence of 1% BSA. Then either 5, 10, 20, 30, 40 or 50 µl of this incubation mixture was replaced by an equal volume of ^{125}I-αBTX solution (365 µCi/10ml H_2O) in the respective wells. After another 4 h incubation at room temperature the incubation mixture was removed and the wells were extensively washed with PBS and counted in a γ-counter.

Results and Discussion

The monoclonal antibodies employed in these studies have been described in detail elsewhere (9). Here it should suffice to point out that they interact with antigenic determinants that exist on both the purified and the membrane-bound receptor from Torpedo marmorata electric organ. Furthermore, antibodies 1-6 (see Table I) are directed against the nicotinic ligand binding regions of Torpedo receptor. All studies described here were performed in parallel with Torpedo membranes (11) and Locusta membranes (12). Since the densities of αBTX-binding sites of the Locusta preparations are much lower than those of the Torpedo preparations, fixation with glutaraldehyde of the former was required in some experiments (15).

Assay Conditions - To determine the minimal concentration of Locusta membranes required to completely coat the cuvettes, increasing aliquots of a Locusta membrane suspension (2 mg/ml) were applied and were assayed with a constant concentration of antibody mAb5. As shown in Fig. 1, the membrane suspension could be diluted 1:100 to still produce maximal assay readings.

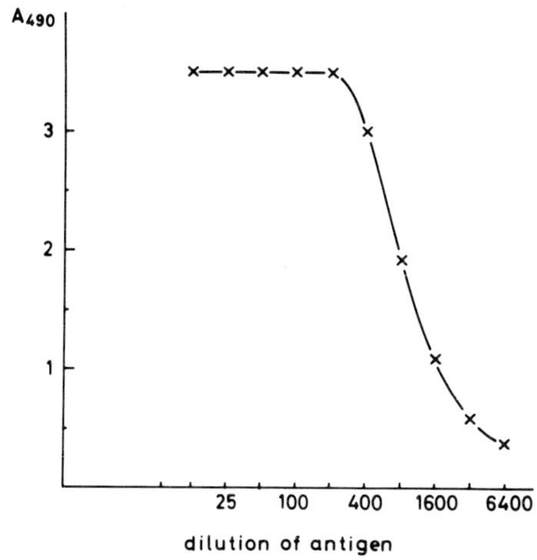

Fig. 1. Determination of the concentration of Locusta ganglionic tissue (2 mg Protein/ml) required to coat the ELISA cuvettes. 200 µl of solutions of the tissue (in 0.1 M $NaHCO_3$, pH 9) at different concentration were allowed to bind over night at 4 °C followed by the ELISA-assay procedure as described in Material and Methods.

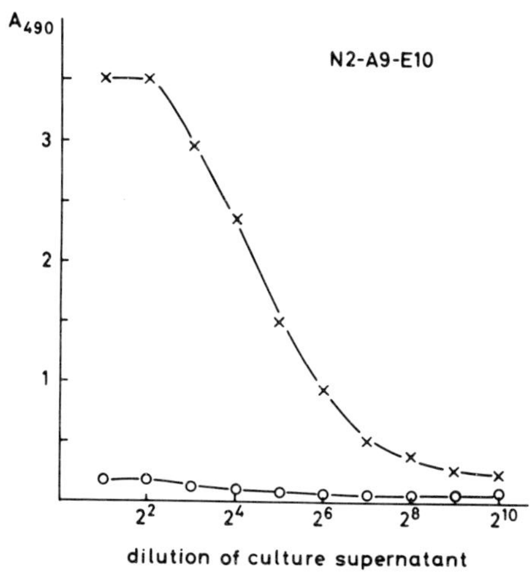

Fig. 2. Serial dilution (1 in 2, dilution in PBS) of 200 µl cell culture supernatant from clones producing antibody mAb5 in ELISA cuvettes coated with Locusta membranes (x) or BSA (1% in PBS) (o), both without fixation.

TABLE I

Cross-reactivity of anti-*Torpedo* receptor antibodies with ganglionic tissue from *Locusta migratoria* as determined by ELISA

Hybridoma clone		ELISA-values with *Locusta* tissue		ELISA-values with *Torpedo* membranes without fixation
		after fixation	without fixation	
mAb	internal code	A_{490} nm	A_{490} nm	A_{490} nm
1	1-A4-C8	0.12	nt	2.7
1a	1-A4-A5	2.2	0.44	1.9
2	2-B4-12	0.39	0.20	3.5
3	5-6/1D6	0.38	0.18	3.5
4	1-C3-C9	0.10	nt	1.9
5	N2-A9-E10	2.0	0.35	2.6
6	XR6-G10-B3	0.33	0.21	3.5
8	NA19-E11	0.25	0.22	3.5
9	1-B6-B10	0.21	0.19	3.5
10	XR33-E7	0.21	0.17	3.5
12	XR64-B2	0.33	nt	3.5
13	XR41-H12	0.30	nt	3.5
14	NB20-D3	0.10	nt	3.5
15	NA23-F10	0.29	nt	3.5
17	XR26-D2	0.28	0.19	3.5
18	XR46-F1	1.1	nt	3.5
20	H5-H12	0.13	nt	3.5
	control	0.13	0.11	0.11

Incubation with cell culture supernatant (200 μl) or ammoniumsulfate precipitated and dialysed monoclonal antibody solutions (20 μl in 180 μl PBS) for 1 h instead of the usual 3 h at 37 °C. (n.t. = not tested).

As shown representatively for a cross-reacting antibody in Fig. 2, ELISA assays of Locusta membranes produced the same dilution pattern previously observed for Torpedo receptor (9,14). At high concentration of antibody (low dilution), the immobilized antigen is saturated and the absorbance readings are maximal. This section of the plot is followed by the dilution range in which decreasing concentrations of antibody will result in a decreasing absorbance. The specificity of the assay is demonstrated by the low background readings (1) when non cross-reacting antibodies were employed (not shown) and (2) when the membrane suspension was replaced by bovine serum albumin.

Table I lists the results of ELISA assays performed with membrane preparations from the two sources. Of 17 antibodies studied, three (mAbs 1a, 5 and 18) clearly reacted also with Locusta membranes. Several other antibodies also showed positive but much smaller assay values. Since the cross-reactivity of monoclonal antibodies can only be complete or zero, the lower assay values as compared to the Torpedo preparation indicate lower affinities of binding of the antibodies to the Locusta membranes. At least for the fixed membranes they are probably not due to a lower concentration of antigenic sites in the cuvettes because comparable readings with both receptor preparations were obtained for at least two antibodies (mAbs 1a, 5). It should be noted, however, that because fixation may alter the antigenic properties of a preparation, the data of Table I only suffice to establish cross-reactivity of a few of the antibodies studied.

Antibody Competition with α-Neurotoxins - Since mAb5 belongs to the antibodies directed against the ligand binding region of Torpedo receptor (9), it was interesting to test whether binding of this antibody to Locusta membranes is also competitive with α-neurotoxins. Two sets of experiments were per-

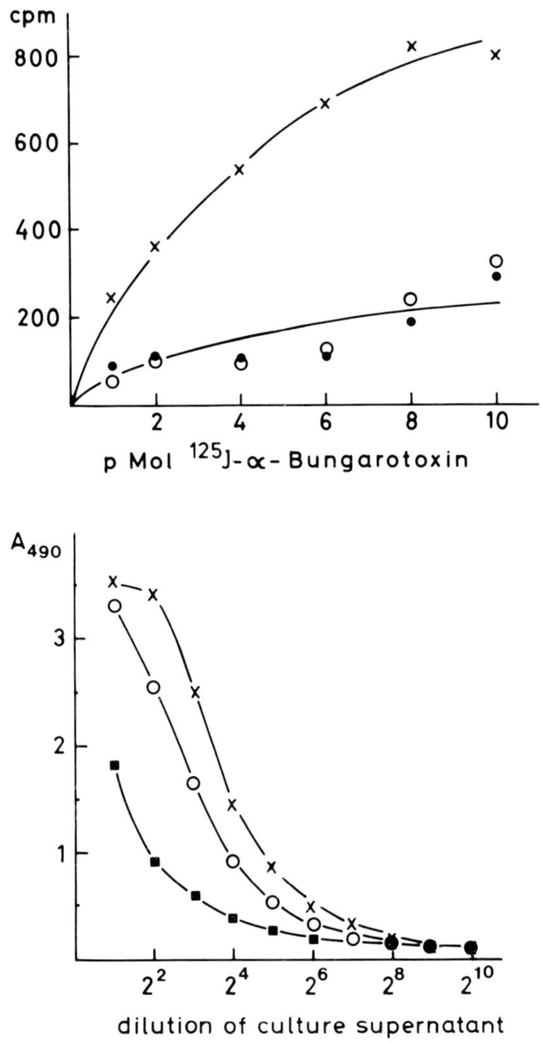

Fig. 3. upper: Binding of ^{125}I-α-bungarotoxin to immobilized *Locusta* ganglionic tissue with (o,●) and without (x) preincubation of antibody.

lower: ELISA assay with mAb5 and fixed *Locusta* membranes: serial dilution of the antibody in buffer (x) or a solution of α-cobratoxin of 5 x 10^{-6}M (o) or 1 x 10^{-4}M (■) in PBS after 2 h preincubation with toxin.

formed (Fig. 3): In the first, binding of ^{125}I-αBTX to immobilized *Locusta* membranes was studied in the absence and presence of mAb5. In the second, the ELISA assay was employed to study binding of mAb5 to *Locusta* membranes in the absence and presence of α-cobratoxin. Both experiments showed competition between antibody and toxin for their sites at the *Locusta* ganglionic tissue.

Similar results were obtained for mAb1a while there was no competition of mAb18 (an antibody not directed against the nicotinic ligand binding regions of *Torpedo* (9)) and α-neurotoxins for binding to the *Locusta* membranes. We conclude that the toxin binding regions of *Torpedo* and *Locusta* membranes must have common structural determinants.

Antibody Competition with Small Cholinergic Ligands - The specificity of monoclonal antibodies directed against the nicotinic ligand binding regions of acetylcholine receptors can be further evaluated by competition experiments employing a variety of small cholinergic and muscarinic ligands (9). We have therefore expanded binding studies with mAb5 and *Locusta* ganglionic tissue to include competition experiments with acetylcholine, tubocurarine, decamethonium and atropin (Fig. 4, Table II). In *Locusta* membranes not fixed with glutaraldehyde all four ligands appear to be of almost equivalent potency at preventing antibody binding, although the effect was only observed at rather high concentrations of ligand (10^{-2}-10^{-3}M). In contrast, local anesthetics at comparably high concentrations (10^{-2}M) were found to be ineffective (data not shown). The experiments were repeated with *Locusta* membranes fixed to the plastic surface with glutaraldehyde. Under these conditions we not only observed the expected higher assay readings but there was also a distinct order of competition for the cholinergic ligands (Fig. 4, Table II). In addition in contrast to the findings with unfixed tissue the muscarinic anta-

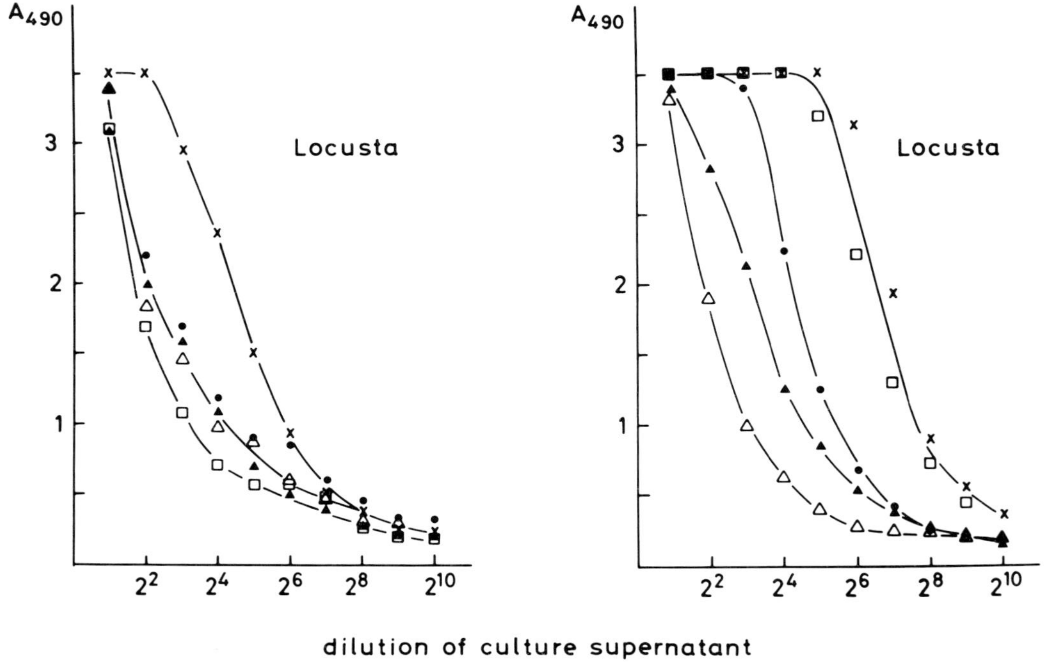

Fig. 4. ELISA assay with mAb5 and (left) unfixed or (right) fixed Locusta membranes: serial dilution of the antibody in buffer (x), 10^{-2}M acetylcholine (●), 10^{-3}M tubocurare (▲), 10^{-2}M decamethonium (△) or 10^{-3}M atropin (□). Ligands were dissolved in PBS and incubated with the immobilized Locusta membranes for 15 min prior to the serial dilution with mAb5.

gonist atropin did not reduce binding of mAb5 to the Locusta membranes.

To test the significance of ligand competition patterns after tissue fixation, a similar set of experiments was also performed with Torpedo membranes (Fig. 5, Table II). Since the receptor concentration of this tissue is high and, thus, does not limit the significance of ELISA assays performed with non-fixed material, the observed difference in the competition pattern with glutaraldehyde fixed material must have other causes: Obviously, membrane fixation by glutaraldehyde does affect the structural integrity of the ligand binding region and, in particular, has a differential effect on the binding of small cholinergic ligands. Thus, although we can demonstrate that the competition patterns between antibodies and small nicotinic ligands are similar for both Locusta and Torpedo membrane preparations, their quantitative assessment must await further studies. In addition, the commonly employed method of antigen fixation by glutaraldehyde should not be overlooked as a possible source of serious errors.

TABLE II

Order of competition of the monoclonal antibody mAb5 with different ligands for binding to Locusta ganglionic tissue and Torpedo membranes, respectively, as deduced from Fig. 4 and Fig. 5.

antigen	Ligands arranged by their order of competition
Locusta	atropin > ACh , deca , tubo > control
Locusta after fixation	deca > tubo > ACh > atropin , control
Torpedo	toxin > ACh > deca > tubo > atropin , choline , control
Torpedo after fixation	toxin > deca > ACh , tubo > atropin , choline , control

ACh = acetylcholine, deca = decamethonium, tubo = tubocurare
Commas separate ligands of equivalent competitive potency.

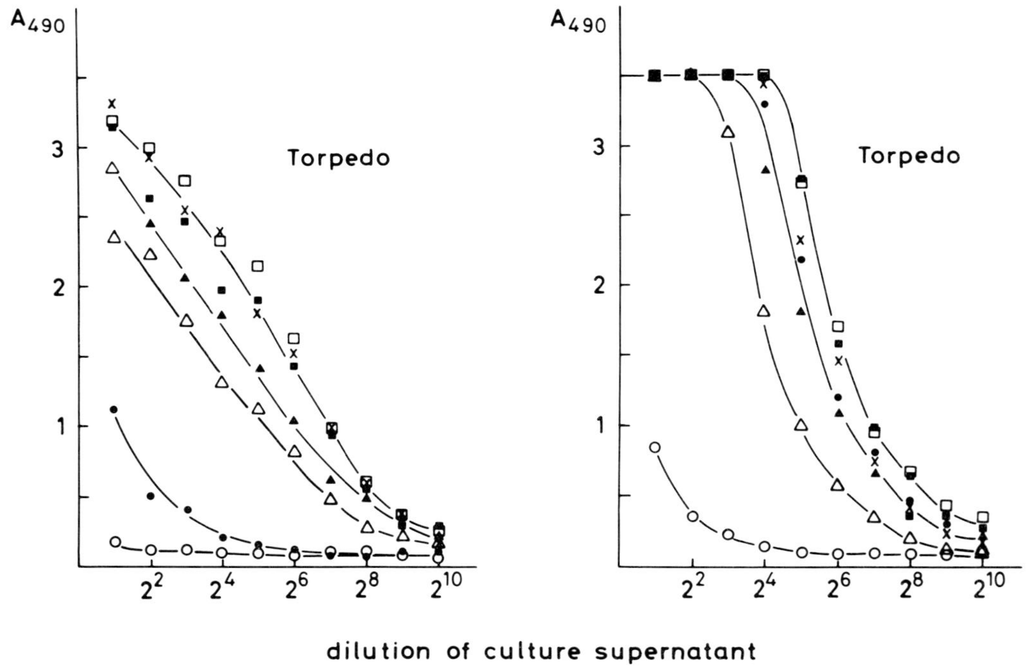

Fig. 5. ELISA assay with mAb5 and (left) unfixed or (right) fixed *Torpedo* membranes: serial dilution of the antibody (1:10 diluted in PBS) in buffer (x), 10^{-2}M acetylcholine (●), 10^{-3}M tubocurare (▲), 10^{-2}M decamethonium (△), 10^{-3}M atropin (□), 10^{-2}M cholin (■) or 10^{-5}M α-cobratoxin (o). Ligands were dissolved in PBS and incubated with the immobilized *Torpedo* membranes for 15 min (with toxin for 2 h) prior to the serial dilution with mAb5.

Concluding Remarks

Although limited in their scope, the present results point to considerable similarities in the structure of the αBTX-binding components in <u>Locusta</u> ganglionic tissue and <u>Torpedo</u> electroplaques: (i) Both tissues bind α-neurotoxins and typically nicotinic ligands in a competitive fashion (10,16). (ii) Several of the monoclonal antibodies directed against <u>Torpedo</u> acetylcholine receptor also bind to <u>Locusta</u> membranes. (iii) Two of these antibodies appear to be directed against the ligand binding regions of both <u>Torpedo</u> and <u>Locusta</u> receptor: Their binding to the antigen preparation is reduced in the presence of α-neurotoxins or small cholinergic ligands. Taken together, these results demonstrate the existence of a site(s) exhibiting specific nicotinic binding characteristic in <u>Locusta</u> head ganglionic tissue.

Acknowledgement

We thank Heike Rimpel for excellent technical assistance. This work was supported by the Deutsche Forschungsgemeinschaft, Schwerpunktprogramm "Biochemie des Nervensystems".

References

1. Salvaterra, P.M., Moore, W.J.: Biochem. Biophys. Res. Commun. <u>55</u>, 1311-1318 (1973).
2. Schleifer, L.S., Eldefrawi, M.E.: Neuropharmacology <u>13</u>, 53-63 (1974).
3. Oswald, R.E., Freeman, J.A.: Neuroscience <u>6</u>, 1-14 (1981).
4. Schmidt, K., Hunt, S., Polz-Tejera, G.: in Neurotransmitters, Receptors and Drug Action (Essman, W.B., ed.) pp. 1-65, Spectrum, New York 1980.
5. Brown, D.: Adv. Cytopharmacol. <u>3</u>, 225-230 (1979).

6. Misgeld, U., Weiler, M.H., Bak, I.J.: Exp. Brain Res. 39, 401-409 (1980).
7. Carbonetto, S.T., Fambrough, D.M., Muller, K.J.: Proc. Natl. Acad. Sci. U.S.A. 75, 1016-1020 (1978).
8. Patrick, J., Stallcup, W.B.: Proc. Natl. Acad. Sci. U.S.A. 74, 4689-4692 (1977).
9. Watters, D., Maelicke, A.: Biochemistry 22, 1811-1819 (1983).
10. Breer, H.: Neurochemistry International 3, 43-52 (1981).
11. Fels, G., Wolff, E.K., Maelicke, A.: Eur. J. Biochem. 127, 31-38 (1982).
12. Breer, H.: Neurochemistry International 3, 155-163 (1981).
13. Rüchel, R., Watters, D., Maelicke, A.: Eur. J. Biochem. 119, 215-223 (1981).
14. Watters, D., Maelicke, A.: Neuroreceptor (Hucho, F., ed.) pp. 329-340, Walter de Gruyter & Co., Berlin 1982
15. Reichlin, M.: Methods in Enzymology (Vunakis, H.V., Langone, J.J., eds.) Vol. 70, pp. 159-165, Acad. Press, New York 1980
16. Maelicke, A., Fulpius, B.W., Reich, E.: Handbook of Physiology, The Nervous System I, 14, 493-519 (1977).

THE SITES OF NEUROTOXICITY IN α-COBRATOXIN

Bakshy A. Chibber[1], Brian M. Martin[1], Malcolm D. Walkinshaw[2], Wolfram Saenger[3] and Alfred Maelicke[1]

[1]Max-Planck-Institut für Ernährungsphysiologie, Rheinlanddamm 201, D-4600 Dortmund, FRG

[2]Department of Chemistry, Edingburgh University, West Mains Road, Edingburgh, Scotland

[3]Institut für Kristallographie, Freie Universität Berlin, Takustr. 6, D-1000 Berlin 33, FRG

Summary

We have chemically modified groups of amino acids in the sequence of α-cobratoxin and have studied the obtained derivatives as to their affinity of binding to the acetylcholine receptor protein from Torpedo marmorata. We find that neither a single positively charged residue nor tyrosine nor the integrity of the disulfides is absolutely essential for toxicity. Furthermore, the single tyrosine and the area around the extra disulfide do not participate in binding to the receptor. Together with previous findings on this interaction, this suggests a multipoint attachment of toxin and receptor involving several locally separate structural elements of the toxin.

Introduction

Soon after the first crystal structure of a curaremimetic snake neurotoxin was solved (1,2), it has been called a "Rosetta Stone" and was believed to provide "a broad understanding of the structure-function relationship" with "the reactive site

Toxins as Tools in Neurochemistry

of the molecule persuasively defined" (3). With two more three-dimensional structures of snake neurotoxins solved (4,5), with more amino acid sequences, toxicity/binding studies of chemically modified toxins and even natural and synthetic peptides with residual toxicity (6-8), there still has not been defined a minimal number and/or region of functional amino acid residues necessary to ellicit toxicity and to permit binding to the receptor.

The failure of previous biochemical studies to provide experimental support to the proposed structure-function relationships (1,3,9) has prompted as to tackle this problem in a more rigorous way than was previously attempted: In contrast to earlier studies of this kind we have modified whole groups of amino acids instead of single amino acid residues. The homogeneity of the derivatives obtained was established by, among other methods, peptide mapping and amino acid analysis. The biological function was assayed by binding studies to the purified receptor protein. Based on the known properties of the toxin-receptor interaction (10,11), our studies focussed on the amino acid residues carrying positive charge and on the cystines. We can show that neither one of the positively charged residues nor tyrosine nor the integrity of the disulfides is absolutely essential for the interaction with the receptor protein.

Materials and Methods

These are described in detail in the full publications on this subject (8,12).

Results

Modification of the Free Amino Groups - The structure of α-cobratoxin contains six free amino groups (N-terminal and the ε-amino groups of the five lysine residues) (Fig. 1). These were modified as follows: (i) All amino groups were converted to the corresponding N.N-dimethyl derivatives by reductive methylation using formaldehyde and sodium borohydride. (ii) α-Cobratoxin was treated with a large excess of acetic anhydride at pH 8 followed by hydroxylaminolysis of the esters formed with tyrosine, serine and threonine. (iii) α-Cobratoxin was exhaustively methylated employing dimethylsulfate at pH 9.

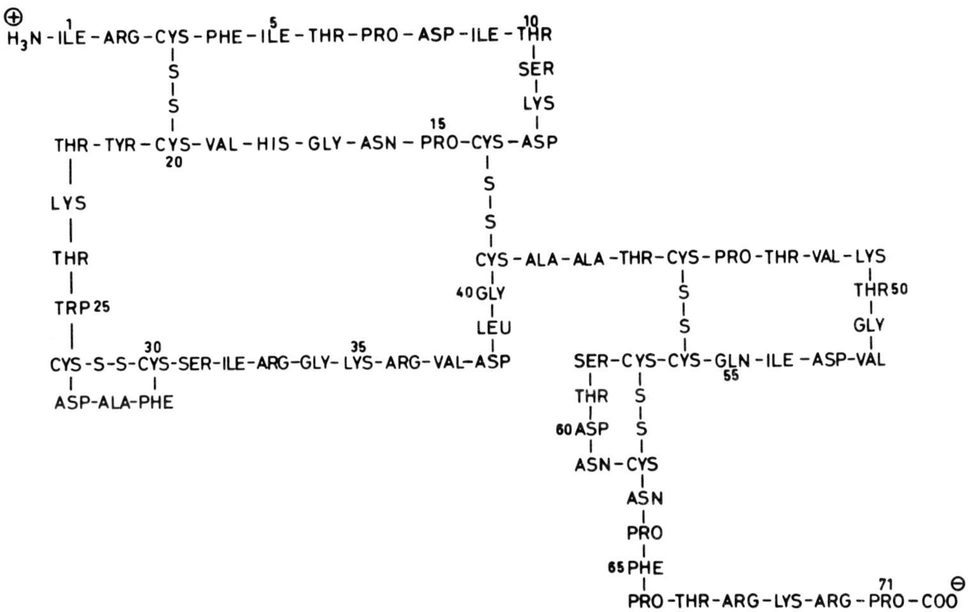

Fig. 1. Amino acid sequence of α-cobratoxin.

Modification of Arginine Residues - α-Cobratoxin contains five arginine residues one of them (Arg-33) being invariant in all known sequences of curaremimetic toxins (13). To derivatise all arginine guanido groups, the toxin was treated with excess cyclohexane-1,2-dione in borate buffer at pH 9. The modified toxin was kept in borate buffer except if it was freeze dried from 20% acetic acid. As repetitively checked by amino acid analysis these conditions maintained complete modification of the arginines.

<u>Modification of the Tyrosine Residues</u> - The single tyrosine residue of α-cobratoxin was nitrated using low concentrations of tetranitromethane at pH 8.

<u>Cleavage and S-alkylation of the Disulfides</u> - Depending on the methods employed complete reduction of all five disulfide or selective reduction of the extra disulfide of "long" neurotoxins was achieved:
(a) Selective reduction of the extra disulfide (Cys26-Cys30) was achieved by treatment of the toxin with dithiothreitol in a molar ratio of 1:1 followed within 20 min by alkylation of the released thiols. The following alkylation reagents were employed: iodoacetamide, vinyl pyridine and iodoacetaminoethylnaphthylamine-5-sulfonic acid.
(b) Complete reduction of disulfides was achieved by several methods the following one proving the mildest with respect to the biological activity of the product: The extra disulfide of α-cobratoxin was first selectively reduced and alkylated as described above. Then the same procedure was repeated five more times.

<u>Peptide Cleavage</u> - Fully S-carboxyamidomethylated α-cobratoxin was treated with Lys-C protease (Boehringer, Mannheim) in 0.1 M $(NH_4)_2CO_3$, pH 8.0 at 37° for 60 min at a substrate: enzyme ratio of 50:1. The reaction was terminated by addition of 0.1 M phosphoric acid to pH 3.0. The resulting peptide mixture was resolved on an RP-8, 10 µm reverse phase column employing a gradient elution system consisting of triethylamine phosphate without and with 80% acetonitril (8). Nine major peptides were obtained and employed in the tests described below.

TABLE

Affinities of Binding of α-Cobratoxin and its Derivatives to the Acetylcholine Receptor from *Torpedo marmorata*.

Toxin Derivative	K_I [M]	Rel.Affinity [%]
native α-toxin	2×10^{-10}	100
hexa-N,N-dimethyl	8×10^{-10}	25
hexa-acetyl	4×10^{-7}	0.05
methylcarbamylated	1×10^{-8}	5
exhaustively methylated	1×10^{-7}	0.1
cyclohexanedione treated	2×10^{-8}	1
tetranitromethane treated	5×10^{-10}	40
di-S-pyridylethylated	1×10^{-10}	200
di-S-carboxyamidomethylated	2×10^{-11}	1000
di-S-AEDANS	3×10^{-9}	7
fully reduced, carboxyamido-methylated	3×10^{-6}	0.02

The binding affinities of the derivatives were determined by competition binding employing [^{3}H]monopyridoxamine phosphate labelled α-cobratoxin (10) and purified receptor (14). The experiments were performed at room temperature with constant incubation times of 300 min.

Binding Studies - The obtained toxin derivatives were subjected to competition binding experiments with tritiated α-cobratoxin (10) and purified receptor from *Torpedo marmorata* (14). The derivatives competitive with the radiolabelled toxin yielded linear double reciprocal plots with extrapolated intersection points at the $1/c_B$-axis identical with that of radioactive toxin. Hence, binding of derivatives and radiolabelled toxin was mutually exclusive with a single class of toxin sites for each derivative at the receptor. The resulting inhibition constants are listed in the table. As shown there, the derivatives vary by orders of magnitude in their affinities of binding. Yet, even the derivatives with comparatively low affinity still bind to the receptor with affinities comparable to those of the natural transmitter. Thus, the related K_D-values must be considered to represent a specific binding interaction of the toxin derivatives and the receptor.

Discussion

Fig. 2 is a view from the front site of α-cobratoxin with some of the residues modified in this study indicated. As shown by the competition binding studies described above, not a single of the modified residues is absolutely essential for receptor binding in the sense that modification of this residue completely abolishes binding. This precludes the classical concept of a narrowly defined "reactive site" for this group of polypeptides. As the binding affinities are modified to various extends depending on the type of modification, it is clear, however, that some of the modified residues must be "involved" in binding. This agrees with previous findings showing that the interaction of toxin and receptor is largely electrostatic (10) and that conversion of the charges renders the neurotoxins inactive (15,16). What alternatives do then exist for possible site(s) responsible for interaction with the receptor?

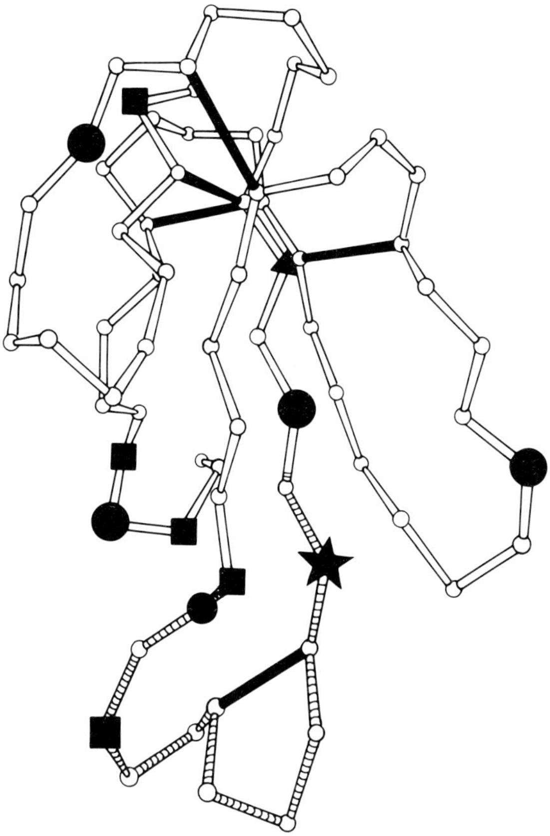

Fig. 2. Front side view of α-cobratoxin. The positions of some of the residues in the structure modified in this study are indicated: ●, lysine, ■ , arginine, ▲, tyrosine, ★, tryptophan. For details concerning the three-dimensional structure see manuscript by W. Saenger et al., this volume.

One possibility not tested in our experiments is that one or more of the free amino groups together with one or more of the arginines form the active region(s). If this applies, these residues must function separately in the sense that modification of one of them does not affect interaction of the other with the receptor. These arguments then lead to the postulation of multiple but independent sites of interaction.

A second possibility for the structure-function relationship of curaremimetic neurotoxins is exposed when the charge patterns of these molecules are considered (Fig. 3). These show a conspicuously redundant distribution of charges within certain regions of the structure. These regions are defined by

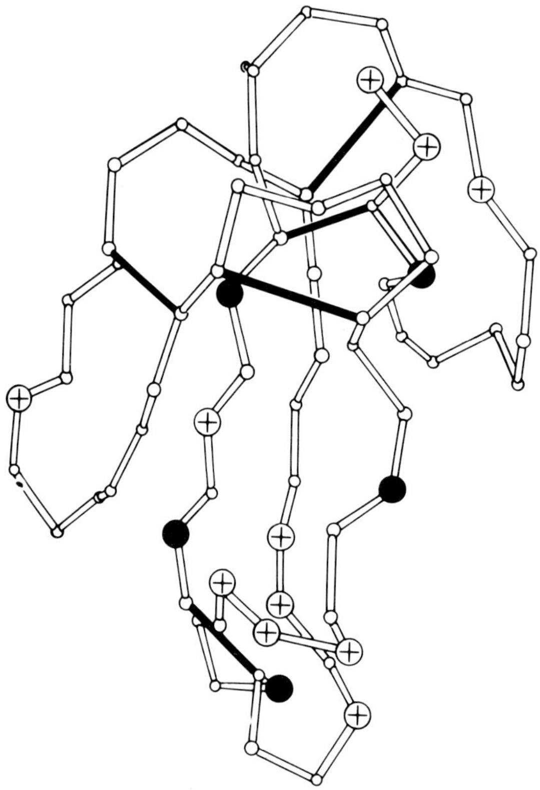

Fig. 3. Back side view of α-cobratoxin. The positions of the amino acid residues carrying positive charge (+) and the aromatic amino acids (●) are indicated.

Ile-1, Arg-2 and Lys-12 (region I), Arg-33, Lys-35 and Arg-36 (region II) and Arg-68, Lys-69 and Arg-70 (region III). These regions may be considered as superstructures providing centers of charge for an initial interaction which then may be followed by conformational adjustments as they are indicated in the slow rates of toxin-receptor complex formation (10) and the changes in fluorescence accompanying some of these interactions (11).

Both possibilities require multipoint recognition patterns of neurotoxins and exclude several of the earlier suggestions as to the organization of the reactive site of these polypeptides: Certainly, pairs of amino acids mimicking the structure of acetylcholine (3,9) can be excluded. Similarly, less than the whole "concavity" (residues 22-55 in α-cobratoxin) (3) is involved in binding to the receptor. Thus, while our studies do not suffice to unambiguously identify the individual elements required for receptor recognition and binding, we can certainly exclude the essence of those suggestions for the reactive site (3) that were put forward in the initial enthusiasm following elucidation of the first three-dimensional structure of a curaremimetic neurotoxin.

Acknowledgements

This work was supported by the Deutsche Forschungsgemeinschaft, Schwerpunktprogramm "Biochemie des Nervensystems".

References

1. Tsernoglou, D., Petsko, G.: FEBS Lett. 68, 1-4 (1976).
2. Low, B.W., Preston, H.S., Sato, A., Rosen, L.S., Searl, J.E., Rudko, A.D., Richardson, J.S.: Proc. Natl. Acad. Sci. USA 73, 2991-2994 (1976).
3. Low, B.W.: Handbook of Experimental Pharmacology, Vol. 52 (Lee, C.Y., ed.) pp. 213-257, Springer Verlag, Berlin 1980
4. Walkinshaw, M.D., Saenger, W., Maelicke, A.: Proc. Natl. Acad. Sci. USA 77, 2400-2404 (1980).
5. Kistler, J., Stroud, R.M. Klymkowsky, M.W., Lalancette, R.A., Fairclough, R.W.: Biophys. J. 37, 371-383 (1982).
6. Tsetlin, V.I., Karlsson, E., Utkin, Yu. N., Pluzhnikov, K.A., Arseniev, A.S., Surin, A.M., Kondakov, V.V., Bystrov, V.F., Ivanov, V.T., Ovchinnikov, Yu.A.: Toxicon 20, 83-93 (1982).
7. Juillerat, M.A., Schwendimann, B., Hanert, S., Fulpius, B.W., Bargetzki, J.P.: J. Biol. Chem. 257, 2901-2907 (1982).
8. Martin, B.M., Chibber, B.A., Maelicke, A.: Toxicon (1983) in press.

9. Walkinshaw,M.D., Saenger,W., Maelicke,A.: Structural Aspects of Recognition and Assembly in Biological Macromolecules (Balaban,M., Sussman,J.L., Traub,W. and Yonath, A., eds.) pp. 313-324 (1981)

1o. Maelicke,A., Fulpius,B.W., Klett,R.P., Reich,E.: J.Biol. Chem.252, pp. 4811-483o (1977)

11. Kang,S., Maelicke,A.: J.Biol.Chem.255, pp.7326-7332, (198o)

12. Martin,B.M., Chibber,B.A., Maelicke,A.: J.Biol.Chem. (1983) in press

13. Karlsson,E.: Handbook of Experimental Pharmacology, Vol.52 (Lee,C.Y., ed.) pp. 159-212, Springer Verlag, Berlin 1979

14. Rüchel,R., Watters,D., Maelicke,A.: Eur.J.Biochem.119, pp. 215-223 (1981)

15. Fulpius,B.W., Maelicke,A., Klett,R.P., Reich,E.: Cholinergic Receptors (Waser,P., ed.) pp. 76-85, Raven Press, New York 1975

16. McConnell,M.G.: Ph.D.thesis, Columbia University, New York 1975

α-COBRATOXIN AND α-BUNGAROTOXIN, TWO MEMBERS OF THE "LONG" NEUROTOXIN FAMILY - A STRUCTURAL COMPARISON

Wolfram Saenger

Institut für Kristallographie, Freie Universität Berlin,
Takustr. 6, D-1000 Berlin 33

Malcolm D. Walkinshaw

Department of Chemistry, University of Edinburgh,
West Mains Rd., Edinburgh EH9 3JJ, Scotland

Alfred Maelicke

Max-Planck Institut für Ernährungsphysiologie,
Rheinlanddamm 201, D-4600 Dortmund 1

Introduction

Snake venom contains a class of curaremimetic neurotoxins which block synapses through their high affinity binding to acetylcholine receptor located in the post-synaptic membrane. The amino acid sequences of about hundred of these neurotoxins have been determined. If they are aligned such that cysteins and invariant amino acids fall into register, two families become apparent (1). In one, called "short", there are 60-62 amino acids in one polypeptide chain whereas in the "long" toxin family, 71-74 amino acids are found. In the "short" toxins, there are eight cystein residues forming four disulfide bridges and the "long" toxins also display cysteins in related positions with two extra cysteins not observed in the "short" family. Besides these invariant cysteins, in both families additional conserved residues are found: Tyr25, Trp29, Asp31, Arg37, Gly38, Gln44, Pro50, Asn69.

Toxins as Tools in Neurochemistry

Because there are so many invariant, conserved amino acids in the "long" and "short" toxin families and chemical studiés indicated the four common disulfide bridges to be formed between corresponding cysteins, structural homology appeared possible. Since now four crystallographic analyses are available which give detailed information on the architecture of "long" and "short" neurotoxins, a structural comparison can be made.

Two independent groups reported the crystal structure analysis of presumably the same "short" toxin, erabutoxin b from the sea snake Laticauda semifasciata (2,3) and two crystal structures of long neurotoxins have been reported (4,5). These are α-cobratoxin from Naja naja siamensis (4) and α-bungarotoxin from Bungarus multicinctus (5,6). In this report, a brief account and structural comparison of the results of these studies is presented.

Structural Comparison

All three toxins exhibit a common overall shape consisting of three loops (Figure 1). Two shorter ones are flanking a central long loop, hanging down from a globular head. The folding of the polypeptide chain in the head region is determined and stabilized by the four disulfide bridges common to "short" and "long" neurotoxins. The fifth disulfide bridge in the "long" neurotoxins is formed between two closely located cysteins in the tip of the long, central loop. Since the loops are arranged in a side-by-side mode, the structure of the toxins is rather flat, with a front and back side. At the back side, the two toxin families are very different, with a long carboxy terminal oligopeptide tail hanging down and protecting the back side in the "long" toxins whereas in the "short", the tail is cut (Figure 1).

From chemical and spectroscopic studies which indicated amino acid side chains essential for binding and from the structural information we have now, it appears that the front side is providing for a binding surface recognized by the acetylcholine receptor. This can be more narrowly specified because there are many evidences which suggest that the long central loop and the lower tip of the loop III (the loop on the right side of the central loop (Figure 1) are actually bound to the receptor. Because this topic has been stressed in other publications (7,8) we do not go into these details here. Instead, a comparison of the two "long" toxin structures, α-cobratoxin and α-bungarotoxin, is of interest.

Stereo views of the back side of these molecules are depicted in Figures 2 and 3. At a first glance, the two molecular structures are very similar. There are two major differences, however. In α-cobratoxin, the long, carboxyterminal tail hangs down over the back side of the molecule whereas it is swung to the side in α-bungarotoxin. This difference is not unexpected, however, because in the electron density map of α-cobratoxin, considerable disorder of the tail portion was apparent. Therefore, the picture given in Figure 2 illustrates only one possible arrangement of the tail. Further refinement of the α-cobratoxin crystal structure is under way and will provide more details about the flexibility of this part of the molecule.

In contrast to the mobility of the tail region, the rest of the toxin molecules appears rather stiff, mainly because the central, long loop is stabilized by ß-pleated sheet hydrogen bonding and is forming a triple sheet with the nearby strand of loop III. Both in erabutoxin b and in α-cobratoxin, there are many (9 and 12 respectively) hydrogen bonds stabilizing this particular structure (2,8). The ß-pleated sheet of the long central loop forces the amino acid side chains into

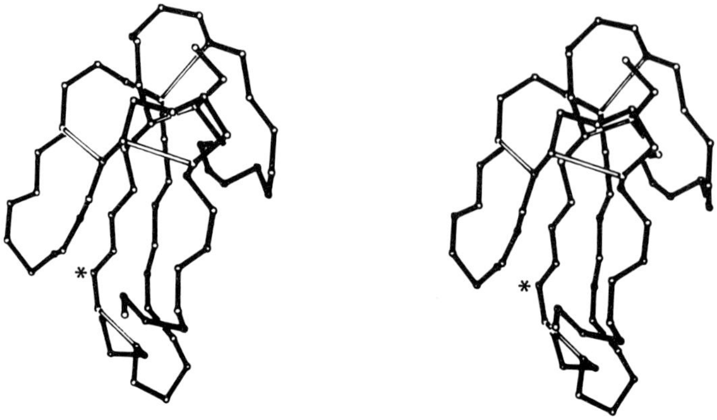

Figure 2.

Stereo view of α-cobratoxin, looking at the back of the molecule. The position of Trp29 is marked by asterisk. Taken from (4)

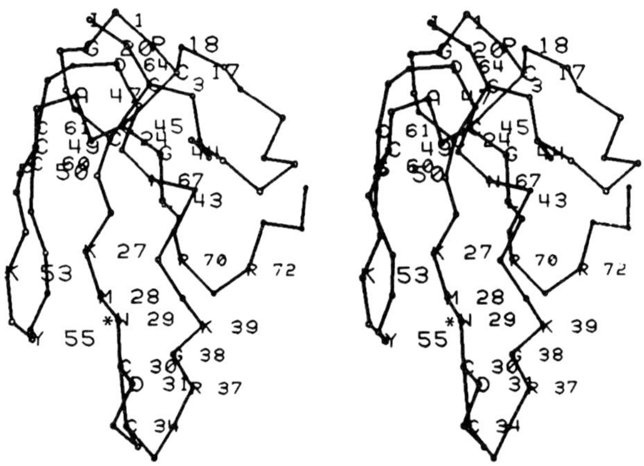

Figure 3.

Stereo view of α-bungarotoxin in the same orientation as Figure 2. Asterisk indicates Trp29. Note similarity in overall shape with α-cobratoxin and differences in carboxy terminus and long central loop. Taken from (6).

well-defined orientations such that they point alternately to the front and back side of the toxin molecules. Most noteworthy, the side chains of invariant, functionally essential Trp29 as well as Tyr25, Asp31, Arg37 are accommodated at the front side which is the presumable acetylcholine receptor binding region.

Surprisingly, this picture is very different in α-bungarotoxin because Trp29 points to the back side of the molecule. This difference in orientation is caused by a disruption of the ß-pleated sheet structure. The stereo view, Figure 3, clearly shows that Tyr25 is still pointing to the front side and residue 26 to the back but then Lys27 is not, as it should be in a regular ß-structure, oriented to the front side. Rather, it is directed sideways, bringing Met28 to the front side and Trp29 to the back. This irregularity is also reflected in the other strand of the long, central loop and, as a consequence, some of the hydrogen bonds stabilizing regular ß-pleated sheets are broken. As described in (5), there remain only four hydrogen bonds in the α-bungarotoxin structure, one in the upper portion of the long, central loop and three between the central loop and loop III.

What is the reason for the structural change in the active site region of the two "long" toxins? There is ample evidence from chemical as well as spectroscopic data that the overall three-dimensional structure of the snake neurotoxins is very rigid and conserved under a wide range of conditions. NMR studies indicated that the ß-pleated sheet configuration is especially preferred and stable in these molecules (Wüthrich and Bystrov, private communication). Since the breaking of a ß-pleated sheet structure is a slow process and crystallization of α-cobratoxin usually occurred over night, this crystal structure probably represents the "long" toxin in its native state, as also corroborated by NMR investigation carried out in Prof. Bystrov's Laboratory under crystallization conditions.

It appears, therefore, that the conformation of α-bungarotoxin does not represent the native state of the molecule. Obviously, a reorientation of the long, central loop region has taken place during crystallization and in this connection it would be of interest to know crystallization conditions and intermolecular hydrogen bonding interactions within the crystal lattice. Since crystallization was very slow (R.M. Stroud, personal communication), aggregation of α-bungarotoxin molecules via hydrogen bonding and refolding of the central loop might have taken place.

Does this mean that the toxin changes its conformation when binding to the acetylcholine receptor, as suggested by Kistler and coworkers(5)? If the α-bungarotoxin structure found in the crystalline state does not represent an artifact, then this might be true. All evidences thus far indicate that snake neurotoxins bind slowly and strongly to the receptor and therefore, a disruption of the ß-stranded structure could well occur. The question remains whether the conformation of α-bungarotoxin as found in the crystalline state is really similar to the one bound to the receptor, a question that can be answered only if a crystalline acetylcholine receptor-toxin complex is available or if toxin is specifically labelled to monitor such conformational changes upon binding using spectroscopic methods.

Acknowledgement

This work was supported by a NATO-grant to A.M. and W.S. and by an Alexander von Humboldt fellowship awarded to M.D.W.

1. Karlsson, E.: in Handbook of Experimental Pharmacology, ed. Lee, C.Y. (Springer, Berlin) 52, 159-212 (1979).

2. Low, B.W., Preston, H.S., Sato, A., Rosen, L.S., Searl, J.E., Rudko, A.D., Richardson, J.S.: Proc. Natl. Acad. Sci. USA 73, 2991-2994 (1976).

3. Tsernoglou, D., Petsko, G.: FEBS Lett. 68, 1-4 (1976).

4. Walkinshaw, M.D., Saenger, W., Maelicke, A.: Proc. Natl. Acad. Sci. USA 77, 2400-2404 (1980).

5. Kistler, J., Stroud, R.M., Klymkowsky, M.W., Lalancette, R.A., Fairclough, R.H.: Biophys. J. 37, 371-383 (1982).

6. Agard, D.A., Stroud, R.M.: Acta Crystallogr. A38, 186-194 (1982).

7. Low, B.: in Handbook of Experimental Pharmacology, ed. Lee, C.Y. (Springer, Berlin), 52, 213-257 (1979).

8. Walkinshaw, M.D., Saenger, W., Maelicke, A.: in Structural Aspects of Recognition and Assembly in Biological Macromolecules, ed. Balaban, M., Sussman, J.L., Traub, W., Yonath, A., pp. 313-324 (1981).

ACETYLCHOLINE RECEPTOR INTERACTION WITH THE NEUROTOXIN II PHOTOACTIVABLE DERIVATIVES

Victor Tsetlin, Kirill Pluzhnikov, Andrej Karelin, Vadim Ivanov

Shemyakin Institute of Bioorganic Chemistry, USSR Academy of Sciences, Moscow, USSR

Introduction

X-ray analysis, NMR, and other physical methods allow to gain an insight into spatial structure of proteins and their complexes with respective receptors. Selective chemical modification considerably expands the possibilities of the above methods. Purely chemical approaches, such as usage of bifunctional reagents, can also provide information on the spatial organization of protein complexes.
Binding of neurotoxins to the nicotinic acetylcholine receptor (AchR) has been earlier characterized relying on the EPR and fluorescence studies of the AchR complexes with the neurotoxin II *Naja naja oxiana* (NT-II) derivatives having a spin or fluorescence label at identified sites (1,2). Herein we report on probing the neurotoxin-AchR interaction by photoinduced crosslinking. To this end, the NT-II singly labeled photoactivable derivatives have been prepared, the label positions identified, and their binding to the purified AchR has been analyzed.

Results

At first we have tried the reaction with 4-fluoro-3-nitro-

Toxins as Tools in Neurochemistry

phenylazide in dimethylsulfoxide under the conditions of α-bungarotoxin modification (3). However, in spite of varying the reagent-toxin ratio or reaction time, we managed to isolate only di- and poly-labeled compounds, rather than mono-labeled ones.

In dimethylformamide, in the presence of triethylamine and 1,6-fold reagent excess, two singly labeled derivatives were isolated (Fig. 1). Since their total yield was only 0.7%, we faced the problem of searching for a more appropriate reagent.

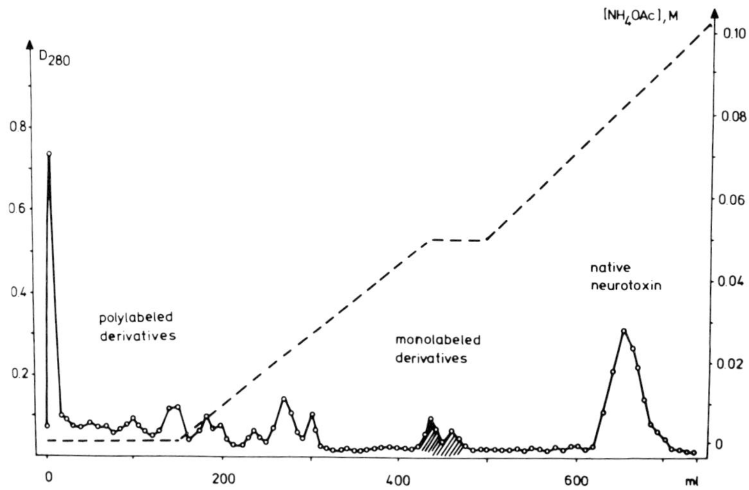

Fig. 1. BioRex 70 column (1 x 15 cm) chromatography of the products of NT-II modification with 4-fluoro-3-nitrophenylazide. Mono-labeled compounds are marked.

We have chosen N-hydroxysuccinimidyl p-azidobenzoate relying on the conditions earlier worked out for the synthesis of NT-II spin labeled derivatives with N-hydroxysuccinimide esters of the piperidine-series spin labels (4). NT-II modification was carried out at the 1,1:1 reagent-protein ratio (pH 8.0, 6 M Gu·HCl, 20 hr, 24°). After desalting, the ion-exchange chromatography (Fig. 2) afforded 6 fractions (total yield 12%) eluting at the position of mono-labeled NT-II derivatives.

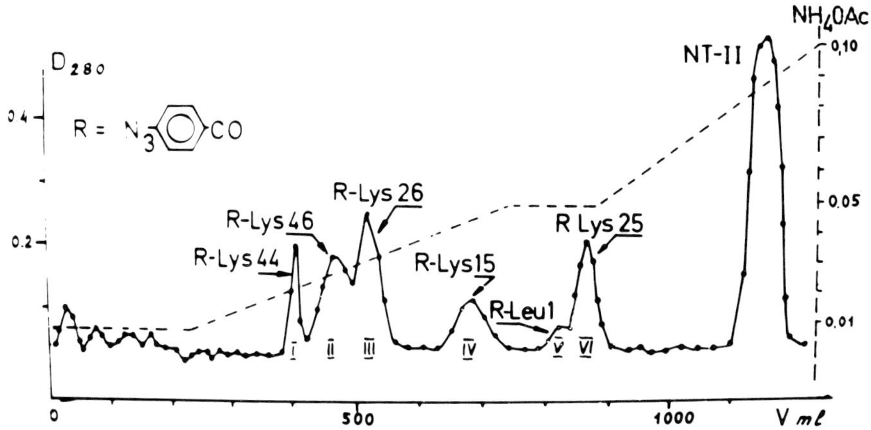

Fig. 2. BioRex 70 column (1 x 13 cm) chromatography of the NT-II p-azidobenzoyl derivatives. (Chromatography and lyophilization were performed in the dark).

Localization of photoactivable groups incorporated into proteins poses a serious problem. We have directly identified the label positions taking into account the possibility of reduction of azide into amino group (5). Having modified the conditions described in (5), namely by lowering pH to 7.5, using high concentration of dithioerythritol and longer reduction time, we succeeded in reducing the azide group and disulfide bonds, still avoiding the intramolecular crosslinking.

The reduced and carboxymethylated derivatives were subjected to trypsinolysis, the hydrolysates being subsequently separated by gel filtration (Fig. 3).

Increase in the D_{280}/D_{206} ratio, as compared with the absorption for the respective peptides of the NT-II hydrolysate, allowed to distinguish the peptides with the additional chromophore in the molecule. Their structure was determined by amino acid analysis and, if necessary, by N-terminal analysis or Edman degradation. For example, chromatography of the fraction III (Fig. 3B) gave the peptide 26-32, which had the D_{280}/D_{206} ratio 0.93 (instead of 0.45 for the respective

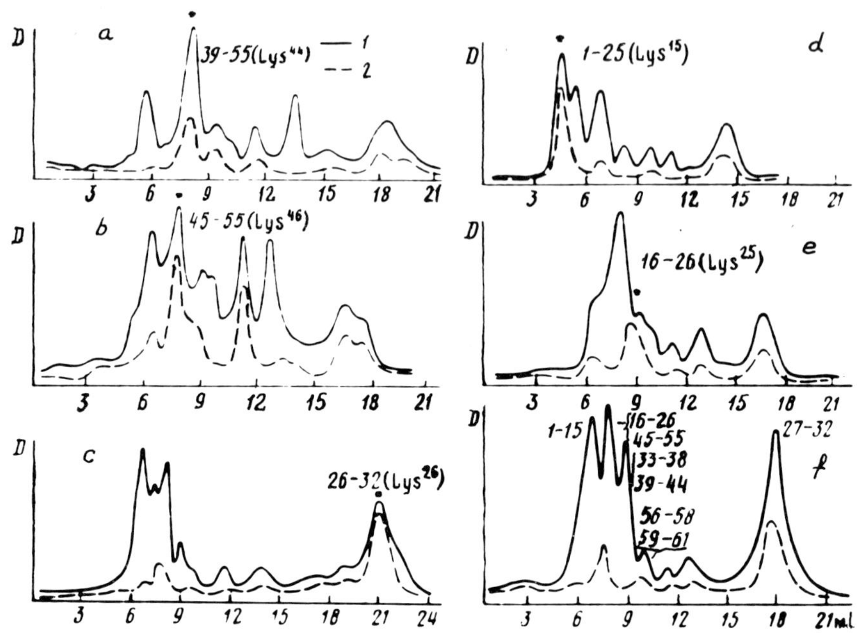

Fig. 3. Sephadex G-25sf column (0.4 x 125 cm) chromatography of the tryptic hydrolysates obtained from the NT-II p-azidobenzoyl derivatives 0.1 M NH_4OAc buffer pH 4.5; peptide detection at 206(1) and 280 nm (2). A-F: hydrolysates of fractions I-IV, VI (see Fig. 2) and NT-II, respectively. Labeled peptides are marked with an asterisk.

unmodified peptide). Since the labeled peptide has only one functional group of the Lys 26, this fact testifies to modification of this residue.

Localization of the labels in fractions I, II and IV was performed in a similar way. Identification of the modified residue was more difficult in case of fraction VI. The tryptic hydrolysate contained two labeled peptides, 1-15 and 16-26, because of incomplete separation of fractions V and VI (Fig. 2). After removal of fraction V admixture from fraction VI by rechromatography, only one labeled peptide 16-26 was found in the tryptic hydrolysate of the latter (Fig. 4B). Therefore, Lys 25 and Leu 1 are the modified residues in the fractions VI and V, respectively.

Binding studies were performed using the Triton X-100 solubilized AchR from the *Torpedo marmorata* electric organs. The purification included affinity chromatography on a NT-II-Sepharose column. The AchR activity was determined from binding $[^3H]$-acetylated NT-II or toxin 3 *Naja naja siamensis*. Molecular weight of AchR was taken as 250 kD (6), and the stoichiometry of the neurotoxin-AchR complex 2:1 (1,7). All the experiments were performed with the AchR of 90% activity, i.e. ~ 7.3 nmol toxin binding sites/mg receptor protein.

The specificity of noncovalent interaction between the AchR and NT-II photoactivable derivatives was proved by the following control experiments.After AchR incubation with a photoactivable derivative (1h in the dark), a 5-fold excess of radioactive neurotoxin was added to the complex. The following Sephadex G-75 chromatography showed 94% displacement of photoactivable neurotoxin by a radioactive one. On the other hand, excess photoactivable neurotoxin almost completely displaces the radioactive neurotoxin.

The crosslinking experiments were performed using a 250 w halogen lamp, the short-wave irradiation being not cut off. The control experiments revealed only a 10% decrease in the AchR activity and no change in its spectral characteristics after 20 min irradiation. No activity loss was observed under the conditions (6 min) adopted for irradiation of the toxin-AchR complexes.

The extent of crosslinking (Table 1) after irradiation was determined from the decrease in the subsequent assay of the radioactive neurotoxin binding. In parallel experiments the AchR activity was determined by its incubation with the same amount of radioactive toxin prior to irradiation.

All the photoactivable derivatives, excluding one with the label at Lys 44 form the crosslinks with AchR (Table 1).

Table 1. Covalent binding of the NT-II photoactivable derivatives to AchR

Fraction number	Modified residue	Extent of crosslinking %
I	Lys 44	0
II	Lys 46	77
III	Lys 26	76, 72*
IV	Lys 15	81, 82*
V	Leu 1	82*
VI	Lys 25	88*

* Displacement by [^{3}H]-acetylated NT-II, displacement by [^{3}H]-acetylated toxin 3 *Naja naja siamensis*.

A high efficiency of crosslinking is remarkable, and may be due to the exceptional stability of the AchR-neurotoxin complexes not severely impaired by the incorporated labels. With the purpose of identifying the AchR subunits involved in neurotoxin binding, we have prepared a series of NT-II derivatives, each of them containing radioactive p-azidobenzoyl group at a different position. Modification of NT-II was carried out with N-hydroxysuccinimidyl [^{14}C]-p-azidobenzoate (5 Ci/mol). Localization of the labels described above holds also for the radioactive derivatives, as follows from the identity of the reaction conditions and close similarity of the profiles given in Figs. 1 and 4.

The fractions used for the AchR binding studies are identified by bars in Fig. 4. The toxins modified at Lys 44 and Lys 46 are not always separable by one chromatographic run. Since the Lys 44-derivative forms no crosslinks with the AchR, we could use the fraction containing both Lys 44- and Lys 46-derivatives to identify the AchR subunit forming crosslinks with the latter. After irradiation, the AchR complexes with the photoactivable neurotoxins were incubated

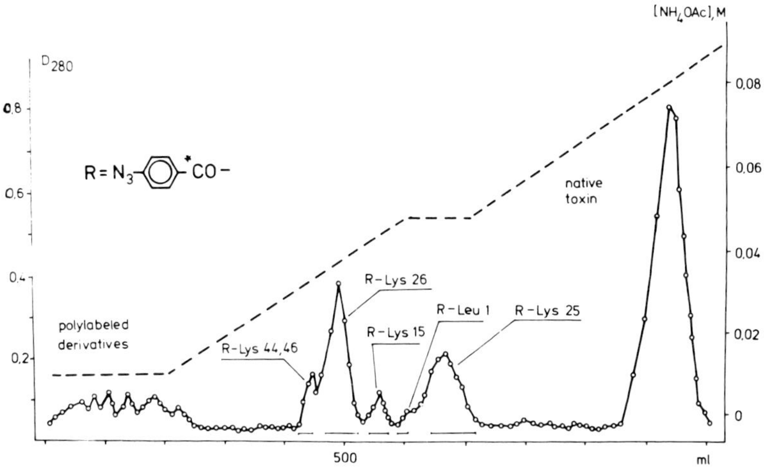

Fig. 4. Separation of the products of NT-II modification with N-hydroxysuccinimidyl [^{14}C]-p-azidobenzoate. For the conditions see the legend to Fig. 2.

in a buffer containing 2% SDS and 5% mercaptoethanol, the subunits being subsequently separated by polyacrylamide gel electrophoresis. The gels were stained, sliced treated with H_2O_2, and the radioactivity was analyzed using a scintillation counter.

The results clearly demonstrated that all 4 types of the AchR subunits (α , β , γ , δ) have contacts with the bound neurotoxins. Table 2 enlists the predominant contacts between the AchR subunits and photoactivable groupings at certain positions of the NT-II molecule. Labeling of only one subunit is characteristic of the derivatives having photoactivable groups either at position Lys 26 or Lys 46. On the other hand, the Leu 1-derivative is attached to the two subunits (α and γ). With the Lys 15-derivative, the radioactivity was found in three subunits (α , γ and δ). Binding of the Lys 15-derivative was also examined in the presence of NT-II hexa(trifluoro)acetylated derivative, which has been earlier shown to displace singly labeled NT-II derivatives only from 50% of the binding sites (1,4). Here-

Table 2. Contacts between NT-II and the AchR subunits

NT-II modified residue	Radioactive subunit
Leu 1	α γ
Lys 25	α β
Lys 26	δ
Lys 46	β
Lys 15	α γ δ
Lys 15*	α

*Binding in the presence of hexa(trifluoroacetyl)-NT-II

with, only α-subunit was found to be radioactively labeled. This experiment means that at one of the two neurotoxin-binding sites the label at Lys 15 is attached to the α-subunit, while at the other site it may contact the γ- or δ-subunits; therefore, the latter two should be in proximity. The studies with the photoactivable derivatives confirmed the multipoint character of the AchR-neurotoxin interaction established earlier by spectral studies on the spin and fluorescence labeled NT-II derivatives (2).

In view of high rigidity of the NT-II spatial structure, essentially preserved upon amino group modification or upon changes in the temperature, pH or polarity of the medium, the identified crosslinks allowed us to propose two most plausible models for the arrangement of the bound neurotoxins in the AchR binding site (Fig. 5). So far no experimental data could be found in literature which would show what region of a neurotoxin contacts a definite AchR subunit. As far as the AchR subunit mutual disposition is concerned, several models are available (8-12),their common feature being the separated position of the two α-subunits. The models are based essentially on usage of bifunctional reagents, which produced several crosslinks but not those between α subunits,and electron microscopy data. The latter are interpreted as evidence against adjacent disposition of α-subunits, relying on detection of the neurotoxins bound to AchR (11-12). However, our results demonstrate that the binding sites targeted in such a way should not be mixed with the α-subunits.

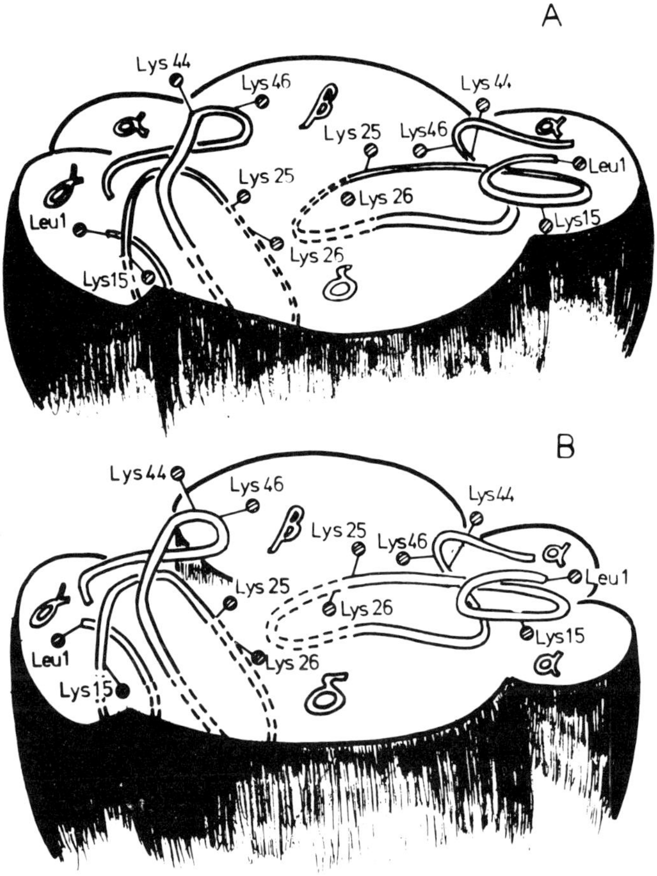

Fig. 5. Models for mutual disposition of neurotoxins and AchR subunits based on photoinduced crosslinks between them.

Model A implies that one α-subunit has no contacts with any of the photoactivable groups of the bound neurotoxins.

Possibly, this subunit is screened by some other regions of the neurotoxin where no labels have been incorporated so far. On the other hand, two α -subunits, being adjacent within the framework of model B, may have less discriminated role in the neurotoxin binding. Apparently, the studies on binding of all photoactivable derivatives under the conditions of one binding site being blocked, might help to delineate in more detail the topography of the AchR and neurotoxin interacting surfaces.

References

1. Tsetlin, V.I., Karlsson, E., Arseniev, A.S., Utkin, Yu.N., Surin, A.M., Pashkov, V.S., Pluzhnikov, K.A., Ivanov, V.T., Bystrov, V.F., Ovchinnikov, Yu.A.: FEBS Lett., 106, 47-52 (1979).

2. Tsetlin, V.I., Karlsson, E., Utkin, Yu.N., Pluzhnikov, K.A., Arseniev, A.S., Surin, A.M., Kondakov, V.I., Bystrov, V.F., Ivanov, V.T., Ovchinnikov, Yu.A.: Toxicon, 20, 83-93, (1982).

3. Hucho,F.: FEBS Lett., 103, 27-32 (1979).

4. Ivanov, V.T., Tsetlin, V.I., Karlsson, E., Arseniev, A.S., Utkin, Yu.N., Pashkov, V.S., Surin, A.M., Pluzhnikov, K.A., Bystrov, V.F.: in Natural toxins (Eaker, D., Wadström, T, eds.), pp. 523-530, Pergamon Press, Oxford and New York (1980).

5. Staros, I.V., Bayley, H., Standring, D., Knowles, J.R.: Biochem. and Biophys. Res. Communs., 80, 568-573 (1978).

6. Reynolds, J.A., Karlin, A.: Biochemistry, 17, 4679-4684 (1978).

7. Damle, V.N., Karlin, A.: Biochemistry, 17, 2039-2041 (1978).

8. Hucho, F.: Trends in Biochem. Sci., 6, 242-245 (1981).

9. Karlin, A.: The cell surface and neuronal function (Gotman, C.W., Paste, G., Nicolson, G.L., eds.) North Holland Publishing Co., Amsterdam, p. 191 (1980).

10. Kistler, J., Stroud, R.M., Klymkowsky, W., Lalancette, R.A., Fairclough, R.H.: Biophys. J., 37, 371-383 (1982).

11. Holtzman, E., Wise, D., Wall, J., Karlin, A.: Proc. Natl. Acad. Sci. USA, 79, 310-314 (1982).

12. Zingsheim, H.P., Barrantes, F.J., Frank, J., Hänicke, W., Neugebauer, D.-Ch.: Nature, 299, 81-84 (1982).

FLUORESCENCE AND CIRCULAR DICHROISM STUDIES ON SNAKE VENOM NEUROTOXINS AND THEIR DERIVATIVES

Alexander Surin, Yuri Utkin, Kirill Pluzhnikov, Victor Tsetlin
Shemyakin Institute of Bioorganic Chemistry, USSR Academy of Sciences, Moscow, USSR

Introduction

A high degree of the primary structure homology for postsynaptic short-chain and long-chain neurotoxins is the core of the hypothesis about similarity of their spatial organization and molecular mechanism of interaction with the acetylcholine receptor (AchR) (1). The X-ray analysis of a short toxin, erabutoxin (2,3), and long toxins, toxin 3 from Naja naja siamensis (4) and α-bungarotoxin (5), shows that the folding of the polypeptide backbone is of the same character for both toxin types. Though recently long-chain neurotoxins have been carefully analyzed by the NMR technique (6-9), their conformations in solutions are still less known than those of short-chain neurotoxins (see (10) and references therein). These studies revealed common features in solution structures of neurotoxins of both types and showed their similarity to crystal structures. Further investigations are still necessary to establish the degree of similarity in static structure or dynamic characteristics of toxins and the reasons underlying different conformational stability and potency of various toxins.

Herein we report on a comparative study of functionally important residues in short and long neurotoxins by spectroscopic methods. The toxins have the limited number of aromatic chromophores (Trp and Tyr residues), therefore in order

Toxins as Tools in Neurochemistry

to procure additional information we inserted dansyl labels at identified positions of the amino acid sequence - in particular, at lysine residues.
The fluorescence method applied to study the interaction of dansyl derivatives of a long neurotoxin allowed to compare the topography of binding for long and short neurotoxins.

Results

1. Chemical modification of neurotoxins.

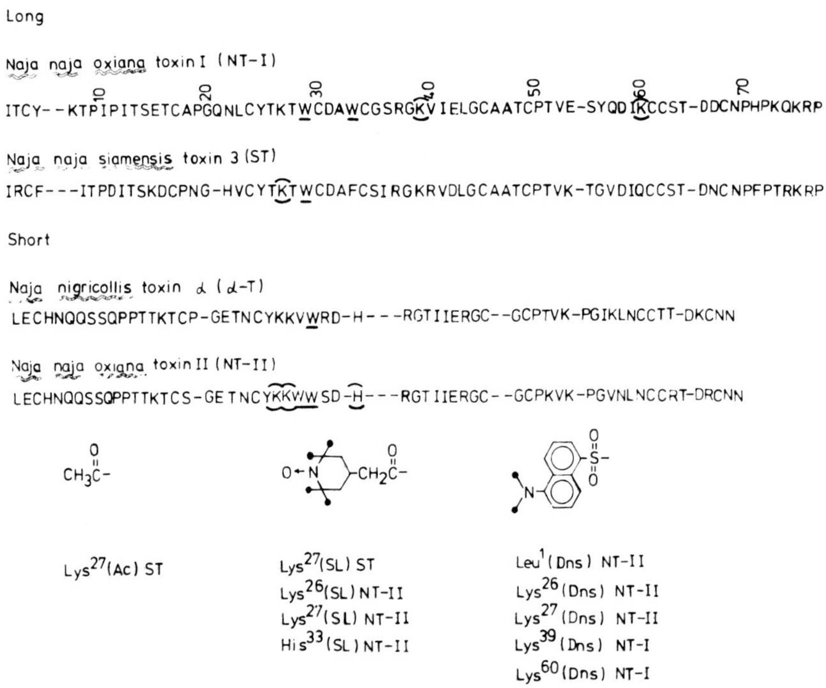

Fig. 1. Amino acid sequence of the short and long-chain neurotoxins and their chemically modified derivatives. Numbering according to (11).

Fig. 1 shows amino acid sequences of neurotoxins and the structure of their derivatives investigated in this work. Synthesis of dansyl derivatives of the *Naja naja oxiana* neurotoxin II, (NT-II), was carried out according to the procedure of Chicheportiche et al. (12) for the *Naja haje* toxin homologous to NT-II. NT-II was treated with the 30-fold ex-

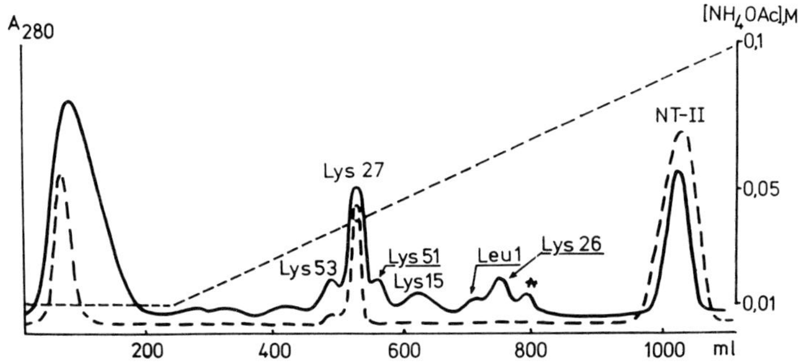

Fig. 2. BioRex 70 chromatography of the products of NT-II modification by DNS-Cl in the absence (- - -) and presence (——) of guanidinium chloride. Asterisk marks the fluorescence labeled peptide 28-37.

cess of DNS-Cl, separation of the resultant mixture (Fig. 2) yielded virtually the only derivative modified at lysine residue 27 (numbering according to (11)).This finding corroborated "superreactivity" of this residue manifested in reaction with DNS-Cl (12). To insert the labels at other positions of the amino acid sequence, we tried a number of procedures. Application of a mild dansylating reagent, dansylglycine N-hydroxysuccinimide ester also yielded the only derivative modified at Lys 27. Reaction of NT-II, having the preliminarily acetylated Lys 27 amino group, with this reagent resulted in a mixture of products. After its separation on CM-32 and BioRex 70, the derivative containing a dansylglycyl group at Lys 53 was isolated, whereas only minor amounts of other derivatives were found. Application of a denaturing agent (6 M guanidinium chloride) appeared to be more effective in reaction with DNS-Cl. Addition of the 30-fold excess of DNS-Cl yielded mainly the derivatives containing two or more labels in the molecule. The 3-fold excess of DNS-Cl per one mole of the toxin was the most optimal for obtaining monolabeled derivatives (Fig. 2). Localization of the labels was carried out as in (13).

Besides amino group modification, the additional incorporation of the label (~ 8%) - removable by NH_2OH - occurred at some nonidentified residue. All lysine modified derivatives contained 15-25% of the label at the Leu 1 α -amino group. That was not the admixture of the Leu1 mono-derivative, as the latter eluted from the column at the definite value of the buffer molarity. Apparently, at rapid removal of DNS-Cl, as upon desalting, the di-dansyl derivative labeled at one of lysine residues and α-amino group underwent some conformational changes, compensating the charge decrease upon the α -amino group blocking, and was eluted together with the corresponding lysine-monodansyl derivative. One of the derivatives, containing 20% of the label at the α -amino group was desalted in 5M Gu·HCl followed at first by slow lowering of the denaturant concentration down to 1M, and then its final removal by gel-filtration. After that we managed to separate di- and mono-derivatives by Bio Rex 70 ion-exchange chromatography.

NT-II modification under various conditions (Table 1) provided information on reactivity of amino groups. Without denaturing agents the reaction with DNS-Cl as well as with 2,2,6,6-tetramethyl-4-carboxymethylpiperidyl-1-oxyl N-succinimide ester (14) preferentially yielded the derivative labeled at Lys 27. On the other hand, in reaction with acetoxysuccinimide, all lysine residues manifested close reactivity (15). It means that "superactivity" of Lys 27 concerns only the reagents containing bulk hydrophobic groups. Increasing of reagent size led to the preferable modification of Lys 27 even under denaturing conditions.

In contrast to NT-II, modification of long-chain neurotoxin I (NT-I) under conditions described in (12) yielded several mono-dansyl derivatives (fractions 2-6, Fig. 3). To establish the positions of labels, the reduced and carboxymethylated

Table 1. Yields of monolabelled NT-II derivatives

Modified residue	Yield (%)		
	Acetoxysuccinimide	DNS-Cl	DNS-Cl + Gu·HCl
Leu 1	-	-	1, 3
Lys 15	6, 6	-	1, 8
Lys 26	2, 8	-	2, 3
Lys 27	6, 4	18, 2	7, 3
Lys 51	4, 2	-	1, 9
Lys 53	6, 6	-	2, 2

derivatives were subjected to tryptic (fractions 3 and 6) or chymotryptic (fraction 5) hydrolyses with the following peptide separation by Sephadex G-25 and DEAE-cellulose DE-52 chromatography. Derivatives from fractions 3 and 5 were established to be labeled at Lys 60 and Lys 39, respectively. Upon hydrolysis of fraction 6, peptide 28-37 was formed, i.e. Lys 27 remained unmodified. Fractions 2 and 4 lost their labels during localization and, consequently, did not contain derivatives labeled at lysine residues. Thus, the NT-I dansylation yielded no derivative labeled at Lys 27, the residue invariant in short and long neurotoxins.

Upon α-bungarotoxin acetylation (16) the derivative modified at Lys 53 was formed with the maximal yield. On the other hand, the reaction of toxin 3 from the Naja naja siamensis venom (ST) with 2,2,6,6-tetramethyl-4-carboxymethylpiperidyl-1-oxyl N-hydroxysuccinimide ester (6) yielded only one monoderivative labeled at Lys 27. Apparently, the microenvironment of Lys 27 and Lys 53 differs more in long than that in short neurotoxins. Interestingly, the available ^{1}H-NMR data show that the mobility of invariant Tyr 25 in α-bungarotoxin and toxin B from Naja naja (9), and Trp 29 in ST (8) is high-

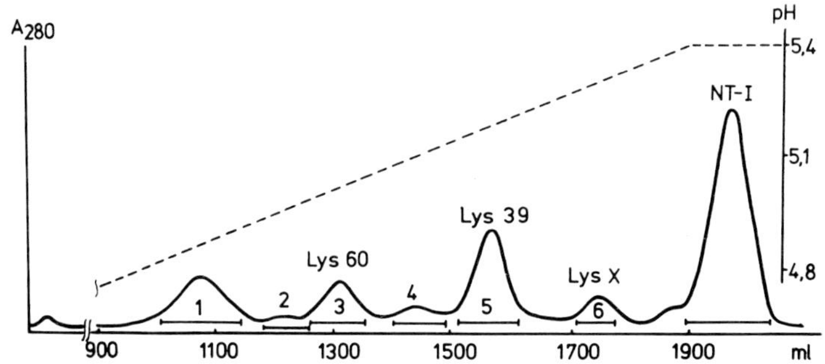

Fig. 3. Separation of dansylated NT-I derivatives by CM-32 ion-exchange chromatography.

er than in short toxins. That evidences in favor of some structural differences in the fragments including Lys 27 and these residues.

2. CD and fluorescence studies of neurotoxins

Incorporation of labels practically did not change the shape of CD spectra of the studied neurotoxins within the absorption region both of peptide chromophores (190-240 nm) and aromatic amino acid residues (240-310 nm) (Fig. 4). Evidently chemical modification virtually did not affect the secondary structure of toxins and did not disturb the environment of aromatic amino acid residues.

For two of five obtained mono-dansyl derivatives, namely, for Lys^{26} (DNS) NT-II and Lys^{27}(DNS) NT-I, the additional negative bands appeared in CD spectra within the 300-400 nm region, where the long-wavelength band of the dansyl chromophore was situated. The dansyl group having a symmetry plane, does not belong to optically active chromophores. The presence of the above-mentioned band could hardly be explained by the influence of asymmetric carbon atoms of Lys 26 and Lys 27 residues

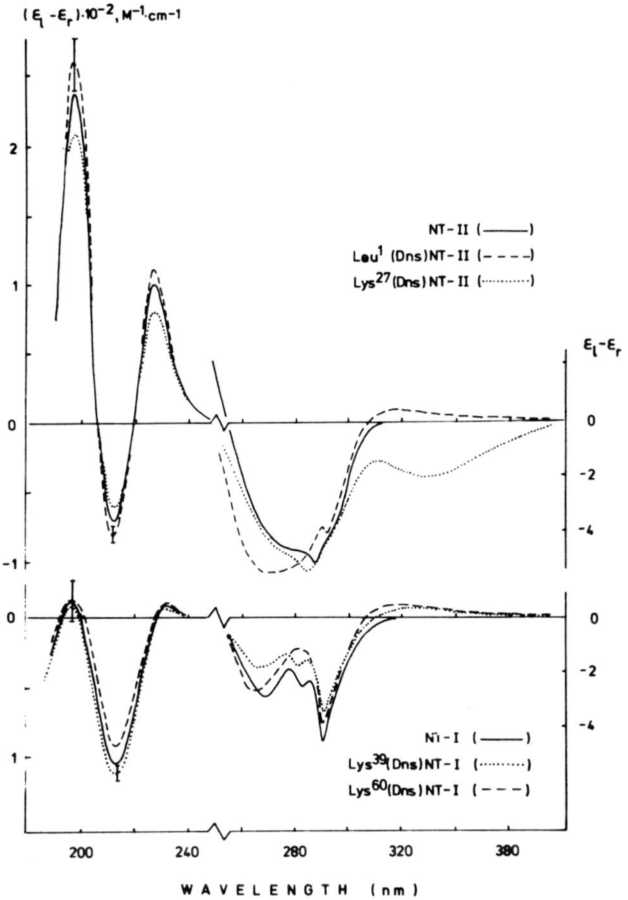

Fig. 4. Circular dichroism spectra of the short and long neurotoxins and their mono-dansylated derivatives.

because in CD spectrum much weaker band was observed within the wavelength region above 300 nm for Leu^{1} (DNS) NT-II in which dansyl was separated with only one bond from the asymmetric carbon atom. Besides, in derivatives of the long neurotoxin NT-I where dansyl labels were also attached to the ε -amino groups of lysine residues the bands within the 300-400 nm region were characterized by small amplitude (Fig. 2B). The presence of considerable Cotton effect within the 300-400 nm region for derivatives Lys^{26}(DNS)NT-II and Lys^{27}(DNS) NT-II depended on asymmetry of the environment of dansyl labels. In Leu^{1}(DNS) NT-II and NT-I mono-dansyl derivatives the environment of dansyl labels was more isotropic and their chiroptical properties were weakly expressed, probably due to the less limited Brownian movement of dansyl groups. This is in line with the EPR data for mono-spin labeled derivatives of NT-II (17): iminoxyl radicals bound to Lys 26 and Lys 27 side chains are less labile comparing with spin labels attached to other amino groups of this toxin. Comparison of fluorescence parameters of native neurotoxins and their derivatives shed light on the reasons of differences in optical properties of dansyl labels. Fluorescence maxima of dansyl labels in NT-II and NT-I derivatives are blue-shifted in comparison with a model compound , DNS-Leu (Table 2). Therefore, the neigbouring amino acid side chains should influence the microenvironment of dansyl groups. However, a correlation between the position of maximum or the fluorescence quantum yield and the presence of Cotton effect in the long wavelength absorption band could not be established (Table 2).

Table 2. Fluorescent and Chiroptical Parameters of the Dansyl Labels in the Neurotoxin II and Neurotoxin I.

Samples	Fluorescence of Dns-labels		Cotton effect at 330 nm ($\varepsilon_L - \varepsilon_R$), $M^{-1}cm^{-1}$
	Emission maximum (nm)	Quantum yield	
Leu^1(Dns) NT-II	560	0.14	+0.3
Lys^{26}(Dns) NT-II	555	0.14	-2.0
Lys^{27}(Dns) NT-II	563	0.12	-2.0
Lys^{39}(Dns) NT-I	555	0.13	+0.4
Lys^{60}(Dns) NT-I	540	0.17	+0.3
Dns-Leu	575	0.09	

Fluorescence of NT-II and toxin α from Naja nigricollis (α-T) is due to exposed tryptophan residues (18). Spectra of the two toxins are similar at neutral pH (Fig. 5). Upon

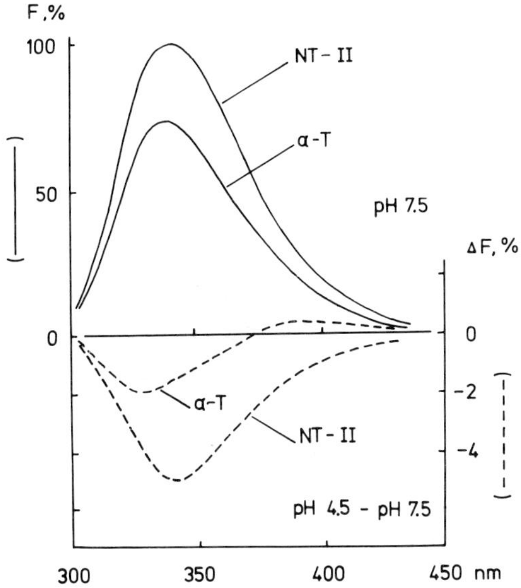

Fig. 5. Fluorescence (F) and difference fluorescence (Δ F) spectra of toxin α and neurotoxin II.

pH decreasing to 4,5 slight red shift in the α-T spectrum was clearly observed by monitoring fluorescence in a differential mode. In case of NT-II the analogous pH decrease resulted only in diminishing the fluorescence quantum yield without changing the band shape. Thus the mobility of local microenvironment differs quite noticeably for tryptophan residues in these highly homologous neurotoxins. A relevant observation was made by Menez et al. (19) at fluorescence studies of erabutoxins a and b (19).
Fig. 6 shows fluorescence pH titration curves for α-T, as well as for NT-II and its derivatives. Data for a model compound, devoid of ionogenic groups are also given. For α-T and NT-II the curves reflect influence of the ionic state of a carboxyl group(s), imidazole ring and one or several amino

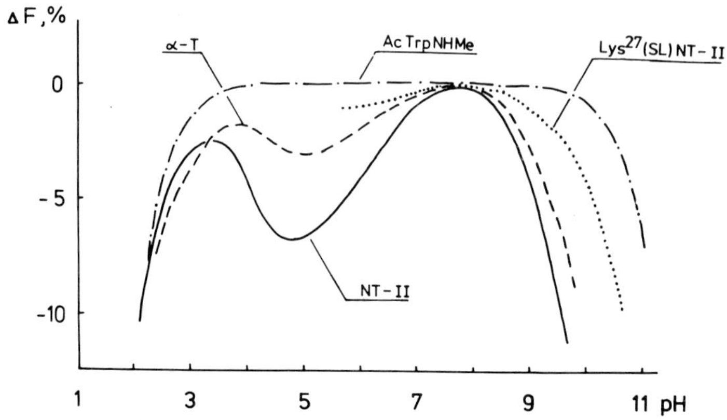

Fig. 6. Differential pH-titration curves of toxin α, neurotoxin II, its mono(spin-labeled) derivative and model compound.

groups. A comparative analysis of NT-II and α-T by NMR and fluorescence methods (20) revealed the influence of His 33 titration on the characteristics of the invariant Trp 29. Thus decreasing fluorescence intensity of α-T and NT-II upon pH fall from 7,5 to 4 (Fig. 6) can be explained by quenching the Trp 29 fluorescence by the charged imidazole ring of His 33. This reason is quite possible. Firstly, in model peptides consisting of Trp and His residues ionization of the His side chain is accompanied with the sharp drop in fluorescence intensity (21). The fluorescence decrease in neurotoxins is quite small comparing with the model systems, probably due to more rigid fixation of indole and imidazole rings in these proteins that prevents their effective contact. Secondly, in the NT-II derivative spin labeled at His 33, the fluorescence quantum yield is 20% lower. The iminoxyl radical is known to be an effective quencher of fluorescence of aromatic chromophores if the distance between the aromatic molecule and radical is 4-6 Å (22). The transition corresponding to His 33 ionization is virtually absent in the pH-titration of the NT derivative spin labeled at Lys 27 ε-amino group (Fig. 6). On the other hand, this transition is preserved with the spin label attached to Lys 26. For both spin labeled derivatives the fluorescence quantum yields decrease 2-fold, whereas the excited state lifetime remains the same as in the native toxin. This finding implies a static mechanism of fluorescence quenching, i.e. the labels form non-fluorescent complexes with the Trp indole rings. Apparently, the iminoxyl radical at Lys 27 amino group is situated not farther than 6 Å from the Trp 29 indole ring and completely quenches its fluorescence; due to this effect His 33 titration can be no longer reflected in the titration curve. The iminoxyl radical at Lys 26 contacts the indole ring of Trp 28 and quenches only its fluorescence. As a result, the region of the titration curve corresponding to the change in fluorescence of invariant Trp 29 is not affected.

It was natural to suggest that dansyl groups at Lys 26 and Lys 27 were located similarly to the spin labels relative to the Trp 28 and Trp 29 indole rings. The mutual arrangement of quite bulky indole and dansyl rings at the distances comparable with their sizes should restrict the mobility. This may be the reason for the anisotropy of the environment of dansyl labels at Lys 26 and Lys 27 that manifested itself as the Cotton effect of the dansyl chromophore long-wavelength band.

The long-wave band of dansyl absorption is overlapped with fluorescence spectra of Trp residues, making possible the radiationless energy transfer from excited indole fluorophores to dansyl labels. This allowed to determine the distance between the Trp 28 indole ring and the label at Lys 27 (23). The analogous experiment was carried out for Lys^{26}(DNS) NT-II, calculations being performed as described in (23). Fig. 7 summarizes the distances between labels attached to Lys 26 or Lys 27 and indole rings of Trp 28 or Trp 29 residues. The

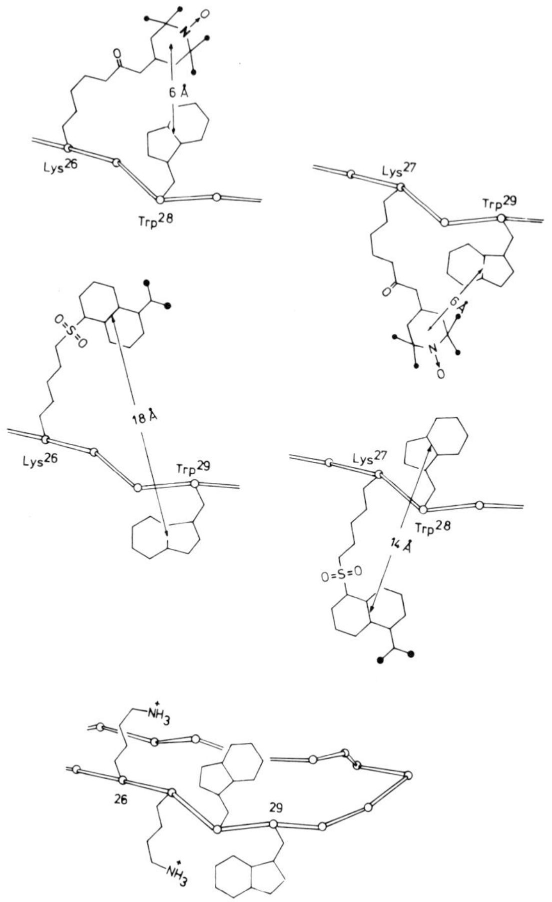

Fig. 7. Scheme of the side chain mutual dispositions for the -Lys[26]-Lys-Trp-Trp[29]-fragment in neurotoxin II.

respective distances between the side chains of the Lys 26-Trp 29 fragment are in accord with the ß-structural conformation of this region (2,3).

Fluorescence spectra show (Fig. 8) that the Trp residues in long toxins NT-I and ST are exposed (18). Acetylation of Lys 27 in ST resulted in a blue shift of the fluorescence maximum and in decreasing the quantum yield by 30%. The CD spectrum of the monoacetyl derivative virtually did not differ from that of native ST, i.e. Lys 27 acetylation did not influence the secondary structure, and the observed fluorescence changes were not induced by the toxin conformation disturbance. In analogous derivative of ST, where the spin la-

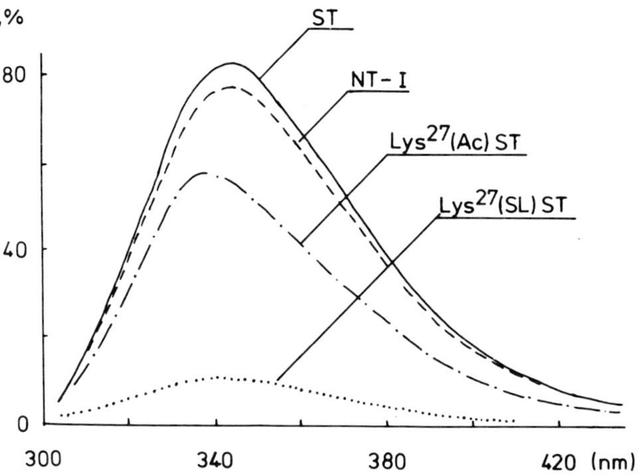

Fig. 8. Fluorescence spectra of long-chain neurotoxins and their derivatives.

bel is attached to the ε-amino group of Lys 27, intensity decreased eight-fold as compared with the native neurotoxin. Probably the Lys 27 side chain is one of the groups, forming the microenvironment of the invariant tryptophan as in case of the short neurotoxin. Data on fluorescence quenching of ST or its monoacetyl derivative confirmed this assumption. The Stern-Volmer plot for fluorescence quenching of Lys^{27}(Ac) ST with sodium iodide (Fig. 9) had the slope three times less than that for the native toxin. So acetylation diminished the accessibility of the indole ring to an iodide ion. The reason might be the elimination of the Lys 27 positive charge since the Stern-Volmer plots for quenching of ST and Lys^{27}(Ac)ST with the neutral quencher, acrylamide, coincide. Modification of Lys 27 affected the fluorescence quenching of the invariant tryptophanyl not only with extrinsic quenchers, iodide and acrylamide,but also with the quenching

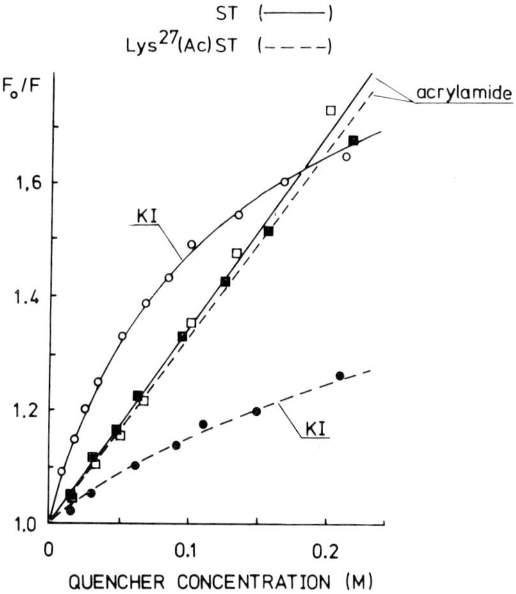

Fig. 9. Stern-Volmer plots for the quenching of the fluorescence of ST and its mono-acetylated derivative.

groups of the protein itself. Comparison of fluorescence pH-titration curves for ST and the monoacetyl derivative (Fig. 10) confirmed that finding. In the alkali region quenching of the derivative occurred at pH values approximately one unit higher than those of ST. Modification of Lys 27 led to elimination of the transition reflecting the change in the ionic state of this residue.

Thus the comparative fluorescence investigations of ST and its monoacetyl as well as mono-spin labeled derivative testified that the Lys 27 side chain was one of the groups forming the microenvironment of the invariant tryptophanyl. The proximity of the Lys 27 and Trp 29 side chains in ST is similar to the arrangement of homologous residues in neurotoxin II and corresponds to the X-ray data (4). Therefore Lys 27 and Trp 29 residues are a part of the ß-structural region of the neurotoxin central loop.

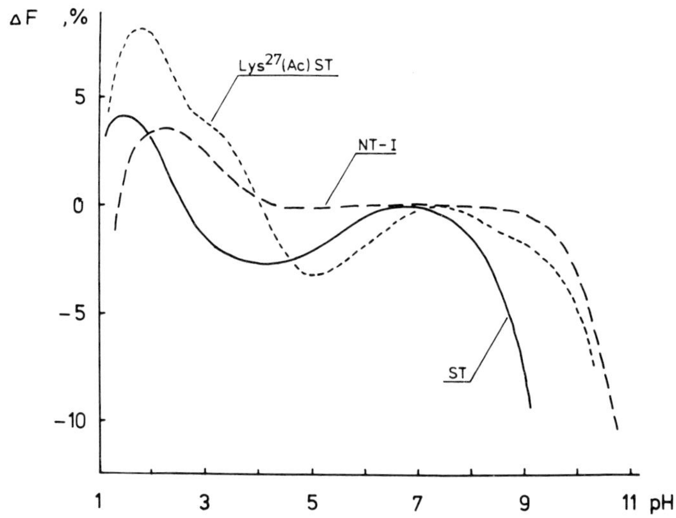

Fig. 10. pH-Dependence of fluorescence intensity for long-chain neurotoxins.

However, some differences exist between the ST and NT-II. Acetylation of Lys 27 in neurotoxin II neither shifted the tryptophan fluorescence maximum, nor decreased the quantum yield or affected the accessibility of Trp residues to the charged quencher (23). Such differences evidence in favor of the definite contribution of the Lys 27 ε-amino group in ST into the total positive charge in the vicinity of the invariant Trp 29, whereas in NT-II the Lys 27 side chain contribution is of lesser importance. Besides, Lys 27 acetylation in ST resulted in the appearance of additional transition with pK about 4,3 (Fig. 10). Acetylation of the homologous residues in NT-II slightly shifted the titration curve into the acidic region (23). Thus mutual arrangement of the charged groups in ST and NT-II differs, at least that of the groups influencing the Trp 29 microenvironment.

The fluorescence pH-titration curve of NT-I differs quite substantially from the curve of ST, especially within the pH region above 4 (Fig. 10). In the alkali region fluorescence quenching of NT-I occurred at pH values one unit higher than in case of ST. Apparently the pK value of the Lys 27 ε-amino group of NT-I is higher than that of ST. Low reactivity of this residue in NT-I as compared with ST and NT-II is probably due to the same reason. According to NMR data (6) the positive charge of Arg 40 affects the Lys 27 state in ST, whereas in NT-I the Val 40 residue occupies this position. Besides, Lys 53 present in the majority of short and long neurotoxins is substituted by the glutamic acid residue in NT-I. Each of these substitutions can decrease pK of the Lys 27 amino group (24).

The region of the ST titration curve within the 4-7.5 pH region corresponds to deprotonation of the His 22 imidazole ring, since according to NMR data only this residue of ST was titrated in the indicated pH range (6, 8). For NT-I, having one histidine residue in the C-terminal fragment of the molecule, no fluorescence change was observed within pH region 4.0 - 7.5.

In addition to fluorescence changes, titration of His 22 was accompanied with small changes in the CD spectra within the region 190-250 nm. pK value equal to 5.4 $\pm$0.1 was calculated from the pH-dependence of ellipticity at 227 nm, which coincided with the pK value found from the fluorescence titration curve. Titration of His 22 apparently resulted in a conformational transition, the secondary structure being altered only slightly. The pK of this transition well agreed with the value recently published by Hider et al. (8). According to (8) the conformational transition in ST affected about one third of amino acid residues spatially close to His 22, but not the direct environment of Trp 29. This suggestion is at variance with our data on fluorescence pH-titration of ST. The appearance of the charge on the His 22 imidazole ring affected the invariant Trp 29 microenvironment, though in the crystal structure these residues were located on the opposite sites of the central loop, at a distance of 20-25 Å (4). Apparently, the differential fluorescence method allowed to detect small changes in the tryptophan microenvironment more accurately than CD and H NMR techniques.

Comparison of the neurotoxin structure and conformational mobility may be relevant for understanding their mode of action. The differences among various neurotoxins in kinetics of their binding to acetylcholine receptor were interpreted in terms of higher conformational stability of the long-chain neurotoxins as compared to short ones (9).

In our opinion, such generalization is not well justified, since the examples are known showing that certain long-chain neurotoxins are more susceptible than short ones to temperature (25, 26), pH or organic solvents (25). Considerable differences in conformational mobility were observed also between the neurotoxins belonging to the same sub-family (27). Investigation of erabutoxins *a* and *b* (19) and the pH titration of ST and NT-I also testified that definite features of the structure (e.g. ionization of the His residue) might bring about the alterations in the conformations of the toxins of one type comparable with those dependent on the difference in the short and long toxin structures.

3. Toxin-receptor interaction

The interaction of fluorescence and spin labeled derivatives of NT-II with solubilized acetylcholine receptor from the electric organ of *Torpedo marmorata* has been earlier studied (6, 17, 28). The side chains of Lys residues, excepting Lys 51 (Lys 44 according to the NT-II sequence), in some or other way interacted with the receptor surface. The largest changes in fluorescence parameters characterizing the interaction of fluorescence labels with the receptor surface (shift of the maxima and increase in the fluorescence quantum yield, energy transfer from the receptor tryptophan residues on the label) were observed for dansyl labels at residues Lys 27 and Lys 53 (6).

According to the X-ray data, the side chains of these residues should be approximately in one direction relative to the plane of the toxin molecule (1). Measurements of distances between spin labels in a series of di-spin labeled NT-II derivatives (28) testified to the same arrangement of Lys 27 and Lys 53 side chains in solution.

The tightest contacts between the receptor and neurotoxin might be expected at those regions where many neurotoxin side chains interact with the receptor surface. According to the above fluorescence studies, the indole ring of invariant Trp 29 is one of the nearest neighbours of the side chain of Lys 27, hence it might be implicated in binding to the receptor molecule.

Earlier, data on complex formation between the acetylcholine receptor and the derivatives of long-chain neurotoxins **with the** labels at identified positions were obtained only for ST and NT-I spin labeled derivatives (17). The aforementioned monodansyl derivatives of NT-I were applied to elucidate the peculiarities of the binding of long toxins to the acetylcholine receptor. Fig. 11 shows fluorescence spectra of dansyl labels for NT-I derivatives in a buffer and in a complex with the solubilized receptor. If fluorescence was excited at 330 nm, the change in the emission band was quite small, i.e. the binding practically did not change the polarity of the environment of labels (29). When excited at 297 nm, i.e. under the conditions when tryptophan residues were also excited, the fluorescence intensity was considerably higher than that for the free toxin. It means that fluorescence of Dns-label was sensitized due to energy transfer from tryptophan residues of the receptor and the labels were located in the vicinity of the receptor surface.

We can conclude that the long-chain neurotoxins share a ge-

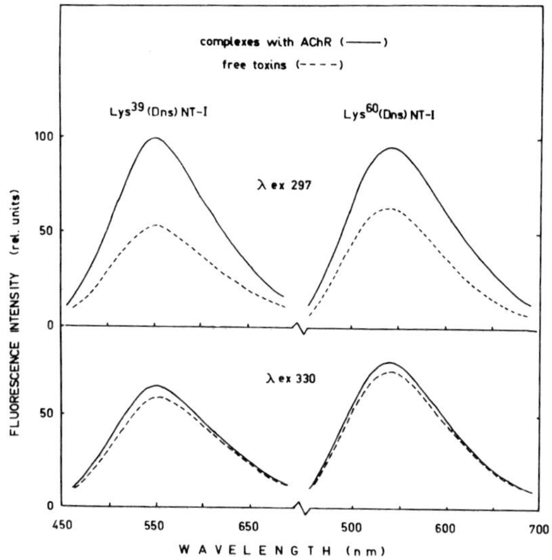

Fig. 11. Fluorescence spectra of mono-dansylated derivatives of NT-I in buffer and in complex with acetylcholine receptor solubilized in Triton X-100.

neral type of binding to the acetylcholine receptor with the short ones. This view is based on the following reasons. (a) In the ST monospin labeled derivative the Lys 27 side chain is in close contact with the receptor (6); (b) Comparative NMR study of NT-I and ST (6) and similarity of the CD spectra of these proteins evidence in favor of similar folding of their polypeptide chains; (c) Side chains of amino acid residues in ST corresponding to Lys 39 and Lys 60 residues in NT-I are oriented in direction opposite to that of the Lys 27 side chain (4).

Using NT-II and ST as examples we demonstrated that there are certain differences in the arrangement of the charged groups relative to invariant residues Lys 27 and Trp 29. These features may be partly responsible for differences in the kinetics of binding of short and long neurotoxins to the receptor. However, more detailed understanding of the factors affecting the kinetics of the toxin-receptor interaction needs further investigation .

Acknowledgements. The authors express their gratitude to Dr. Evert Karlsson for many helpful discussions and for making available the toxin 3 acetylated derivatives.

References

1. Low B.W. In: Handbook of experimental Pharmacology (Lee, C.Y. ed.) vol. 52, pp. 213-257, Springer Verlag, Berlin (1979).

2. Tsernoglow, M.D., Petsko, G. FEBS Lett. 68, 1-4 (1976).

3. Low,B.W., Preston, H.S., Sato, A., Rosen, L.S., Searl, J.E., Rudko, A.D., Richardson, J.S. Proc. Natl. Acad. Sci. U.S.A. 73, 2991-2994 (1976).

4. Walkinshow, M.D., Saenger, W., Maelicke, A. Proc. Natl. Acad. Sci. USA, 77, 2400-2404 (1980).

5. Kistler, J., Stroud, R.M., Klymkowsky, M.W., Lalancette, R.A., Fairelough, R.H. Biophys. J. 31, 371-383 (1982).

6. Tsetlin, V.I., Karlsson, E., Utkin, Yu.M., Pluzhnikov, K.A., Arseniev, A.S., Surin, A.M., Kondakov, V.V., Bystrov, V.F., Ivanov, V.T., Ovchinnikov, Yu.A. Toxicon 20, 83-93 (1982).

7. Inagaki, F., Claydon, N.J., Tamiya, N., Williams, R.J.P., Eur. J. Biochem. 120, 313-322 (1981).

8. Hider, R.C., Drake, A.F., Inagaki, F., Williams, R.J.P., Endo, T., Miyazawa, T. J. Mol. Biol. 158, 275-291 (1982).

9. Endo, T., Inagaki, F., Hayashi, K., Miyazawa, T. Eur. J. Biochem. 120, 117-124 (1981).

10. Pashkov, V.S., Arseniev, A.S., Utkin, Yu.N., Tsetlin, V.I., Bystrov, V.F. Bioorgan. Khim. 8, 588-615 (1982).

11. Karlsson, E. In: Handbook of Experimental Pharmacology (Lee C.Y. ed.) vol. 52, p. 159-212, Springer-Verlag, Berlin (1979).

12. Chicheportiche, R., Vincent, J.-P., Kopeyan, C., Schweitz, H., Lazdunski, N. Biochemistry 14, 2081-2091 (1975).

13. Pluzhnikov, K.A., Karelin, A.A., Utkin, Yu.N., Tsetlin, V.I., Ivanov, V.T., Bioorgan. Khim. 8, 905-913 (1982).

14. Utkin, Yu.N., Pashkov, V.S., Pluzhnikov, K.A., Kurjatov, A.B., Arseniev, A.S., Tsetlin, V.I., Bystrov, V.F., Ivanov, V.T. Bioorgan. Khim. 9, 437-449 (1983).

15. Tsetlin, V.I., Arseniev, A.S., Utkin, Yu.N., Gurevich, A.Z., Senyavina, L.B., Bystrov, V.F., Ivanov, V.T., Ovchinnikov, Yu.A. Eur. J. Biochem. 94, 337-346 (1979).

16. Dolly, J.O., Nockles, E.A.V., Lo, M.M.S., Bernard, E.A. Biochem. J. 193, 919-923 (1981).

17. Tsetlin, V.I., Karlsson, E., Arseniev, A.S., Utkin, Yu.N., Surin, A.M., Pashkov, V.S., Pluzhnikov, K.A., Ivanov, V.T., Bystrov, V.F., Ovchinnikov, Yu.A., FEBS Lett. 108, 47-52 (1979).

18. Burstein, E.A., Vedenkina, N.S., Ivkova, M.N., Photochem. Photobiol. 18, 263-279 (1973).

19. Menez, A., Montinay-Garenstier, Th., Fromageot, P., Helene, C. Biochemistry 19, 5202-5208 (1980).

20. Arseniev, A.S., Surin, A.M., Utkin, Yu.N., Tsetlin, V.I., Bystrov, V.F., Ivanov, V.T., Ovchinnikov, Yu.A. Bioorgan. Khim. 4, 197-207 (1978).

21. Shinitzky, M., Goldman, R. Eur. J. Biochem. 3, 139-144 (1967).

22. Green, J.A., Singler, L.A., Parks, J.H. J. Chem. Phys. 58, 2690-2695 (1973).

23. Surin, A.M., Utkin, Yu.N., Pluzhnikov, K.A., Efremov, E.S., Tsetlin, V.I., Bioorgan. Khim. 7, 342-356 (1981).

24. Tanford, C., Roxby, R. Biochemistry 11, 2192-2198 (1972).

25. Tsetlin, V.I., Mikhaleva, I.I., Myagkova, M.A., Senyavina, L.B., Arseniev, A.S., Ivanov, V.T., Ovchinnikov, Yu.A. In: Peptides Chemistry, Structure and Biology (Walter R., Meinhofer J., eds.) Ann. Arbor.: Ann Arbor Sci. pp. 935-941 (1975).

26. Drake, A.F., Dafton, M.J., Hider, R.C. Eur. J. Biochem. 105, 623-630 (1980).

27. Nebdryk-Viala, F., Thiery, C., Menez, A., Tamiya, N., Thiery, J.M. Biochim. Biophys. Acta 626, 321-331 (1980).

28. Tsetlin, V., Karlsson, E., Utkin, Yu., Arseniev, A., Pluzhnikov, K., Surin, A., Pashkov, V., Bystrov, V., Ivanov, V. In: Chemistry of Peptides and Proteins (Woelter W., Wünsch, E., Ovchinnikov Yu.A., Ivanov V.T., eds.) vol. 1, pp. 389-398, Walter de Gruter & Co., Berlin (1982).

29. Chen, R.F. Arch. Biochem. and Biophys. 120, 609-620 (1967).

MAGNETIC RESONANCE EVALUATION OF SNAKE NEUROTOXIN STRUCTURE-FUNCTION RELATIONSHIP

V.F.Bystrov[+], V.I.Tsetlin[+], E.Karlsson[&], V.S.Pashkov[+], Yu. Utkin[+], V.I.Kondakov[+], K.A.Pluzhnikov[+], A.S.Arseniev[+], V.T. Ivanov[+], Yu.A.Ovchinnikov[+]

[+] M.M.Shemyakin Institute of Bioorganic Chemistry, USSR Academy of Sciences, Moscow, USSR

[&] Institute of Biochemistry, Uppsala, Sweden

Summary

The crucial details of the spatial structure in solution and of interactions of polypeptide neurotoxins are revealed by a combination of spectroscopic techniques (NMR, EPR, CD, laser Raman, fluorescence) and selective chemical modification of native molecules. High field NMR 2D-spectroscopy was extensively used as a very informative method in conformational analysis.

Series of homologous *Naja* snake venom "short" and "long"type neurotoxins as well as their fluorine and spin labeled derivatives were studied by NMR and EPR to reveal intramolecular nonbonding contacts and consequently to establish the polypeptide backbone folding and conformational dynamic of the molecules in different solvents.

By EPR study of selectively spin labeled "short" and "long" type neurotoxins interacting with *Torpedo marmorata* nicotinic acetylcholine receptor the topography of the binding site and the type of interactions were evaluated. The results are compared with data obtained by fluorescence spectroscopy and photoaffinity cross-linking.

Toxins as Tools in Neurochemistry

Introduction

Molecular conformation represents a crucial factor in the functioning of physiologically active compounds. To this one must add that it is becoming more and more recognized that the dynamic aspects, the degree of flexibility of the molecular structure under corresponding physiological conditions is of equal importance.

As far as tertiary structure of peptides and proteins are concerned, one must admit that even minor conformational transitions could play an important role in their biological mechanism of action. Thus, it is vitaly significant to evaluate not only the binding site in, for instance, the protein ligand-receptor interaction, but also to present data on conformational changes, if any, that could have occured in the course of binding.

Of the various spectroscopic techniques available for a molecule in solution magnetic resonance is the choice for

- delineation of dynamic conformations and intermolecular interactions under various environmental including those resembling the physiological conditions,
- sheding light on the specific intramolecular interactions leading to given conformational properties,
- identification of particular groups participating in interactions with biological targets, for instance with ligands and receptors,
- revealing the role of the molecular spatial structure and its dynamical properties in a particular biological process performed by the molecule; that is to answer the question of just why has nature constructed such a molecule for a given task and not another one.

This report summarizes the results obtained recently in the Shemyakin Institue of Bioorganic Chemistry on the structure and the conformational transitions of snake venom neurotoxins in solution and on the topography of their binding with the target of the neurotoxin action - the nicotinic acetylcholine receptor.

There is a large set of homologous snake neurotoxins; among them there are two major types - "short" neurotoxins with 6o-62 amino acid residues and four disulfide bonds, and "long" toxins with 71-74 residues and five disulfide bridges (1). Additionally there is a set of nonneurotoxic polypeptides - cytotoxins and cardiotoxins - with close homology to the "short" neurotoxins (1). The degree of overall conformational stability has been studied by circular dichroism spectra at different pH, temperature and solvent composition (2). It was shown that "short" neurotoxins are most resistant to environmental changes, while cardiotoxins and cytotoxins are less resistant.

Materials and Instrumentation

Naja naja oxiana neurotoxin I and II are prepared according to (65,66). Naja mossambica mossambica neurotoxin III is obtained as in (12). Naja nigricollis α-toxin and toxin 3 were kindly provided by Prof.P.Boquet and Dr.E.Karlsson. Preparation of selectively trifluoroacetylated, spin- and dansyl-labeled derivatives of neurotoxin I and II and toxin-3 is described in (7,8,11,13-18,34,51).

Acetylcholine receptor protein from the electric organ of Torpedo marmorata was isolated in Triton X-1oo by affinity chromatography on neurotoxin-Sepharose, gel- and ion-exchange chromatography and resulted in solubilized preparations with 7-1o nM α-toxin binding sites per 1 mg of receptor protein (67). In fluorescence and ESR experiments respectively 2.5-8.o · $1o^{-7}$ M and 3.5 · $1o^{-6}$ M aqueous solutions of receptor (o.3 M sodium phosphate buffer, pH 7.5, o.o25 % Triton X-1oo o.1 M NaCl, o.o2 % NaN_3) were used.

The NMR spectra were recorded on Varian SC-3oo or Bruker WM-5oo spectrometers. The pH measurements were carried out directly in the 5 mm o.d. NMR tubes using an 18o-mm Ingolf 4o5 M3

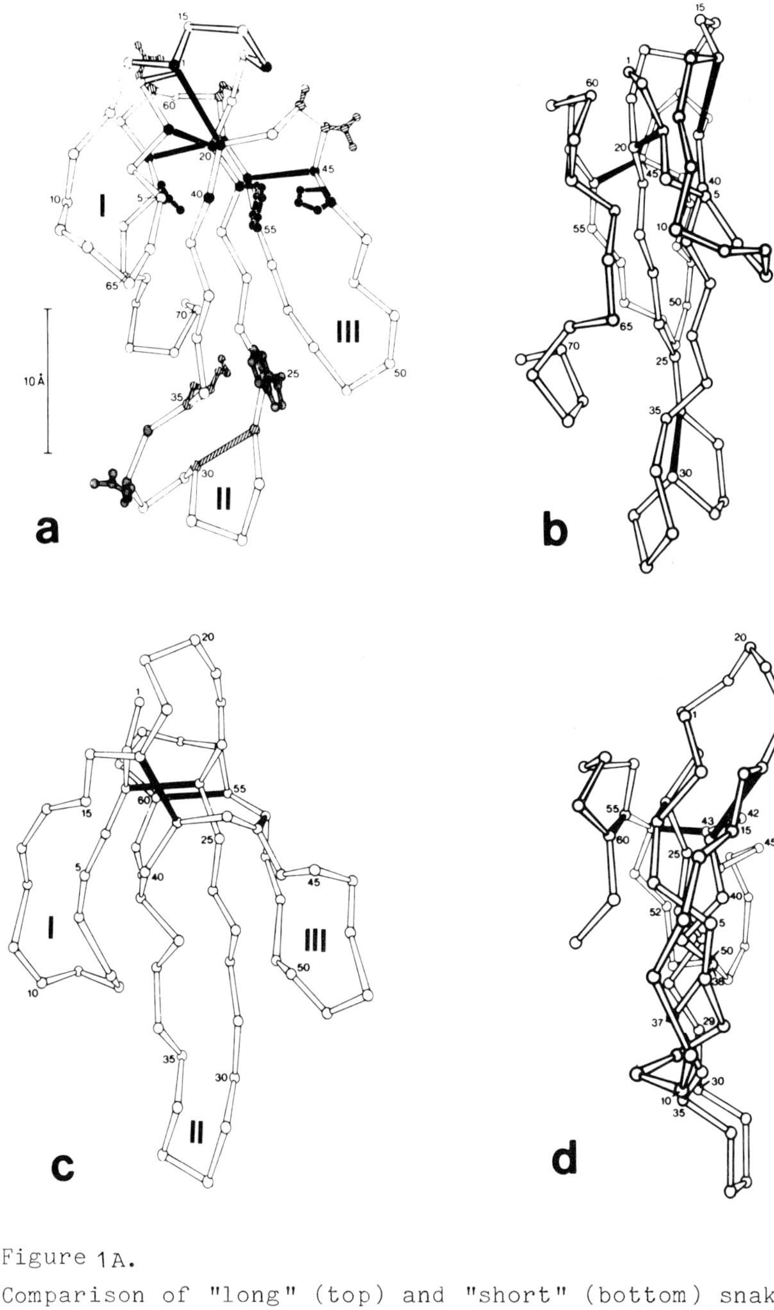

Figure 1A.

Comparison of "long" (top) and "short" (bottom) snake neurotoxins. Shown are views of the front side (left) and from the side (right). Loops I, II, III marked by Roman numbers. In (a), invariant residues are indicated. Numbering is sequential, Trp25 in (a) called Trp29 in the text, according to Karlsson's nomenclature (1).

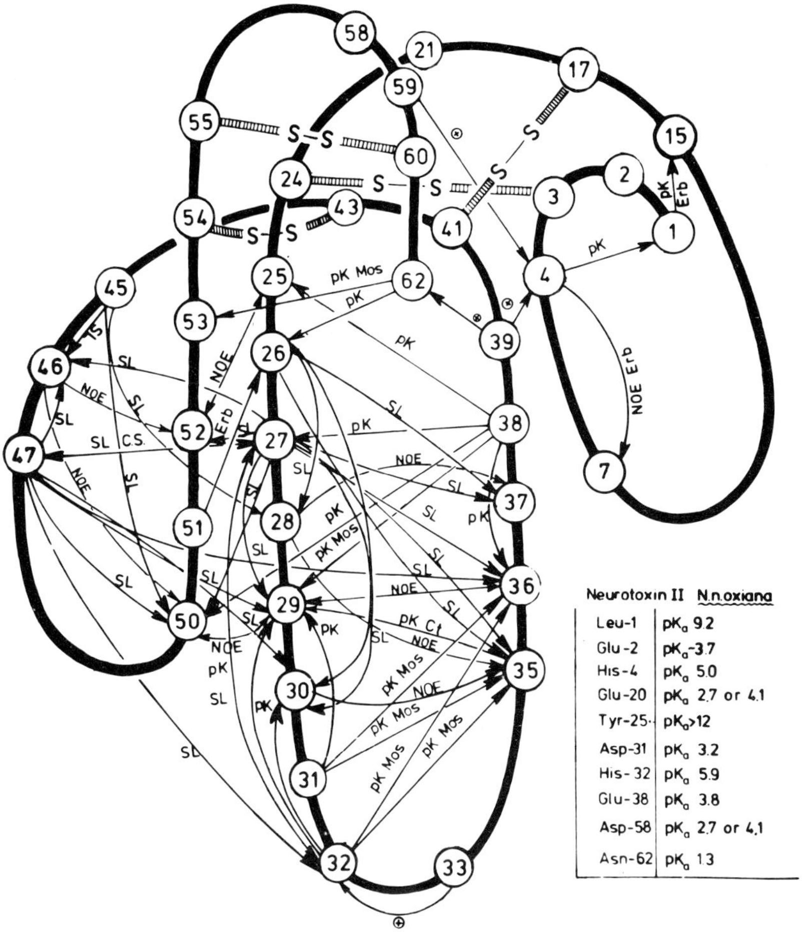

Neurotoxin II	N.n.oxiana
Leu-1	pK_a 9.2
Glu-2	pK_a-3.7
His-4	pK_a 5.0
Glu-20	pK_a 2.7 or 4.1
Tyr-25	pK_a>12
Asp-31	pK_a 3.2
His-32	pK_a 5.9
Glu-38	pK_a 3.8
Asp-58	pK_a 2.7 or 4.1
Asn-62	pK_a 1.3

Fig.1B Backbone folding of "short" neurotoxin molecule as derived from NMR study of the Naja naja oxiana neurotoxin II (4). Additional contacts deduced from NMR studies of other "short" neurotoxins are indicated as follows: Mos-neurotoxin III Naja mossambica mossambica, Erb - erabutoxin b Laticauda semifasciata (26-28) and Ct - cobratoxin Naja naja atra (23). Abbreviations: pK - protonation of ionogenic group influence on chemical shifts, (+) - charge effect on pK values, NOE - nuclear Overhauser effect, CS - chemical shift changes upon selective modification, SL - spin label broadening effect.

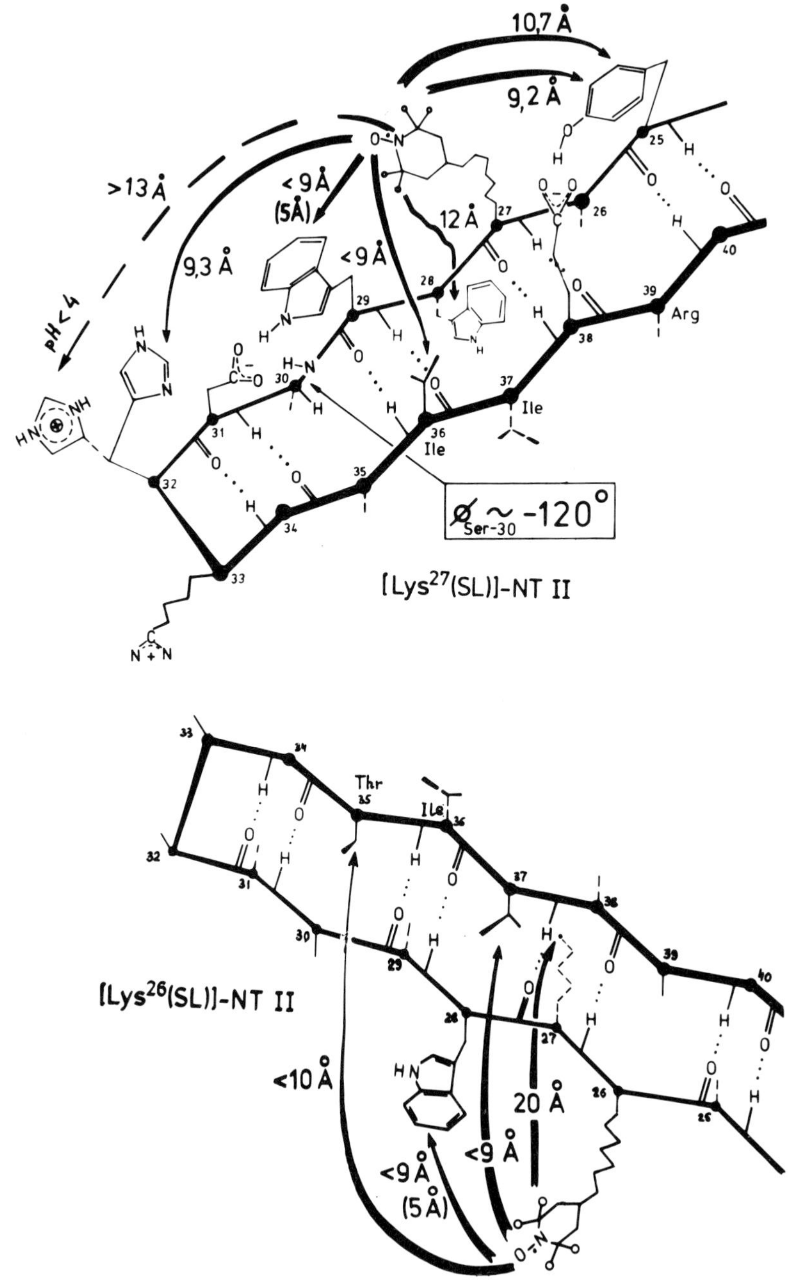

Fig.2 The β-antiparallel sheet conformation of the central loop of the Naja naja oxiana neurotoxin II molecule. A- Upper view. The spin label-proton distances were obtained from the NMR spectrum of 1-oxyl-2,2,6,6-tetramethylpiperidine-4-carboxy(ε-amino-Lys^{27}) neurotoxin II derivative. B - Bottom view. Distances were obtained for the SL-Lys^{26} derivative (4,1o,11,14-16). The Lys^{26}-Lys^{27} interlabel distance was evaluated from ESR spectrum of the dispin-labeled neurotoxin II (4,8,9,11-15,17).

combination microelectrode. For D_2O solution the values used are the direct reading of an Orion Research Model 6o1 digital pH-meter. Analysis of experimental pH dependencies of chemical shift were performed by interactive curve fitting software (68).

The ESR spectra were obtained on a Varian =-1o9 X-band spectrometer equipped with a Varian E-9oo Data Aquisition System.

Fluorescence measurements were made using an Aminco SPF-1ooo instrument. CD spectra were taken on a Jobin-Yvon III Dichrograph. Laser Raman spectra were obtained on a Ramanor HG-2D spectrometer.

Results

"Short" neurotoxins: Extensive NMR studies (3-18) demonstrated that the backbone folding of the "short" neurotoxins (Table 1) in solution is very similar to the X-ray crystal structure (19-21).Fig.1B shows the intramolecular contacts revealed by proton NOE (16,18), spin label induced paramagnetic relaxation (4,8,9,11,13-16,18) and the influence of functional group ionization on the chemical shift of spatially close protons (3-12, 15,16). The general features of this backbone folding are preserved over a wide range of pH and temperature, at least for terrestrial snake neurotoxins (17).

Side chains of course could be subject to considerable movements, especially if they are not in a strictly constrained environment. For instance, a reorientation of the His^{32} imidazole ring is observed when it is protonated (4,17) as shown in Fig 2. For constructing the antiparallel β-structure of the *Naja naja oxiana* neurotoxin II central loop we used the data obtained from NMR spectra of the two monospin-labeled derivatives (4,1o,11,14-16). It is supported also by pH titration of the connectivities (3-5,12,16) (see Fig.1) and by NOE between spatially close groups (16). As an example, Fig.3 shows the truncated NOE difference spectrum where the downward directed signals correspond to the result of irradiation of the

Table 1: Amino acid sequences of snake "short" neurotoxins (1) studied by NMR: 1, (3-18), 2 (6), 3 (12), 4 (22), 5 (23-25), and 6-8 (26-28). The standard IUPAC one-letter code is used for amino acid residue notation.

	1				5					1o					15					2o					25
1) Neurotoxin II, <u>N.n.oxiana</u>	L	E	C	H	N	Q	Q	S	S	Q	P	P	T	T	K	T	C	S	-	G	E	T	N	C	Y
2) α-Toxin, <u>N.nigricollis</u>																		P	-						
3) Neurotoxin III, <u>N.mossambica m.</u>		N					M		A						T	R			R	W					
4) Neurotoxin I, <u>N.mossambica m.</u>										E					T	R			G						
5) Cobrotoxin, <u>N.n.atra</u>											T				T	G			G						
6) Erabutoxin a, <u>L.semifasciata</u>	R	I		F			H					Q						P	S			S	S		
7) Erabutoxin b, <u>L.semifasciata</u>	R	I		F			H					Q						P	S			S	S		
8) Erabutoxin c, <u>L.semifasciata</u>	R	I		F			H					Q						P	S			S	S		

	26				3o					35					4o					45					5o					55					6o		62
1) Nt-II	K	K	W	W	S	D	H	R	G	T	I	I	E	R	G	C	G	C	P	K	V	K	P	G	V	N	L	N	C	C	R	T	D	R	C	N	N
2) a-T			V		R															T					I	K					T			K			
3) Nt-III			R		R					Y	R	T								T			K		I	Q		H			T	S		N			
4) Nt-I			R		R					Y	R	T								T			K		I	E					T						
5) Ct			R		R					Y	R	T								S			N		I	E	I				T						
6) Er-a	N		Q				F													T					I	K		S			E	S	E	V			
7) Er-b	H		Q				F													T					I	K		S			E	S	E	V			
8) Er-c	H		Q				F													T					I	N		S			E	S	E	V			

γ-methylproton resonance of Ile^{36}. Those signals demonstrate the cross-loop NOE with the aromatic Trp^{29} protons as well as trivial side chain NOE with α-, β- and γ-CH protons of Ile^{36}. Similarly many other NOE contacts were detected (16) to confirm the backbone folding shown in Fig.1 at different pH of aqueous solution.

The disadvantage of the NOE experiments is the limited range of controlled distance, usually not more than 5 Å (29). Spin labeling studies complement the NOE data in that the proton

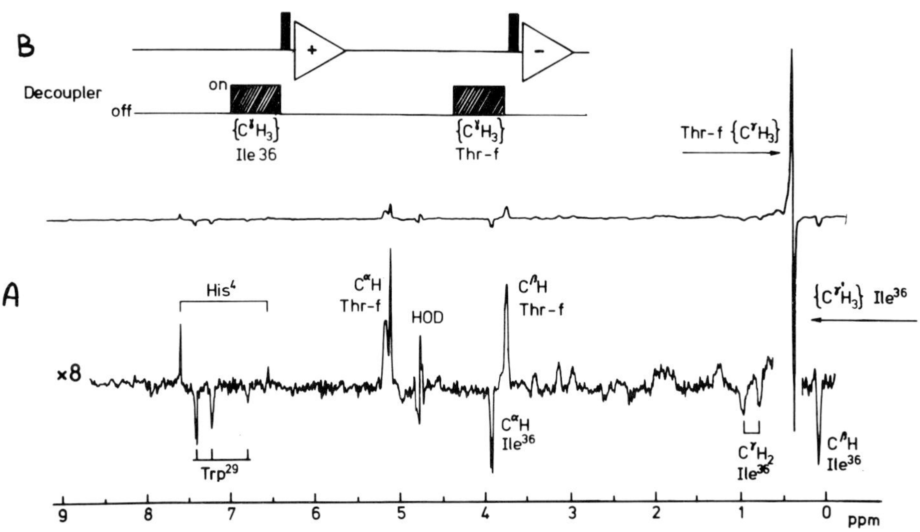

Fig.3 A - Truncated NOE difference 5oo MHz spectrum of <u>Naja naja oxiana</u> neurotoxin II (9 mM solution in D_2O, pH 8.o at 25°C) for irradiation of $C^{\gamma}H_3$ resonance of Ile^{36} residue (downward signal) and of $C^{\gamma}H_3$ resonance of one of threonine residue (upward signal). B - Diagram of pulse sequence used in obtaining the spectrum. Irradiation time and field strength are o.3 sec and 4 Hz, respectively.

Table 2: Selected inter side-chain distances (Å) in the Naja naja oxiana neurotoxin II molecule in solution as measured by reporter group techniques (4), and in the crystal (in brackets) as estimated from X-ray data on erabutoxin b (19-21).

Residue	Lys^{26}	Lys^{27}	Trp^{28}	Trp^{29}	Lys^{45}	Lys^{47}
Lys^{15}	26[a] (29)	18[a] (15)			23[a] (15)	21[a] (19)
Tyr^{25}	13[b] (18)	10[b] (10)		13[c] (9)	10[b] (16)	12[b] (14)
Lys^{26}	-	20[a] (22)	6[d] (9)	12[b,e] (16)	23[a] (29)	26[a] (26)
Lys^{27}		-	12[b,e] (18)	6[d] (10)	19[a] (8)	15[a] (6)
Trp^{28}			-	9[c] (8)	9[b] (25)	11[b] (19)
Trp^{29}				-	10[b] (17)	8[b] (11)
Lys^{45}					-	23[a] (8)

[a] From ESR spectra of dispin-labeled compounds (4,9)

[b] From proton NMR spectra of monospin-labeled compounds (4,10,14-16,18)

[c] From energy transfer measurements by fluorescence spectroscopy of the Naja nigricollis toxin α (33)

[d] From quenched fluorescence spectra of monospin-labeled derivatives (34, 35)

[e] 18 and 14 Å from the fluorescence spectrum of the monodansyl-labeled compound (34,35)

relaxation, line width in particular, is perturbed in the range up to 12-15 Å from the nitroxyl radical (3o). For Naja naja oxiana neurotoxin II monospin-labeled derivatives were obtained for His^{32}, Glu^{2}, the α-amino group of Leu^{1} and γ-amino groups of lysine residues 15,26,27,45 and 47 (8,9,11,31). Proton NMR spectra of those derivatives were used for evaluating the apparent distances of spin label-protons (4,1o,11,14-16, 18). One example of this technique was shown in Fig.2. Some of the other results are included in the general picture of the molecular topography (Fig.1).

Another use of spin labels is illustrated by double labeled neurotoxin derivatives that help to overcome the problem of signal resolution and assignment. The simultaneously fluorine and monospin-labeled Naja naja oxiana neurotoxin II derivatives provide apparent distances deduced from fluorine resonance line widths (13). All ten possible ε-lysine dispin-labeled compounds were obtained by careful purification of corresponding reaction products and interspin distances were evaluated (4,8,15) from dipole-dipole interaction effects in ESR spectra at $77^{o}K$ (32). Conservation of the intact conformation of the chemically modified derivatives was carefully controlled by CD and NMR. The most representative apparent inter-side-chain distances obtained using double-labeled compounds are compiled in Table 2. Tryptophan fluorescence spectra of monospin-labeled and of dansyl-Lys^{27} derivatives provide additional interresidue distances (34,35). Their overall agreement with distances estimated from the published X-ray crystal structure of erabutoxin b (19-21) proves the adequacy of the utilized approach and the stability of the "short" neurotoxin backbone folding. Some discrepancies in Table 2 could be easily rationalized either by differences in the primary structure of the Naja naja oxiana neurotoxin II and the Laticauda semifasciata erabutoxin b (there is, for instance a deletion in position 19 in the neurotoxin II sequence), or by motional averaging of side chain position in solution.

Although the gross structure of "short" neurotoxins is generally very resistant to denaturation by a variety of conditions, individual representatives of this class of snake neurotoxins do vary in the precise degree of this stability. For instance the sea snake erabutoxins are highly vulnerable to denaturation conditions and their spatial structure is more flexible as was shown by NMR and optical spectroscopy (28,36,37).

"Long" neurotoxins: This (15) as well as other (38-43) NMR studies of snake "long" neurotoxins (Table 3) indicate the general agreement of the molecular backbone folding in solution shown in Fig.4 with the model obtained by X-ray structure analysis of crystals(44-47). Moreover comparison of spatial organization (45) of the neurotoxins demonstrates that they possess a common core structure schematically presented in Fig.5. It consists of antiparallel β-strands shown by thick solid lines in Fig.5.

The particular conformational stability of the core structure for "long" neurotoxins was revealed by 2-D NMR (NOE) at different pH values of an aqueous toxin solution. NOESY spectra at pH 2.2 and 7.5 show specific cross-peaks between α-proton resonances (Fig.6) and between methyl and aromatic protons (Fig.7), which serve as a basis for constructing the antiparallel β-structure of "long" neurotoxins shown in Fig.8. The NOE connectivities for backbone amide NH protons are best observed at low pH values (Fig.9). Due to the increased rate of deuterium exchange at pH 7.5 only one connectivity between the backbone NH of Tyr^{25} and the α-proton of Cys^{45} is observed in the NOESY spectrum.

If one could consider the core antiparallel β-structure shown in Fig.5 as relatively stable towards pH changes, the other regions of "short" and "long" neurotoxins may contain certain fragments predisposed to conformational transitions. For "short" neurotoxins as mentioned above conformational changes occurred at the tip of the central loop (4,1o,15,16) and in the N-termi-

Table 3: Amino acid sequences of snake "long" neurotoxins (1) studied by NMR: 1,2 and 4 this paper and (15), 1 (43), 3 (41,42), 5 (39), and 6 (42). The standard 1UPAC one-letter code is used for amino acid residue notation.

```
                                             1   5    10   15   20   25

1) Toxin 3, Naja naja siamensis              IRCF---ITPDITSKDCPNG-HVCY

2) Toxin 3, Naja naja naja

3) Toxin B, Naja naja naja

4) Neurotoxin I, Naja naja oxiana            T Y  KTPIP   ET AP QNL

5) α-Bungarotoxin, Bungarus multicinctus      V H TTATIPSSAVT PP ENL

6) Toxin Ls III, Laticauda semifasciata      RE Y   -LNPNDTQT  S QEI

                   26  30   35   40   45   50   55   60   65   70   75
1) N.S. Tx-3       TKTWCDAFCSIRGKRVDLGCAATCPTVK-TGVDIQCCST-DNCNPFPTRKRP

2) N.n. Tx-3             G                    R

3) N.n. Tx-B             G   S                R

4) N.o. NT-I              W GS   VIE          E SYQ  K      D  H KQ

5) α-BTX           R M       S   V E        SK  PYEEVT      K HP K QPG

6) Tx-Ls III       V S  N W  S   VLEF       S N   TE K   A  K TY
```

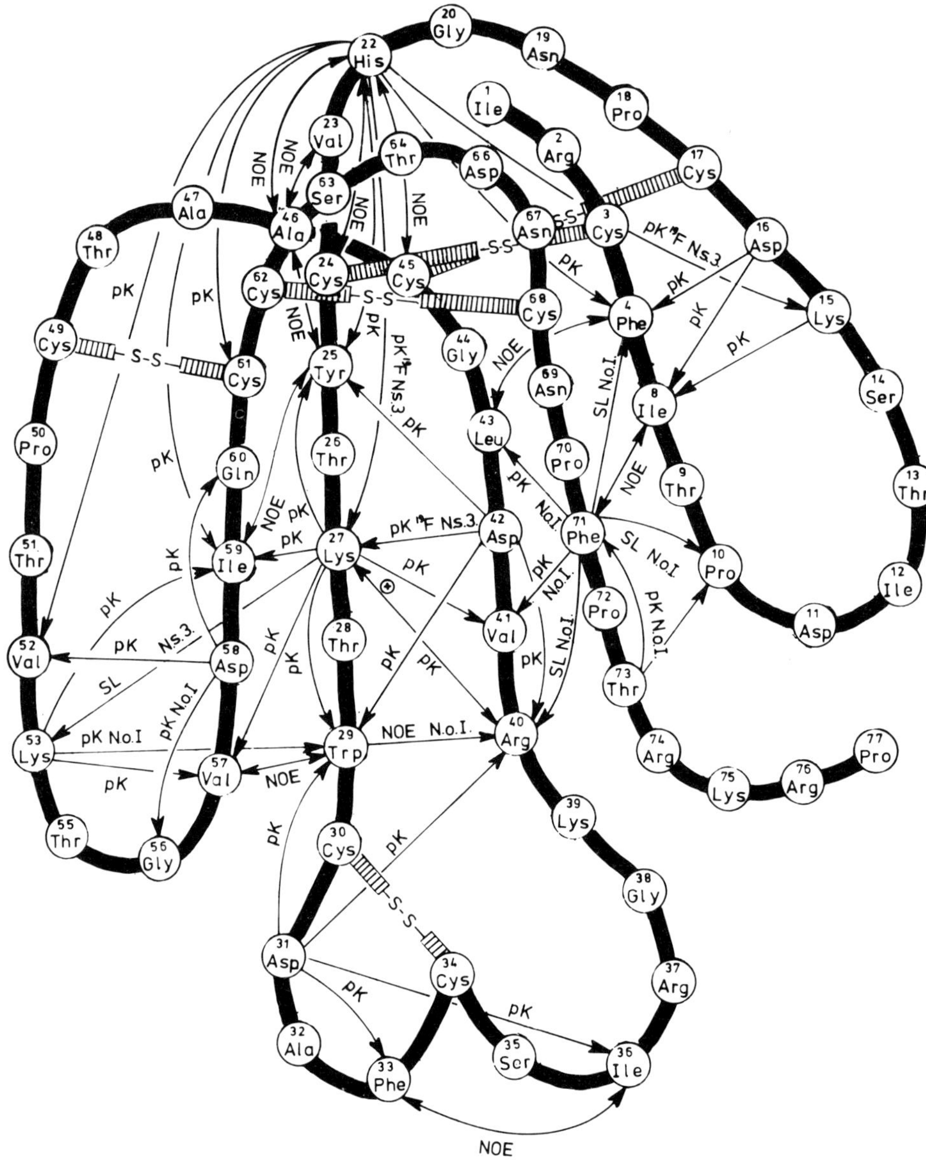

Fig.4 Backbone folding of "long" neurotoxin molecule as derived from NMR study of the Naja naja siamensis neurotoxin 3 (15). Additional contacts are deduced from NMR studies of the neurotoxin I Naja naja oxiana (N.o.I) (15). Other abbreviations see capture to Fig.1.

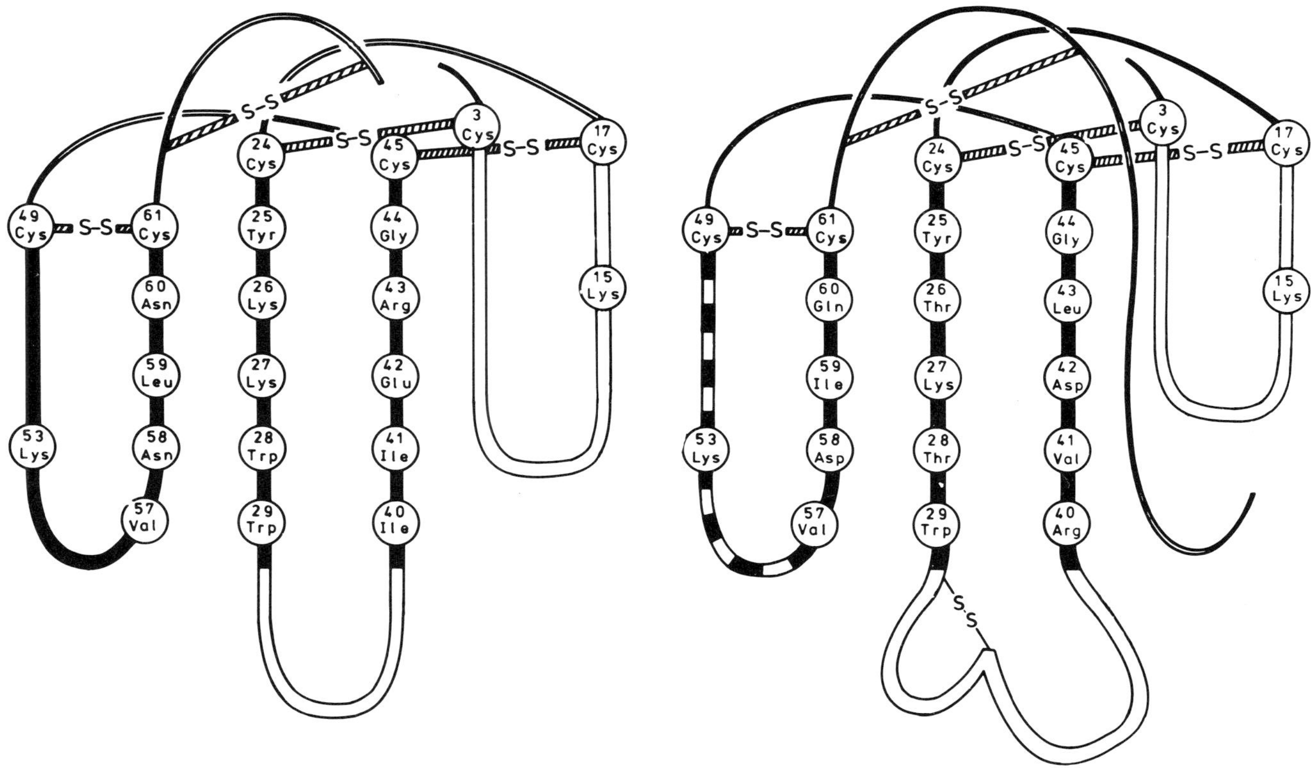

NT II (Naja n. oxiana) NT 3 (Naja n. siamensis)

Fig.5 The common core sturcture in the spatial organization of "short" and "long" types of snake neurotoxins

TOXIN 3 *Naja naja siamensis*

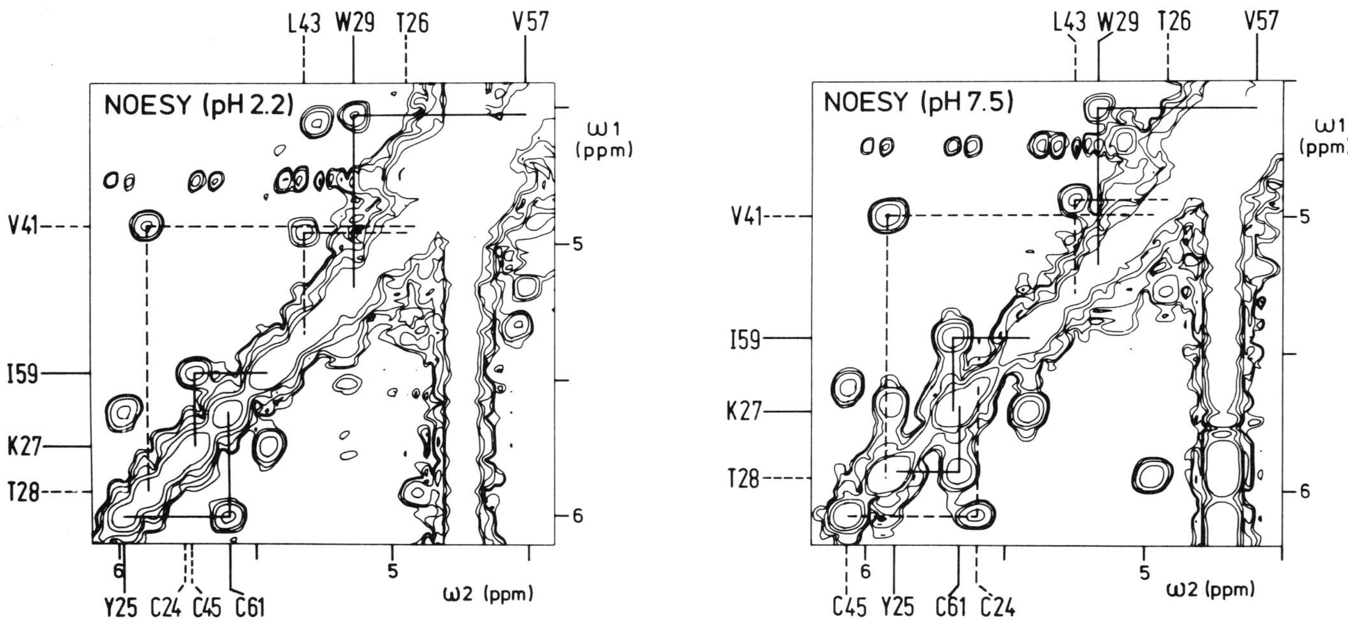

Fig.6 Contour plot of 500 MHz proton NOESY spectra (τ_m 0.1 sec) of 7 mM solution of the *Naja naja siamensis* toxin 3 in D_2O, 32°C at pH 2.2 (ω_1 and ω_2 4.4 - 6.1 ppm) and at pH 7.5 (ω_1 and ω_2 4.5 - 6.2 ppm). The $C^{\alpha}H$ connectivities are indicated for the following amino acid residues: Trp^{29} - Val^{57} (——), Thr^{28} - Val^{41} (---), Thr^{26} - Leu^{43} (---), Lys^{27} - Ile^{59} (——), Tyr^{25} - Cys^{61} (---) and Cys^{24} - Cys^{45} (---). The positions of the $C^{\alpha}H$ resonances of the Cys^{24} and Cys^{45} residues coincide at pH 2.2.

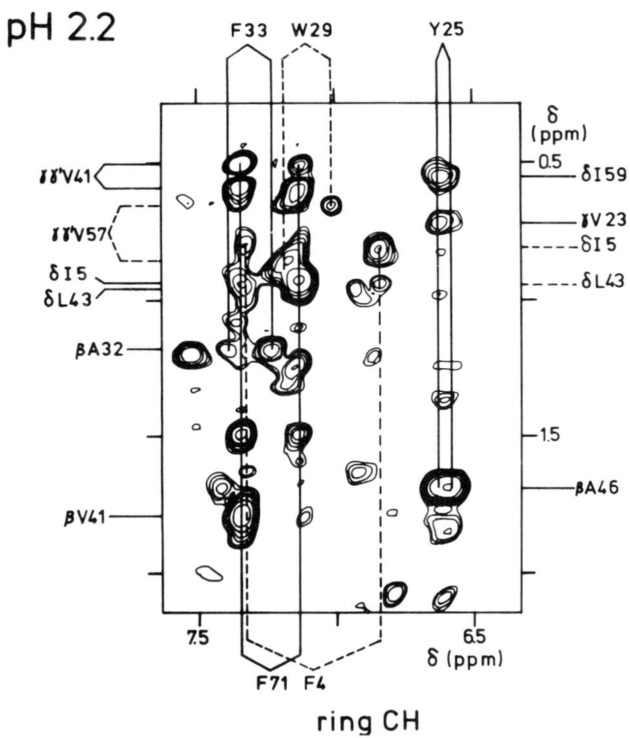

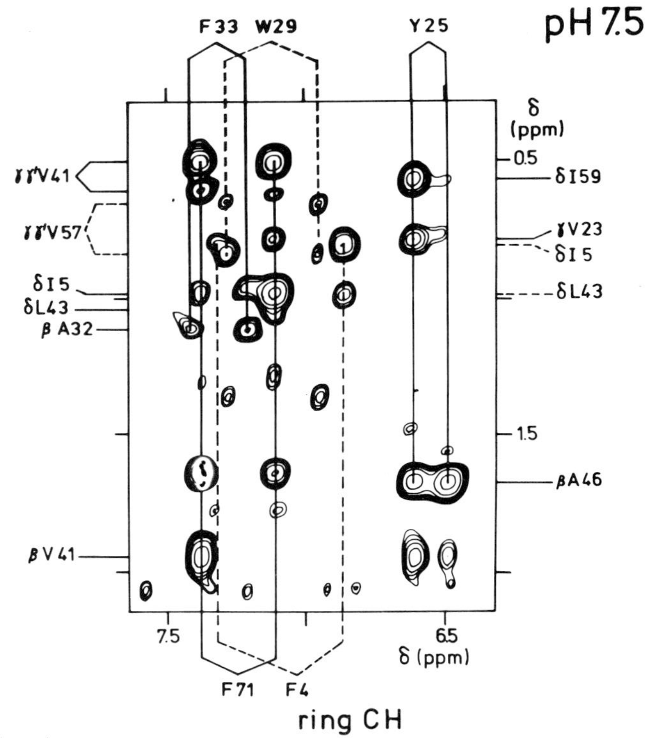

Fig.7 Contour plot of 500 MHz proton NOESY spectra (τ_m 0.1 sec, ω_1 0.4 - 1.8 ppm and ω_2 6.4 - 7.6 ppm) of 7 mM solution of the <u>Naja naja siamensis</u> toxin 3 in D_2O, pH 2.2 and 7.5 at 32°C. The NOE connectivities between aromatic and methyl side chain protons are indicated for the following amino acid residues: Phe^{4} - Ile^{5}, Leu^{43} (---), Tyr^{25} - Val^{23}, Ala^{46}, Ile^{59} (——), Trp^{29} - Val^{57} (---), Phe^{33} - Ala^{32} (——) and Phe^{71} - Ile^{5}, Val^{41}, Leu^{43} (——).

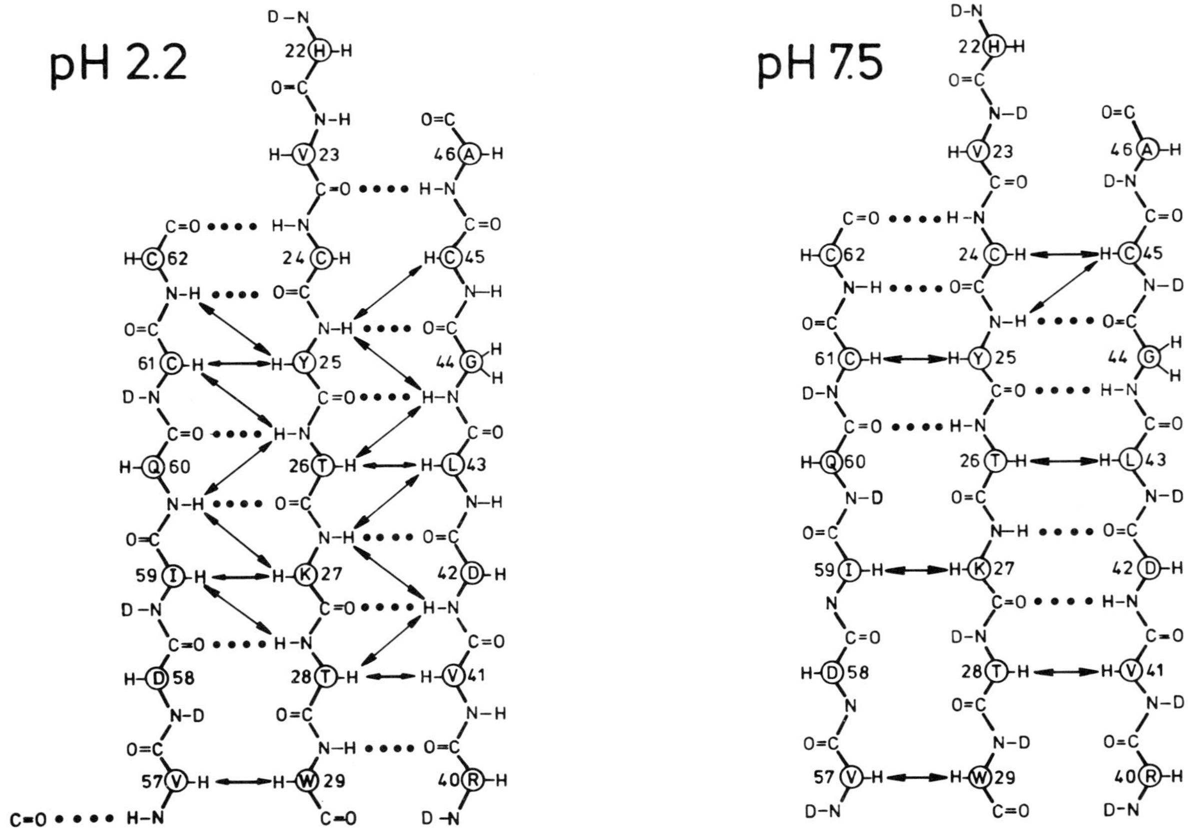

Fig.8 Schematic presentation of three strands antiparallel β-structure of the *Naja naja siamensis* toxin 3. The NOE connectivities between α-proton resonances (Fig.6), between backbone amide NH protons (Fig.9) and between α-proton and backbone amide NH

Fig.8 (contin.): resonances are shown by arrows as observed in NOESY spectra at pH 2.2 and 7.5. The hydrogen bonds are deduced from the deuterium exchange rates of corresponding backbone amide NH hydrogens ($t_{1/2}$ > 2o hours at pH 2.2 and 6-5o hours at pH 7.5).

nal fragment (13,17,27). Concerning "long" neurotoxins, two major changes are revealed. First, the structural transition in the upper loop, initiated by protonation of the His^{22} imidazole ring in the Naja naja siamensis toxin 3, and second in the C-terminal fragment connected with protonation of His^{71} in the Naja naja oxiana neurotoxin I.

It must be pointed out that all of these conformational transitions were detected in connection with protonation of histidine residues and because none of those histidines are conservative in the primary structure of homologous snake neurotoxins, details of the transitions presumably have no direct physiological significance. They are just emphasizing the conformational flexibility of corresponding regions of the molecules.

When the His^{22} imidazole ring of Naja naja siamensis toxin 3 is titrated (pK 5.6), two forms of the molecule are clearly seen by the splitting of the ring C_4H and C_2H signals (Fig.1o) as well as by other signal shapes. In the aromatic proton resonance region two conformational forms are expressed in the difference spectrum by Phe^4, Tyr^{25} and Phe^{71} signals (Fig.11). In the high field region this transition is revealed by one of the Val^{23} methyl resonances (Fig.12). Table 4 compiles the resonance data that respond to the His^{22} protonation either by signal splitting or by inflection in their chemical shift pH-dependencies. Similar observations were made for α-cobratoxin (43) and toxin B (41).

The conformational transition is also demonstrated by the mono- and dispin-labeled derivatives of the Naja naja siamensis toxin.3. For instance ESR spectra indicate that the Lys^{53} side

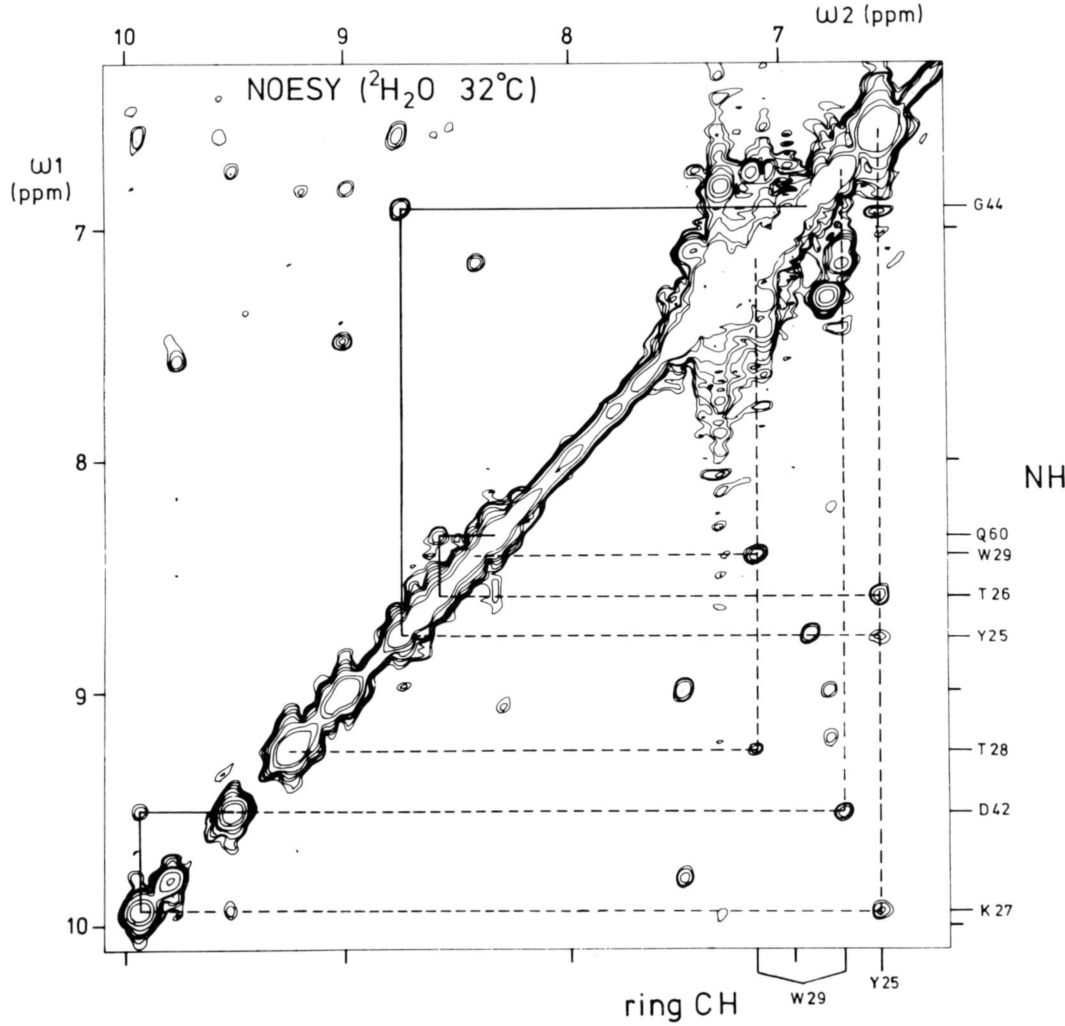

Fig.9 Contour plot of 5oo MHz proton NOESY spectrum (τ_m 0.1 sec ω_1 and ω_2 6.5 - 1o.1 ppm) of 7 mM solution of the *Naja naja siamensis* toxin 3 in D_2O, pH 2.2 at 32°C. The NOE connectivities between backbone amide NH protons (——) are indicated for the following residues: Tyr^{25} - Gly^{44}, Thr^{26} - Gln^{60} and Lys^{27} - Asp^{42}. Connectivities between aromatic CH and backbone NH protons (- - -) are indicated for Tyr^{25} - Tyr^{25}, Thr^{26}, Lys^{27} and Trp^{29} - Thr^{28}, Trp^{29}, Asp^{42}.

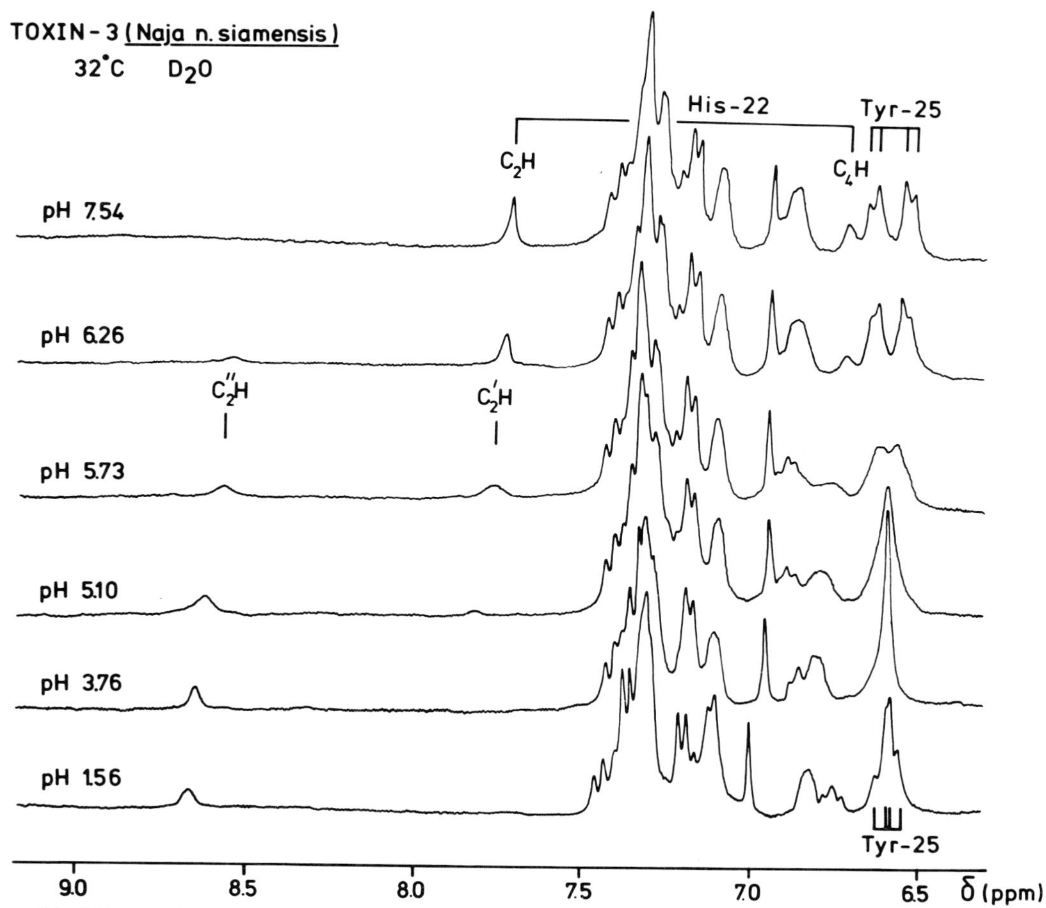

Fig.1o Low field region of 3oo MHz proton NMR spectra of the <u>Naja naja siamensis</u> toxin 3 at different pH of D_2O solution (5 mM, 32°C)

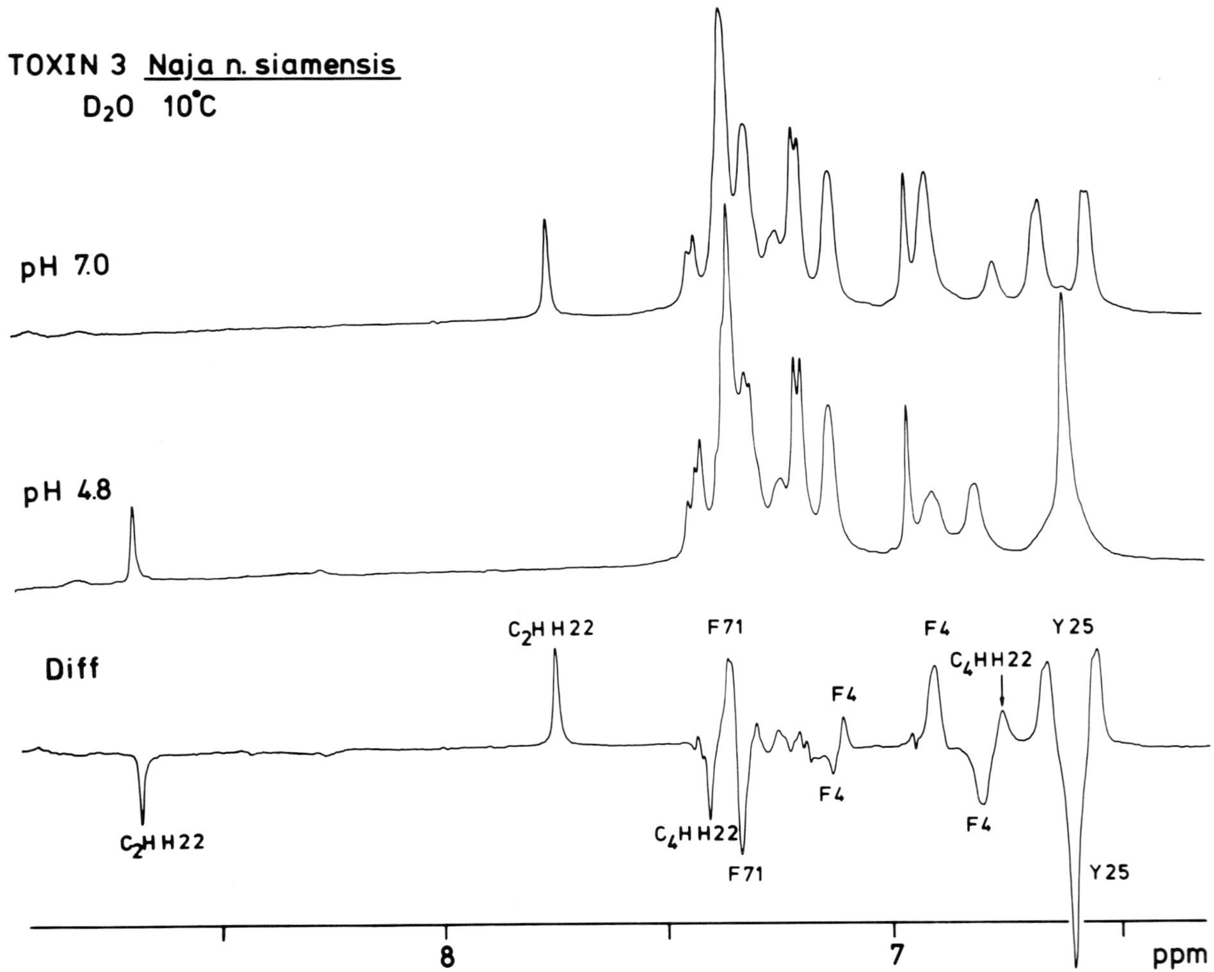
TOXIN 3 Naja n. siamensis
D2O 10°C
pH 7.0
pH 4.8
Diff
C2HH22
F71
F4
Y25
C4HH22
F4
F4
C2HH22
C4HH22
F71
F4
Y25
8
7
ppm

Fig.11

Fig.11 Aromatic region of 5oo MHz proton NMR spectra of a o.o1 M solution of the *Naja naja siamensis* toxin 3 in D_2O, 1o°C at pH 7.o and 4.8. The difference spectrum (bottom) indicates the conformational transition due to protonation of His^{22}.

chain changes its distance relative to the Lys^{27} spin-label from 23 Å in acidic solution (pH 3.5) to 18 Å at physiological pH 7.5 (17). It should be noticed that the Lys^{27} and Lys^{53} residues in "long" neurotoxins correspond to Lys^{27} and Lys^{47} in "short" neurotoxins (1). As was shown above the Lys^{27}-Lys^{47} spin-label distance (15 Å) in the "short" neurotoxin is pH-independent (Table 2). Thus one might speculate that this conformational change could be related to converting the "long" neurotoxin into the physiologically active state by shortening the Lys^{27}-Lys^{53} distance to 18 Å .

Another conformational transition is caused by protonation of His^{71} (p*K* 4.5) in the C-terminal tail of the *Naja naja oxiana* neurotoxin I. It can be detected by the long-range p*K*-effect of His^{71} on chemical shifts of the Ile^{1o}, Ile^{41} and Leu^{43} methyl protons as well as by the pH-dependence of paramagnetic relaxation effects in the NMR spectrum of the selectively spin-labeled His^{71} derivative. The general nature of C-terminal flexibility in "long" neurotoxins is emphasized by NMR study of homologous α-bungarotoxins (39) where the position and steric environment of the neighbouring His^{7o} side chain were found to be temperature dependent and subject to conformational perturbation by His^{7o} protonation. It will be shown in the next section that the C-terminal "tail" undergoes a conformational change on "long" neurotoxin binding to acetylcholine receptor.

Localization and range of two conformational transitions observed in snake "long" neurotoxins are presented in Fig.13. As can be seen a relatively large portion of the molecule is

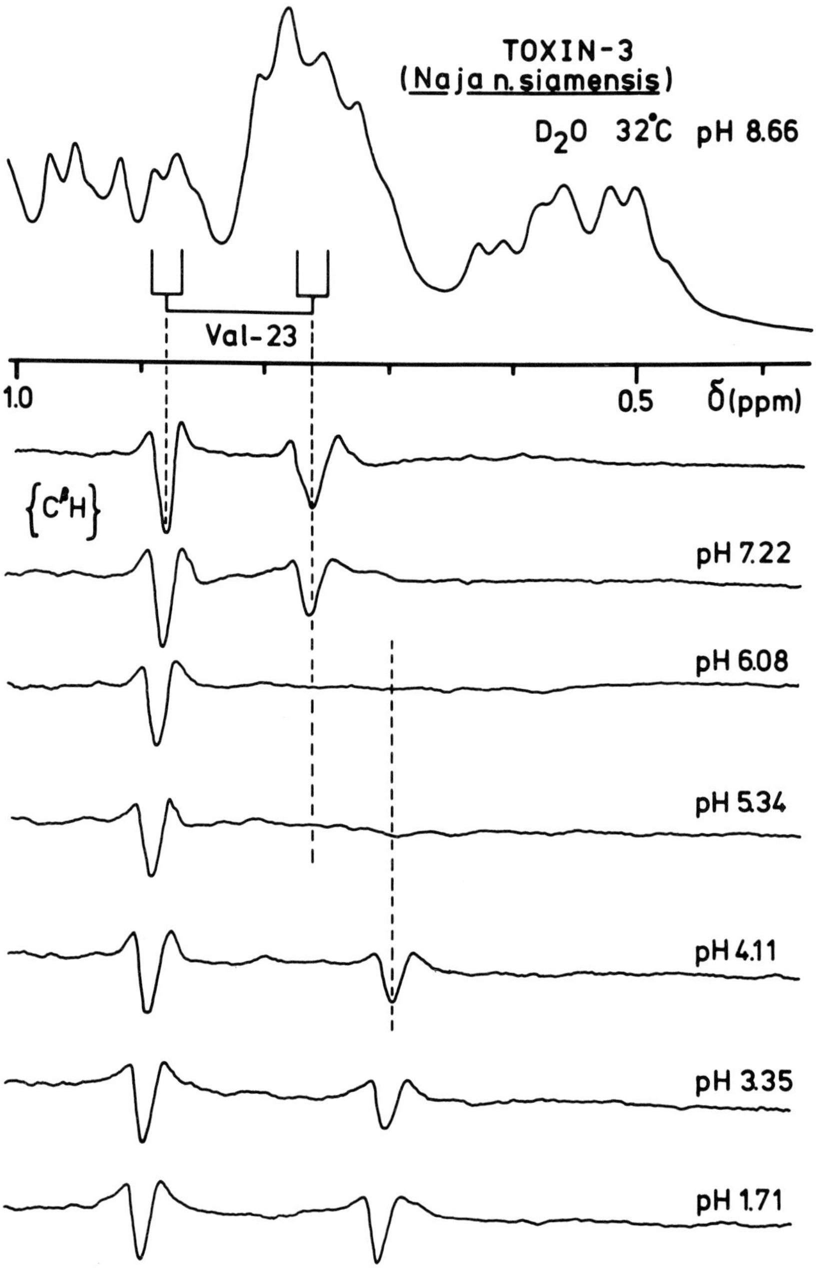

Fig.12 High field region of a 3oo MHz proton NMR spectrum of a 7 mM solution of the <u>Naja naja siamensis</u> toxin 3 in D_2O, pH 8.7 at 32°C and decoupling difference spectra

Fig.12 (contin.): corresponding to different pH values. The $C^{\beta}H$ resonance position of the Val[23] residue is irradiated for decoupling the corresponding γ-methyl resonances.

Table 4 <u>Naja naja siamensis</u> toxin 3 proton resonance response (Δδ) to protonation of the His[22] imidazole ring

Residue	Proton	δ, ppm pH 7.5	Δδ, ppm[a]
His[22]	C_2H	7.71	0.77[b]
Val[23]	γ-CH_3	0.76	-0.10[b]
	γ'-CH_3	0.88	-0.02
Ile[1]	γ-CH_3	0.73	-0.08[b]
Tyr[25]	$C_{2',6}H$	6.65	-0.05
	$C_{3',5}H$	6.55	0.05
Phe[4]	$C_{2',6}H$	6.87	-0.07
	C_4H	7.10	0.03
Ile[59]	γ-CH_3	0.87	0.02
	δ-CH_3	0.56	-0.03
Val[41]	γ-CH_3	0.57	0.02
	γ'-CH_3	0.51	-0.01
Val[52]	γ-CH_3	0.72	-0.02
	γ'-CH_3	0.81	-0.02
Phe[71]	$C_{3',5}H$	7.33	0.02
Lys[15]	ε-CH_2	2.94	0.02

<u>a</u> Positive sign corresponds to low field shift at acidic pH;

<u>b</u> Signal splits at His[22] p<u>K</u> 5.6;

affected but nevertheless the core of the spatial structure is not touched as has been shown here by NMR and as was also confirmed by laser Raman spectroscopy (48) which revealed that β-structure content doesn't change with pH variation and with crystallization.

As compared to the "short" type of snake neurotoxins the "long" neurotoxin molecules have larger fractions predisposed to conformational transitions when environmental conditions are changed. This differential property might be important for the binding of these two neurotoxin types to their target - the nicotinic acetylcholine receptors in postsynaptic membranes.

Binding of snake "short" and "long" neurotoxins to acetylcholine receptor: Despite extensive studies on the acetylcholine receptor (47,49,5o) direct evidence is lacking as to by which region the snake venom neurotoxin molecule binds to the receptor. Availability of series of neurotoxin derivatives with spin (SL) or fluorescence labels in known positions provides the means for outlining the topography of a neurotoxin binding site (4,8,9,11,14,15,51).

Binding of "short" neurotoxins: Labeled derivatives of *Naja naja oxiana* neurotoxin II specifically bind to the solubilized nicotinic acetylcholine receptor protein (AChR) isolated from *Torpedo marmorata* electric organ with dissociation constants ranging from 3 to 8o nM and a 2:1 stoichiometry (51). For comparison the dissociation constant of the natural neurotransmitter acetylcholine is 1oo nM and the stoichiometry is also 2:1 (5o). The two AChR-neurotoxin binding sites were found to be independent and indiscernible by dissociation constants and label's micro-environment but not identical in that the nontoxic hexa(trifluoroacetyl)-neurotoxin II derivative competes for only one AChR binding site (51).

When the monospin-labeled neurotoxin II derivatives combine with AChR their ESR signals broadened to a different extent - considerable for instance for a spin label grafted on the His^{32}

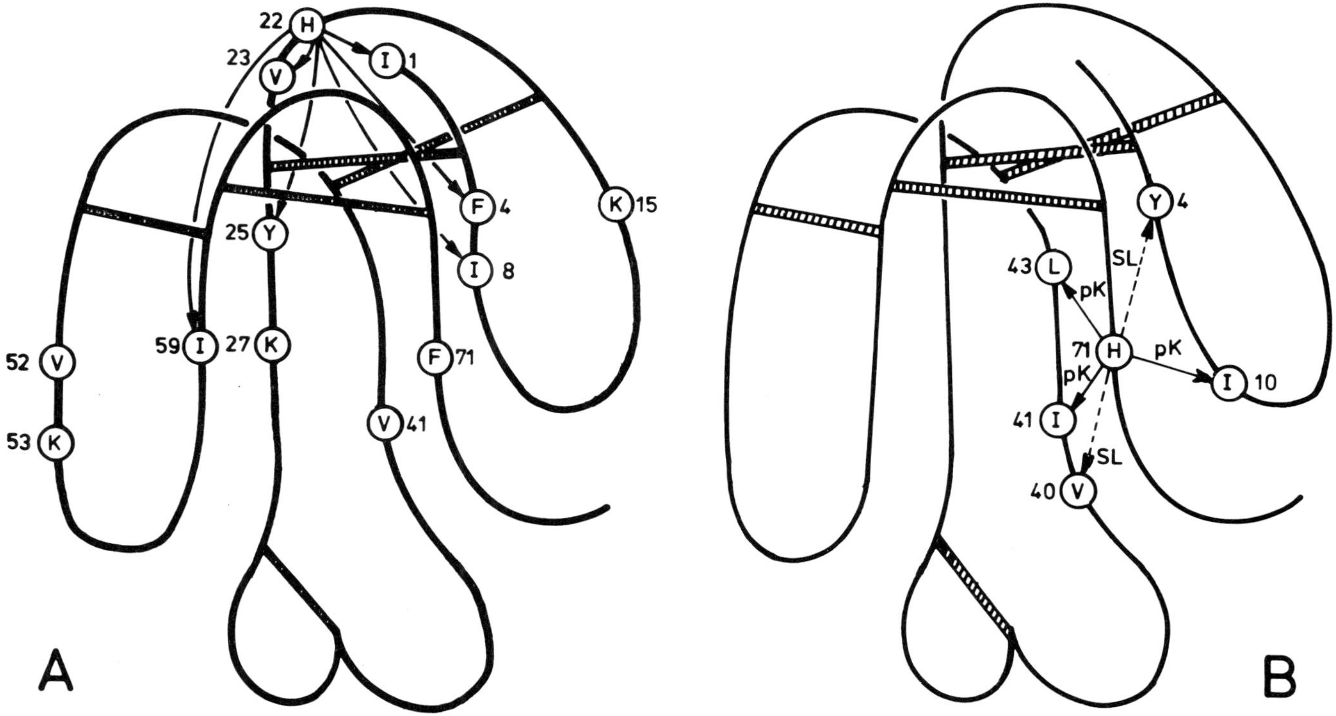

Fig.13 Localization and detected range of two conformational transitions in snake "long" neurotoxins. A - Naja naja siamensis toxin 3 transition initiated by His^{22} protonation (p$\underline{K}$ 5.6); arrows indicate residues whose signals are splitted by the imidazole ring protonation (Table 4). B - Naja naja oxiana neurotoxin I transition with p$\underline{K}$ 4.5 of His^{71}.

residue and only marginal for SL-Lys^{45} (4,15,51). As a first approximation the ESR spectra were treated by the isotropic rotation model to compare the intrinsic mobility by use of the correlation time τ in free and bound states (Table 5). The degree of solvent exposure of spin labels in the toxin-AChR complex were estimated by the paramagnetic effect of the nonspecific probes $K_3Fe(CN)_6$ and $Ni(CH_3COO)_2$ on the ESR central line width H_o (52). The relative value $(\Delta H_o)_{bound}/(\Delta H_o)_{free}$ for AChR-bound and free neurotoxin refers to the same concentration of paramagnetic probe which was chosen much larger than the neurotoxin concentration and was determined by least square fitting of concentration dependence of the line width (51). Table 8 shows that depending on the label position and charge of the probes $Fe(CN)_6^{-3}$ and Ni^{2+} the spin-label accessibility in the complex may remain almost the same as with the free neurotoxin derivative (for instance the labels at the N-terminal and Lys^{45} residues as monitored with ferricyanide or the labels at Glu^2, Lys^{15} and Lys^{45} with the nickel ion), or considerably diminished (Lys^{47} and Lys^{27}).

In summary the ESR data supported by fluorescence measurments (9,15,35,51) provide evidence for the previously proposed (4, 9,15,51) multipoint topography of the "short" neurotoxin binding to AChR. The schematic model of binding is shown in Fig.14 taking into account the discussed close similarity of the spatial structure of "short" neurotoxins in solution to that of the X-ray structure of erabutoxin b (19-21,53). In this model according to the experimental results:
i) both the Lys^{27} and Lys^{47} side chains are immersed into the receptor, but the Lys^{47} label still has enough room for considerable mobility in the corresponding spacious pocket. (Indeed chemical modification of the two highly conservative lysine residues has a most drastic effect on the neurotoxic activity).

ii) neighbour labels at the N-terminal and Glu^2 residues as well as at His^{32} are also in close contact with the receptor surface, presumably surrounded by positive (N-terminal) or negative charges (Glu^2 and His^{32}).

Table 5: Toxicity and ESR data on monospin-labeled (4,15,51) and fluorescence spectroscopy data on monodansyl-labeled (9,15,35,51) derivatives of the *Naja naja oxiana* neurotoxin II interaction with solubilized *Torpedo marmorata* acetylcholine receptor protein, 21 °C.

Label location	Spin-labeled derivatives			
	Toxicity, LD_{5o} µg/Kg[a]	τ bound / τ free	$(\Delta H_o)_{bound}/(\Delta H_o)_{free}$ $Fe(CN)_6^{3-}$	$Ni(Ac)_2^{2+}$
Lys-45	152	1.8	o.8	1.4
Lys-15	11o	2.5	o.3	o.9
Lys-26	85	1o	o.5	o.5
N-term.		> 9	o.7	o.2
Glu-2	3oo	>16	<o.1	o.8
His-32	1oo	>27	<o.o5	o.4
Lys-47	24o	2.2	o.1	o.1
Lys-27	3oo	1o	<o.3	o.2

Label location	Dansyl-labeled derivatives		
	Emission shift $\Delta\lambda^{max}$, nm	Relative intensity, excitation at 298 nm	Displacement by native neurotoxin II %
Lys-45			
Lys-15	-1	1.3	34
Lys-26	-7	1.6	18
N-Term.	o	1.3	22
Glu-2			
His-32			
Lys-47	-25	5.3	76
Lys-27	-19	2.6	7o

[a] LD_{5o} for native neurotoxin II is 6o µg/Kg.

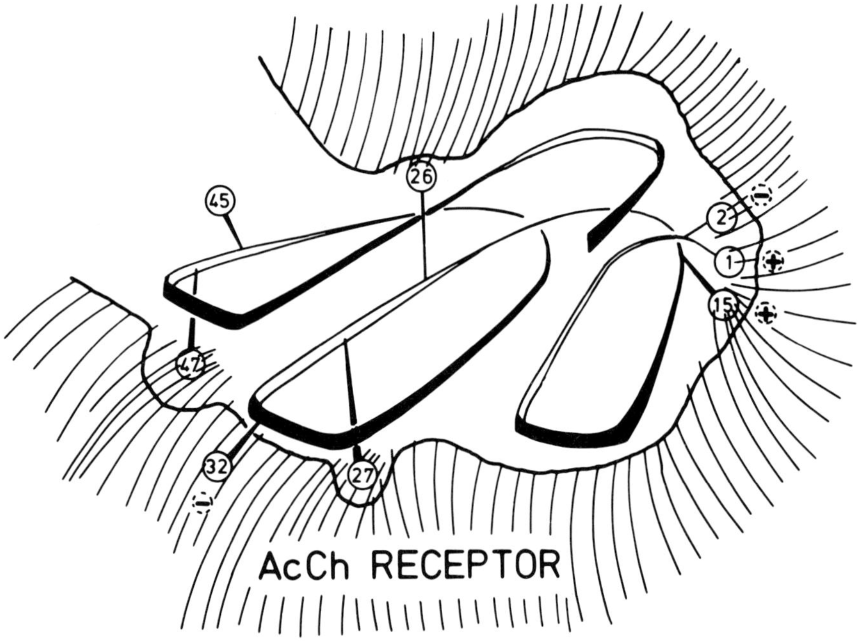

Fig.14 Schematic presentation of the topography of the snake "short" neurotoxin binding with acetylcholine receptor as inferred from the ESR, fluorescence and X-ray data.

iii) the Lys^{26} and Lys^{15} side chains are only in marginal contact with the receptor surface,
iv) while Lys^{45} is not involved in the contact at all.

Increase of fluorescence intensity and a blue shift of the emission maximum observed on binding of the Lys^{27} and the Lys^{47} dansyl-labeled derivatives with AChR correspond to a transfer of the labels into a more hydrophobic environment in the complex (15,35).

Thus there is direct spectroscopic evidence that the "short" neurotoxin binding site is quite extended and that the neurotoxin-receptor interaction is of a multipoint nature, as was postulated before on the basis of the effect of chemical modifications on biological activity (1) and of the erabutoxin b X-ray structure (53). It was also shown by competitive binding experiments that the distance between the two AChR-bound spin-labeled neurotoxin derivatives is over 25 Å .

Binding of "long" neurotoxins: Interactions of mono- and dispin-labeled "long" neurotoxins with AChR were monitored by ESR spectroscopy (Fig.15). As with "short" neurotoxins a 2:1 stoichiometry of the toxin-AChR complexes was found (this study (15) and reports by others (54,55)). The changes of line shape show that binding to AChR leads to a decrease of spin label mobility which is most prominent for the nitroxyl radical grafted on Lys^{27} of Naja naja siamensis toxin 3. Moreover this label experiences an anisotropic type (56) of rotational diffusion (order parameter S ∿ o.5) in the bound state as could be judged from the corresponding ESR spectra.

An interesting feature exhibits the Cys^{30}, Cys^{34} -dispin-labeled derivative of Naja naja siamensis toxin 3 with reduced Cys^{30}-Cys^{34} disulfide bond. For the unbound state of this derivative the ESR spectra do not show any interaction of the spin-labels even at $77^{o}K$ indicating that the interlabel distance is more than 2o-25 Å (32,56). For the AChR-bound state on the other hand the shape of the ESR spectrum (Fig.15) provides clear evidence for an electron spin-exchange interaction, which corresponds to direct motional collision of the two spin labels at a distance of 6-8 Å (57). Thus while the toxin 3 with reduced Cys^{30}-Cys^{34} disulfide bond has a distorted conformation in the region of this bond (central loop of Fig.4) in the bound state the dispin-labeled "long" neurotoxin most likely returns to the native conformation due to the tight binding with the AChR active site. Similar results were recently obtained by Rousselet et al. (58) for corresponding dispin-labeled α-bungarotoxin and its complex with AChR rich membranes.

The solvent accessibility of the spin-labeled "long" neurotoxins bound to AChR is shown in Table 6. There are some similarities in these data with corresponding results for "short"

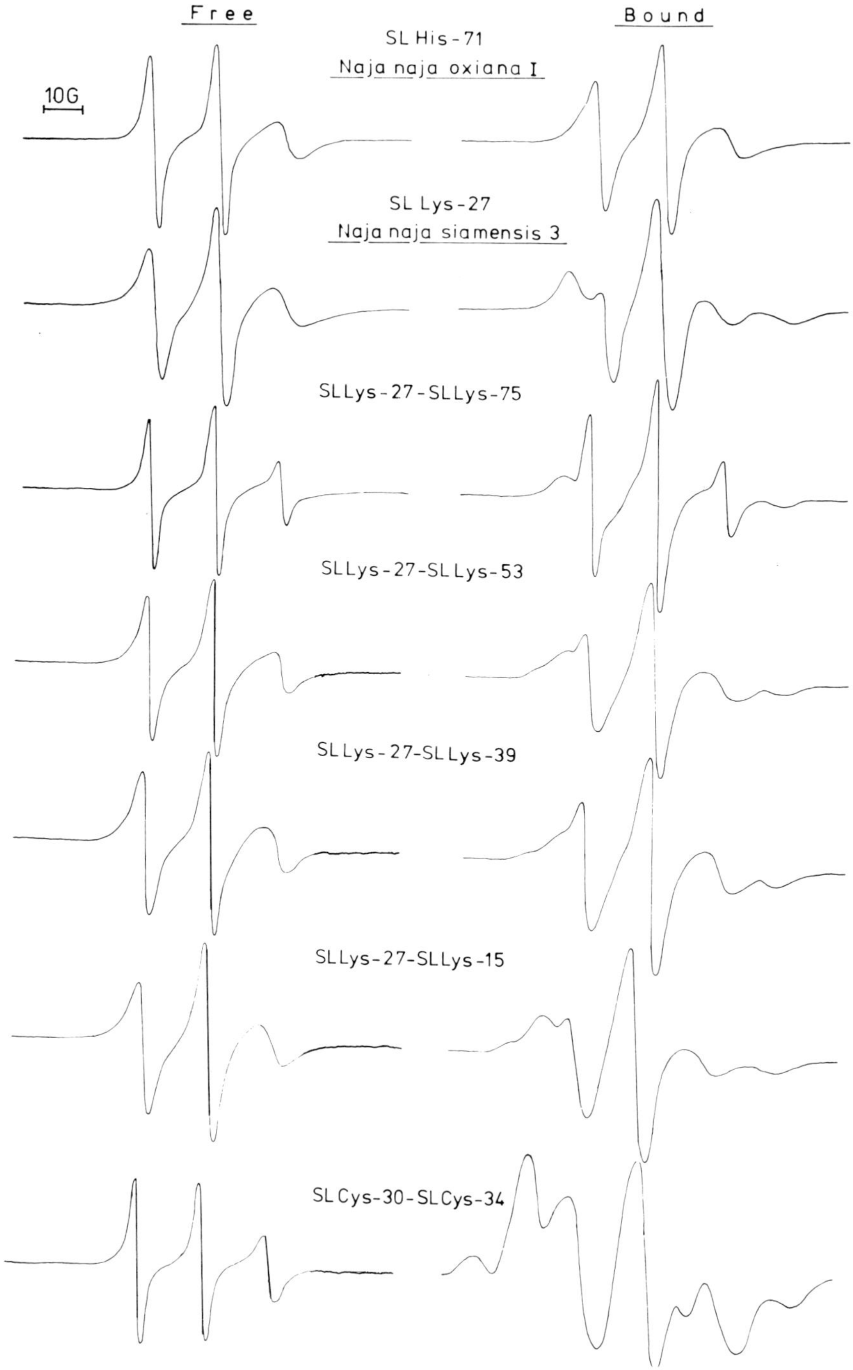
Free
Bound
SL His-71
Naja naja oxiana I
10G
SL Lys-27
Naja naja siamensis 3
SLLys-27-SLLys-75
SLLys-27-SLLys-53
SLLys-27-SLLys-39
SLLys-27-SLLys-15
SLCys-30-SLCys-34

Fig. 15

Fig.15 ESR spectra of mono-and dispin-labeled "long" neurotoxins - Naja naja oxiana neurotoxin I and Naja naja siamensis toxin 3 in buffer solution and bound to Torpedo marmorata acetylcholine receptor (AChR concentration $3 \cdot 10^{-6}$M, Triton X-100 0.025%, phosphate or Tris-HCl buffer pH7.4, 21°C).

neurotoxins. The most buried Lys^{27} spin label of toxin 3 corresponds to Lys^{27} in "short" neurotoxins which is also shielded by the receptor binding site.

Somewhat less pronounced binding is observed for the Lys^{53} spin label (Fig.15) which corresponds to the well bounded Lys^{47} in "short" neurotoxins (Fig.14). The difference is that in "long" neurotoxins the orientation of the Lys^{53} side chain is vulnerable to conformational transition as was shown here by the pH-dependence of the distance between spin labels grafted on Lys^{53} and Lys^{27}. Other spin-labeled "long" neurotoxin residues do not reveal any contact with AChR in agreement with a lack of almost any effect of their modification on the toxicity (1). For instance the accessibility of SL-His^{71} of neuro-

Table 6 Relative accessibility of spin labels in free and AChR-bound snake "long" neurotoxins (15)

Neurotoxin	Spin label position	$(\Delta H_o)_{bound}/(\Delta H_o)_{free}$	
		$Fe(CN)_6^{3-}$:	$Ni(Ac)_2^{2+}$
Neurotoxin I	His-71	1.2	1.3
N.N.oxiana			
Toxin 3	Lys-27	0.2	0.3
N.N.siamensis	Lys-27/Lys-15	0.4/2.2	0.5/0.8
	Lys-27/Lys-39	0.2/2.9	0.2/0.4
	Lys-27/Lys-53	0.2/0.9	0.2/0.6
	Lys-27/Lys-75	0.3/1.2	0.5/1.1
	Cys-30/Cys-34	0.7	1.0

toxin I increased even somewhat in the AChR complex. Thus the C-terminal fragment of "long" neurotoxins does not participate in AChR binding and probably changes its orientation with respect to the rest of the polypeptide backbone. This is in agreement with the "tail" vulnerability to conformational transition (see Fig.13_B).

The available although limited data are combined in a schematic diagram of "long" neurotoxin binding topography shown in Fig.16 which is based also on the similarity of solution and crystalline structure of "long" neurotoxins. Obviously, these preliminary results on spin-labeled "long" neurotoxins cannot give a self-sufficient picture of the binding to the receptor. However it seems highly probable that "long" and "short" neurotoxins share not only common conformational features (45) but also a similar topography of receptor binding. One could speculate that in the process of binding with AChR the "long" neurotoxin molecule is more subject to conformational transition for a better fit to the receptor surface. This might be responsible for the lower rate of association and dissociation of "long" neurotoxins (1) in comparison to "short" toxins; it makes the neurotoxin-receptor complex more quasi-irreversible. That binding involves a conformational change in the "long" neurotoxins is confirmed also by the fact, that the rate of association of *Naja naja siamensis* α-toxin with acetylcholine receptor diminishes rapidly with decreasing temperature, whereas the rate of dissociation of the complex is hardly affected (1,59).

Conclusion

By ESR study of spin labels incorporated into various sites of snake neurotoxin molecules direct information was obtained on the neurotoxin regions involved in binding with its target - the acetylcholine receptor of postsynaptic membranes. It is evident that the molecular recognition of the two proteins is a result of many contributions, non of which are ab-

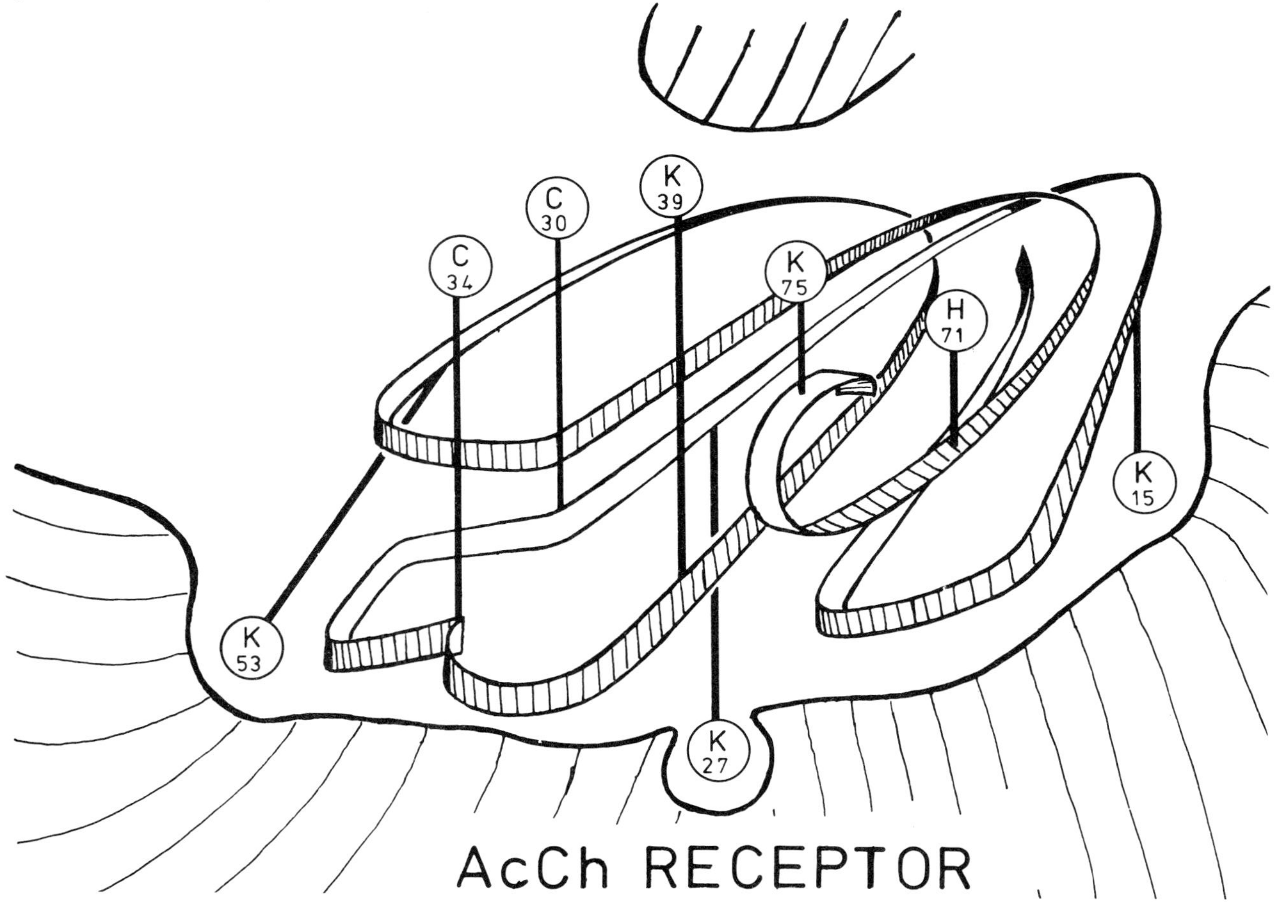

Fig.16 Schematic presentation of the topography of the snake "long" neurotoxin binding with AChR.

solutely essential. The dimension of the neurotoxin-AChR multipoint binding region might be 3o x 2o Å (for instance the distance between Leu^1 and His^{32} side chains is 3o Å according to the X-ray analysis of erabutoxin b (21,53)). Indeed recent results obtained by photoaffinity crosslinking (15,6o-63) and electron microscopy (64) indicate that the neurotoxin molecules when bound to AChR are in direct contact with several subunits of the receptor, not only with the α-subunits the function of which seems to be binding the natural neurotransmitter acetylcholine (49,5o). Thus the revealed multipoint topography of neurotoxin binding which occupies several subunits of the AChR might serve for locking the receptor quaternary structure with normally closed channel and therefore prevent the large structural changes which are believed to be necessary for opening the channel and induction of nerve impulse transmission (47).

References

1) Karlsson,E.: in Snake Venoms (Handbook of Experimental Pharmacology, vol 52) (C.-Y.Lee, ed.) p.159, Springer Verlag, West-Berlin (1978).

2) Tsetlin,V.I., Mikhaleva,I.I., Myagkova,M.A., Senyavina,L. B., Arseniev,A.S., Ivanov,V.I., and Ovchinnikov,Yu.A.,: in Peptides: Chemistry, Structure and Biology (R.Walter and J. Meienhofer, eds.) p.935, Ann Arbor Science, Ann Arbor, Michigan (1975).

3) Bystrov,V.F., Arseniev,A.S., Gavrilov,Yu.D.: J.Magn.Res. 3o, p.151 (1978).

4) Bystrov,V.F., Ivanov,V.T., Okhanov,V.V., Miroshnikov,A.I., Arseniev,A.S., Tsetlin,V.I., Pashkov,V.S., and Karlsson, E.: in Advances in Solution Chemistry (I.Bertini, L.Lunazzi and A. Dei,eds.) p.231, Plenum Press, New York,(1981).

5) Arseniev,A.S., Balashova,T.A., Utkin,Yu.N., Tsetlin,V.I., Bystrov,V.F., Ivanov,V.T., and Ovchinnikov,Yu.A.: Eur.J. Biochem. 71, p.595 (1976).

6) Arseniev,A.S., Surin,A.M., Utkin,Yu.N., Tsetlin,V.I., Bystrov,V.F., Ivanov,V.T., and Ovchinnikov,Yu.A.: Bioorgan. Khim. (Soviet J.Bioorg.Chem.) 4, p.197 (1978).

7) Tsetlin,V.I., Arseniev,A.S., Utkin,Yu.N., Gurevich,A.Z., Senyavina,L.B., Bystrov,V.F., Ivanov,V.T., and Ovchinnikov, Yu.A.: Eur.J.Biochem. 94, p.337 (1979).

8) Ivanov,V.T., Tsetlin,V.I., Karlsson,E., Arseniev,A.S., Utkin,Yu.N., Pashkov,V.S., Surin,A.M., Pluzhnikov,K.A., and Bystrov,V.F.: Toxicon, Suppl.1, p.77 (1979).

9) Ivanov,V.T., Tsetlin,V.I., Karlsson,E., Arseniev,A.S., Utkin,Yu.N., Pashkov,V.S., Surin,A.M., Pluzhnikov,K.A., and Bystrov,V.F.: in Natural Toxins (D.Eaker and T. Wadstöm, eds.) p.523, Pergamon Press, Oxford (198o).

1o) Bystrov,V.F.: in Biomolecular Structure, Conformation, Function and Evolution (R.Srinivasan, E.Subramanian and N.Yathindra, eds.), Vol.2, p.3, Pergamon Press, Oxford (198o).

11) Ivanov,V.T., Tsetlin,V.I., Mikhaleva,I.I., Volpina,O.M., Nuriddinov,A.R., Utkin,Yu.N., Arseniev,A.S., Pashkov,V.S., Karlsson,E., Surin,A.M., and Bystrov,V.F.: in Peptides 1978 (G.Kupryszewski, I.Z.Siemion, eds.), p.41, Wroclaw University Press, Poland (1979).

12) Arseniev,A.S., Pashkov,V.S., Pluzhnikov,K.A., Rochat,H., and Bystrov,V.F.: Eur.J.Biochem 118, p. 453 (1981).

13) Arseniev,A.S., Utkin,Yu.N., Pashkov,V.S., Tsetlin,V.I., Ivanov,V.T., Bystrov,V.F., and Ovchinnikov,Yu.A.: FEBS Letters 136, p.269 (1981).

14) Tsetlin,V.I., Karlsson,E., Utkin,Yu., Arseniev,A., Pluzhnikov,K., Surin,A., Pashkov,V., Bystrov,V., and Ivanov,V.: in Chemistry of Peptides and Proteins, Vol. 1 (W. Voelter, E. Wünsch, Yu.Ovchinnikov and V. Ivanov, eds.) p. 389, Walter de Gruyter, West Berlin (1982).

15) Tsetlin,V.I., Karlsson,E., Utkin,Yu.N., Pluzhnikov,K.A., Arseniev,A.S., Surin,A.M., Kondakov,V.V., Bystrov.V.F., Ivanov,V.T., and Ovchinnikov,Yu.A.: Toxicon 2o, p.83 (1982).

16) Pashkov,V.S., Arseniev,A.S., Utkin,Yu.N., Tsetlin,V.I., and Bystrov,V.F.: Bioorg.Khim (Soviet J.Bioorg.Chem.) 8, p.588 (1982).

17) Utkin,Yu.N., Pashkov,V.S., Pluzhnikov,K.A., Kuryatov,A.B., Arseniev,A.S., Tsetlin,V.I., Bystrov,V.F., and Ivanov,V.T.: Bioorg.Khim. (Soviet J.Bioorg.Chem.) 9, (1983),in press.

18) Pashkov,V.S., Pluzhnikov,K.A., Utkin,Yu.N., Hintshe,R., Arseniev,A.S., Tsetlin,V.I., Ivanov,V.T., and Bystrov,V.F.: Bioorg.Khim. (Soviet J. Bioorg. Chem.) 9 (1983)in press.

19) Tsernoglou,D., and Petsko,G.A.: FEBS Letters 68, p.1 (1976)

2o) Low,B.W., Preston,H.S., Sato,A., Rosen,L.S., Searl,J.E., Rudko,A.D., and Richardson,J.S.: Proc.Natl.Acad.Sci.USA 73 p. 2991 (1976).

21) Kimball,M.R., Sato,A., Richardson,J.S., Rosen,L.S., and Low,B.W.: Biochem.Biophys.Res.Commun. 88, p.95o (1979).

22) Lauterwein,J., Lazdunski,M. and Wüthrich,K.: Eur.J.Biochem. 92, p.361 (1978)..

23) Endo,T., Inagaki,F., Hayashi,K., and Miyazawa,T.: Eur.J. Biochem. 1o2, p.417 (1979).

24) Fung,C.H., Chang,C.C., and Gupta,R.K.: Biochemistry 18, p. 457 (1979).

25) Galat,A., Degelaen,J.P., Yang,C.C., and Blout,E.R.: Biochemistry 2o, p.7415 (1981).

26) Inagaki,F., Miyazawa,T., Hori,H., Tamiya,N.: Eur.J.Biochem. 89, p.433 (1978).

27) Inagaki,F., Tamiya, N. and Miyazawa,T.: Eur.J.Biochem.1o9, p.129 (198o).

28) Inagaki,F., Tamiya,N., Miyazawa,T., and Williams,R.J.P.: Eur.J.Biochem. 118, p.621 (1981).

29) Poulsen,F.M., Hoch,J.C., and Dobson,C.M.: Biochemistry 19, p.2597 (198o).

3o) Morris,A.T., and Dwek,R.A.: Quart.Rev.Biophys.1o, p.421 (1977).

31) Utkin,Yu.N., Pluzhnikov,K.A., and Tsetlin,V.I.: this issue.

32) Likhtenstein, G.I.: Spin Labeling Methods in Molecular Biology, Wiley, New York (1976).

33) Menez,A., Montenay-Garestier,T., Fromageot,P., and Hélène, C.: Biochemistry 19, p.52o2 (198o).

34) Surin, A.M., Utkin,Yu.N., Pluzhnikov,K.A., Efremov,E.S., and Tsetlin,V.I.: Bioorgan.Khim. (Soviet J.Bioorg.Chem.), 7, p.342 (1981).

35) Surin,A.M., Utkin,Yu.N., Pluzhnikov,K.A., and Tsetlin,V.I.: this issue.

36) Thiery,C., Nabedryk-Viala,E., Menez,A., Fromageot,P., and Thiery,J.M.: Biochem.Biophys. Res.Commun. 93,p.889 (198o).

37) Nabedryk-Viala,K., Thiery,C., Menez,N., Tamiya,N., and Thiery,J.M.: Biochim.Biophys.Acta 626, p.321 (198o).

38) Inagaki,F., Miyazawa,T., and Williams,R.J.P.: Bioscience Rep.1, p.743 (1981).

39) Endo,T., Inagaki,F., Hayashi,K., and Miyazawa,T.: Eur.J. Biochem. 12o, p.117 (1981).

4o) Inagaki,F., Clayden,N.J., Tamiya,N., and Williams,R.J.P.: Eur.J.Biochem.12o, p.313 (1981).

41) Endo,T., Inagaki,F., Hayashi,K., and Miyazawa,T.: Eur.J. Biochem. 122, p.541 (1982).

42) Inagaki,F., Clayden,N.J., Tamiya,N., and Williams,R.J.P.: Eur.J.Biochem. 123, p.99 (1982).

43) Hider,R.C., Drake,A.F., Inagaki,F., Williams,R.J.P., Endo, T. and Miyazawa,T: J.Mol.Biol. 158, p.275 (1982).

44) Walkinshaw,M.D., Saenger,W., and Maelicke,A.: Proc.Natl.Acad. Sci.USA 77, p.24oo (198o).

45) Walkinshaw,M.D., Saenger,W., and Maelicke,A.: in Structural Aspects of Recognition and Assembly in Biological Macromolecules (M.Balaban, J.S.Sussman, W. Traub and A.Yonath, eds.) Vol.1, p.313, Balaban ISS, Rehovot and Philadelphia, (1981).

46) Agard,D.A., and Stroud,R.: Acta Cryst. A38, p.186 (1982).

47) Kistler,J., Stroud,R.M., Klimowsky,M.W., Lalancette,R.A., and Fairclough,R.H.: Biophys.J.37, p.371 (1982).

48) Nabiev,I.R., Pluzhnikov,K.A., Trakhanov,S.D., Efremov,E.S., and Tsetlin,V.I.: Bioorg.Khim. (Soviet J.Bioorg.Chem.) 7, p.836 (1981).

49) Hucho,F.: Trends in Biochem.Sci.6, p.242 (1981).

5o) Conti-Tronconi,B.M., and Raftery,M.A.: Ann. Rev. Biochem. 51, p.491 (1982).

51) Tsetlin,V.I., Karlsson,E., Arseniev,A.S., Utkin,Yu.N., Surin,A.M., Pashkov,V.S., Pluzhnikov,K.A., Ivanov,V.T., Bystrov,V.F., and Ovchinnikov,Yu.A.: FEBS Letters 1o6, p.47 (1979).

52) Salikhov,K.M., Dohktorov, A.B., and Molin,Yu.N.: J.Magn. Res. 5, p.189 (1971).

53) Low,B.L.: in Advances in Cytopharmacology (F.Cecarelli and F.Clementi, eds.) Vol. 3, p.141, Raven Press, New York (1979).

54) Ellena,J.F., and McNamee,M.G.: FEBS Letters 11o, p.2oo (198o).

55) Johnson,D.A., and Taylor,P.: J.Biol.Chem 257, p.5632 (1982).

56) Smith,I.C.P., Schreir-Mucillo,S., and Marsh,D.: in Free Radicals in Biology (W.A.Pryor, ed.) Vol.1, Chapter 4, Academic Press, New York (1976).

57) Lackhurst,G.R.: in Spin Labeling, Theory and Applications (L.J. Berliner, ed.) Vol.1, Chapter 4, Academic Press, New York (1975).

58) Rousselet,A., Cartaud,J., Devaux,P.F., and Changeux,J.-P.: EMBO J. 1, p.439 (1982).

59) Lester,H.: J.Gen.Physiol. 57, p.255 (1971).

6o) Hucho,F.: FEBS Letters 1o3, p.27 (1979).

61) Karlin,A., Damle,V., Hamilton,S., McLaughlin,M., Valderamma,R., and Wise,D.: in Advances in Cytopharmacology (F.Cecarelli and F.Clementi, eds.) Vol.3, p.183, Raven Press, New York (1979).

62) Oswald,R.E. and Changeux,J.-P.: FEBS Letters 139, p.225 (1982).

63) Pluzhnikov, K.A., Karelin,A.A., and Tsetlin,V.I.: this issue.

64) Zingsheim,H.P., Barrantes,F.J., Frank,J., Hanicke,W., and Neugebauer,D.-Ch.: Nature 299, p.81 (1982).

65) Grishin,E.V., Sukhikh,A.P., Slobodyan,L.N., Ovchinnikov, Yu.A., and Sorokin, V.M.: FEBS Letters 45, p.118 (1974).

66) Grishin,E.V., Sukhikh,A.P.,Lukyanchuk, N.N., Slobodyan,L. N., Lipkin,V.M., Ovchinnikov,Yu.A., and Sorokin,V.M.: FEBS Letters 36, p.77 (1973).

67) Karlsson,E., Fohlman,J., and Groth,M.: Bull.Inst.Pasteur 74, p.11 (1976).

68) Gurevich,A.Z.:Bioorgan.Khim. (Soviet J.Bioorg.Chem.) 6, p.197 (198o).

PHOTOLABELING OF ACETYLCHOLINE RECEPTOR STATES WITH TRI - PHENYLMETHYLPHOSPHONIUM

Peter Muhn, Leander Lauffer and Ferdinand Hucho
Institut für Biochemie der Freien Universität Berlin,
Fabeckstr.34-36, D-1ooo Berlin 33

Introduction

In the nicotinic acetylcholine receptor-complex (AChR) the transmitter-binding site is coupled with an ion channel which is closed in the postsynaptic membrane at rest. Interaction with two molecules of the agonist acetylcholine regulates the permeability of the channel in an as yet unknown way. Binding site and channel are located on different parts of the receptor protein and appear to interact by mechanisms similar to the interaction between regulatory and catalytic sites in allosteric enzymes (1).

To elucidate the molecular properties of the ion channel one can make use of effectors of the cholinergic synapse which appear to interact specifically with various states of the ion channel. These effectors have been termed non-competitive antagonists. At present the most plausible model of the AChR pictures it as existing in three interconvertible functional states: R, resting (channel closed), A, active (channel open) and D, desensitized (channel closed, affinity for agonists high). A forth state, I, with intermediate affinity has been added recently (2). The model postulates a minimum of four distinct conformations with different affinities for acetylcholine with dissociation constants in the mM, μM an 1o nM range for A,I, and D, respectively.

We have introduced recently (3,4) a ligand which appears to be very useful for the characterisation of various functional states of the AChR. ^{3}H-Triphenylmethylphosphonium (^{3}H-TPMP) binds specifically to the resting receptor (R) but its affini-

Toxins as Tools in Neurochemistry

ty to the AChR is markedly enhanced by cholinergic effectors and most interestingly it blocks agonist induced ion flux through receptor-rich membranes. In these properties it resembles other known non-competitive antagonists e.g. histrionicotoxin, phencyclidine and others (5-7).

Moreover irradiation of the AChR-^{3}H-TPMP-complex with UV-light causes covalent binding of the ligand to the receptor protein (4). This allows to identify the site of binding and presumably the site of action of this effector. In this report we describe in more detail the characteristics of this reaction. We present evidence that the labeling pattern monitors changes of the receptor conformation. For the functional model of the receptor it is of inerest, that there are indications for a conformational state induced by agonists and antagonists as well. This state appears not to be identical with any of the states known so far.

Materials and Methods

(^{3}H)-TPMP^{+}(NEN) was 36 Ci/mmol, all other reagents were of the highest purity commercially available. Experimental conditions concerning sample preparation, UV-irradiation, electrophoresis, fluorography and chemicals were essentially as described (4).

Phospholipase A_2 from Crotalus durissus terrificus venom (Sigma, 17o U/mg protein) was used in Na-Ringer. Incubation time was 3o to 6o min at room temperature.

For the laser-experiments a Quanta Ray Nd:YAG type DCR-2-A pulsed laser at 4-folded frequency (2oo nm) was used. The sample was placed in a 1 ml-cuvette at 8o cm distance without any cooling and flashed with the appropriate amount of pulses.

The effect of carbamoylcholine on the covalent ^{3}H-TPMP-labeling was quantified by cutting the δ-subunit out of the polyacrylamide gels, processed in the usual manner and counted in a liquid scintillation counter with Omnifluor/Toluol/3% Protosol after shaking for 24 hours at room temperature.

Results and Discussion:

Action Spectrum of the Photoreaction

In previous experiments we had observed that there is a significant difference in the photolabeling patterns obtained with AChR from different species of electric fish. The observed incorporation of ^3H-TPMP$^+$ into the δ-polypeptide chain of the receptor complex stimulated by carbamoylcholine was reproducibly obtained with receptor from Torpedo marmorata but not with preparations from Torpedo californica. This indicated that it was not truely the label (^3H-TPMP$^+$) which was activa-

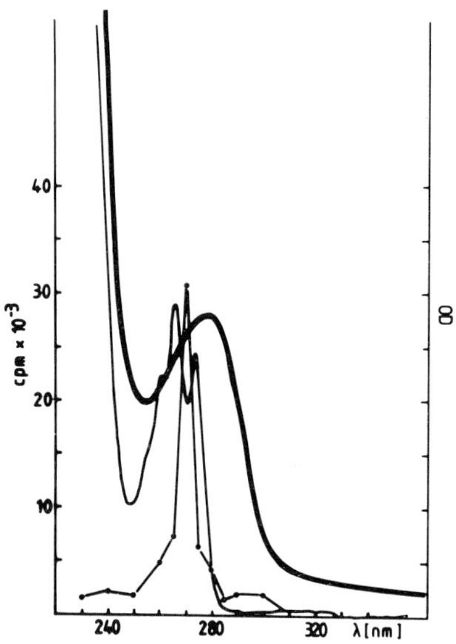

Fig 1 Action spectrum of the photoreaction; 28 µl of Torpedo marmorata-microsacs were incubated for 5 min at room temperature with 7.5 µl ^3H-TPMP$^+$ (o.33 µCi/µl) and 15 µl 1o^{-4} M carbamoylcholine, placed in a 1 ml Quartz cuvette and positioned into a Zeiss MQ4 photometer at the appropriate height, i.e. that the whole volume was in the light beam. The wavelength was afjusted with the monochromator. Irradiation lasted 5 min. The radioactivity incorporated was analysed as described in "Materials and Methods".

ted by light but some component of the protein. This component then would be absent or not active in AChR from *Torpedo californica*.

To investigate this possibility we performed the photolabeling with UV-light of specific wavelengths. Fig.1 shows that light of λ = 27o nm was most efficient in this reaction. Comparison of the action spectrum with the absorption spectra of $TPMP^+$ and AChR, respectively, shows no similarity with the $TPMP^+$ spectrum. Maximum efficiency is obtained close to the maximum of the protein spectrum. This observation and the small-side maximum around 29o nm may hint at the involvement of aromatic aminoacid side chains in the photoreaction.

Ligand specificity of the observed shift of the labeling from alpha to the delta polypeptide chain

The stimulation and the change in the labeling pattern caused by carbamoylcholine are similar to effects obtained by others (6,7) with UV-labeling of AChR with trimethisoquin, chlorpromazin etc. and which have been interpreted as indicative of conformational changes underlying receptor desensitizytion. But this interpretation would be premature in the case of our ^3H-$TPMP^+$-labeling experiments for the following reasons:
1) The change in the labeling pattern induced by carbamoylcholine appears to be slower than desensitization (4).
2) Not only agonists as e.g. carbamoylcholine and acetylcholine but also antagonists as D-tubocurarine, flaxedil, hexamethonium and decamethonium (acting *in vitro* also as an antagonist (8)) stimulate $TPMP^+$ binding and UV-induced incorporation into the δ-polypeptide chain. According to the present understanding antagonists are not capable of desensitizing AChR (perhaps with the exception of some antagonists termed "metaphilic" (9)).

A special problem was posed by the irreversible antagonist α-bungarotoxin. It also stimulated ^3H-$TPMP^+$ labeling of the δ-polypeptide chain (Fig.2). But we observed that this effect

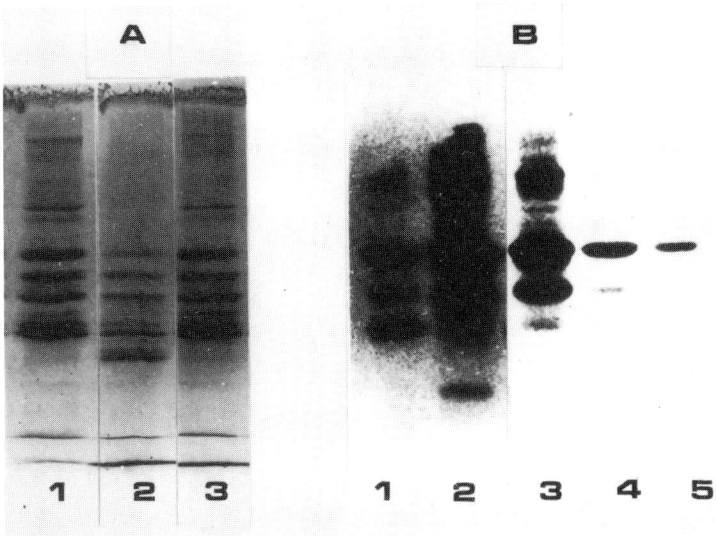

Fig.2 Stimulation of receptor labeling with ^3H-TPMP$^+$ by different α-bungarotoxin preparations; α-bungarotoxin from Boehringer (Mannheim) stimulates (as other agonists and antagonists) preferential labeling of δ-polypeptide chains (at the expense of α-chain-labeling, the lower bands) as is shown in the autoradiograms (B) of SDS-polyacrylamide electrophoresis gels (compare track 3 with the control, track 1 of Fig.2, B). But this effect is considerable less with the toxin from Sigma (track 4) and almost absent with a sample purified in our laboratory from the crude venom (track 5). The Boehringer toxin contains considerable amounts of proteolytic activity, as can be seen on the Coomassie-Blue stained SDS gel (A) of receptor incubated with this toxin (track 2) as compared to native receptor (track 1) and receptor incubated with Sigma toxin (track 3).

varied with toxin from Boehringer, far less with the Sigma sample and virtually non-existent with toxin purified by us from the venom. We suspect that an enzymatic impurity in some of the toxin smaples is responsible for the structural change monitored by the ^3H-TPMP$^+$ labeling. This interpretation is supported by the finding that commercially available toxin indeed contains considerable proteolytic activity (Fig.2 A, track 2) and that under certain circumstances trypsin too enhances labeling of the larger AChR-chains (β and δ) (Fig.2 B, track 2).

Another enzymatic activity which could affect the receptor-rich membranes is lipolysis. Fig.2c shows that photolipase A_2

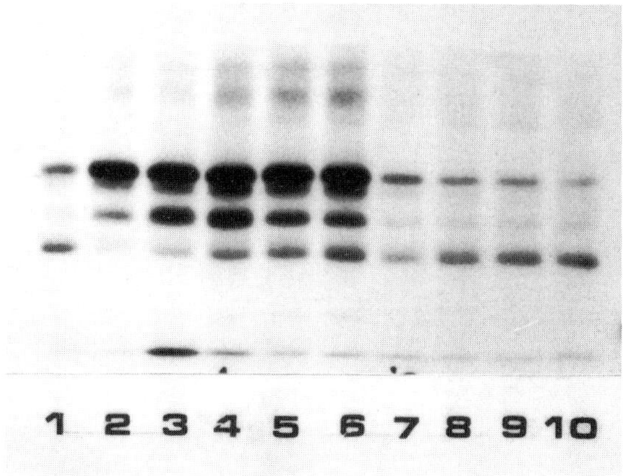

Fig. 2c Effect of phospholipases A_2 and D on the covalent labeling of AChR-receptor; usual samples were incubated with different dilutions of the lipase in Na-Ringer for 60 min at room temperature and then processed as described. The lipase-activities applied per sample were for track number 1) 2 x 1o^{-1} U, 2) 2 x 1o^{-2} U, 3) 2 x 1o^{-3} U, 4) 2 x 1o^{-4} U, 5) 2 x 1o^{-5} U, 6) 2 x 1o^{-6} U, 7) 2 x 1o^{-7} U.

promotes UV-labeling of the δ-chain considerably. The toxin sample used in the experiment shown in Fig.2, A,B, also contained a considerable amount of lipolytic activity as can be shown by thin layer chromatography of receptor-rich membrane. After 6o min incubation of the membrane with 1o^{-5} M α-bungarotoxin (Boehringer) virtually all of the membrane phospholipid was cleaved. Lipolysis was shown previously (1o) to alter AChR in a way similar to receptor desensitization.

Stimulation of $TPMP^+$ binding and covalent incorporation into the δ-polypeptide chain caused by carbamoylcholine is reversible; after removal of the ligand by dialysis the original labeling pattern is obtained again, i.e. most of the radioactivity is found in the α-polypeptide chain.

Removal of peripheral membrane protein by treatment of the receptor-rich membranes at pH 11 (11) does not alter the labeling pattern in the presence or absence of cholinergic effectors. This treatment removes among others a protein of molecular weight 43 ooo the function of which is still unknown. Pretreatment of the AChR-rich membranes with reducing agents, shown to alter e.g. the affinity and cooperativity of the binding sites, has no significant effect either.

Carbamoyl-concentration dependence of the covalent ^{3}H-TPMP^{+}-incorporation.

To investigate further the nature of the conformational change visualized by the covalent TPMP^{+}-labeling we tried to determine the dose dependence of the carbamoylcholine stimulated labeling (Fig.3). The experiment was carried out using different receptor-concentrations, leaving the other parameters unchanged. The lower protein-limit was restricted by the cpm incorporated, i.e. the number of cpm on the gel to be high enough to yield a darkening of the X-ray film within a reasonable time (about 5oo - 1ooo cpm).

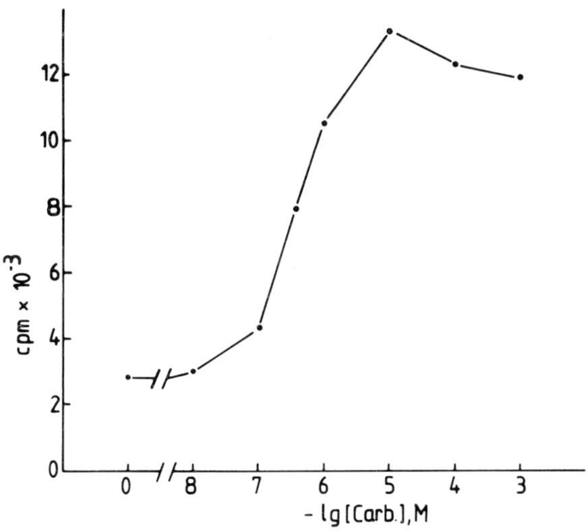

Fig.3 Dose-response curve of the carbamoyl-stimulated covalent labeling with ^{3}H-TPMP^{+}; 15 µg receptor-rich membranes were incubated as usual with increasing amounts of carbamoylcholine. After irradiation the samples were subjected to SDS electrophoresis and fluorographed. The δ-subunits were then sliced out and the ^{3}H-cpm incorporated were counted after dissolving the slices in Toluol/Omnifluor/3% Protosol for 24 hours.

Thus we augmented the protein-concentration up to tenfold. A typical experiment is shown in Fig.3. The value of the EC_{50}, i.e. the apparent K_D, was in the range of the carbamoylcholine affinity of the receptor but it increases with the protein-concentration (not shown). We therefore think the curve shown in Fig.3 represents a titration curve of the ligand binding site, the half-maximal saturation being the equivalence point rather than the "apparent K_D". This would mean that with one agonist molecule bound to the AChR maximum labeling with $TPMP^+$ is obtained. The conformational change underlying the stimulation of the labeling therefore appears to be achieved with one of the two binding sites on the receptor being occupied.

Pulse labeling of AChR at very brief irradiation times

As mentioned in the introduction we believe that UV induced labeling with ^3H-$TPMP^+$ may be a useful tool for monitoring

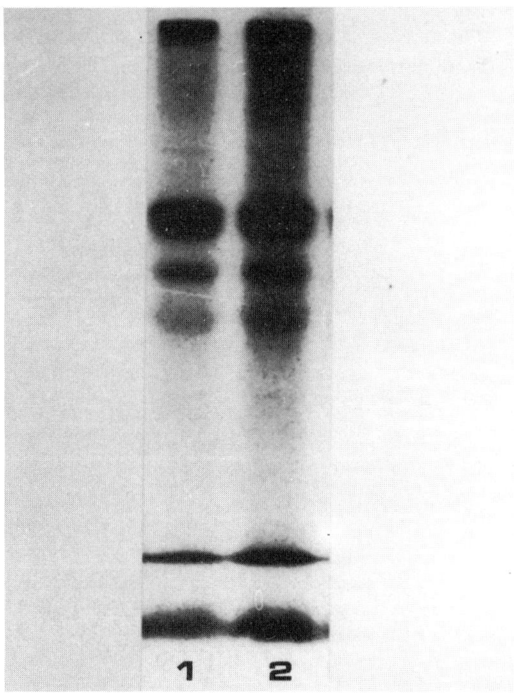

Fig.4 Pulse-labeling of AChR with laser-light; samples (36 µl) in quartz cuvettes, containing the usual amounts of protein, carbamoylcholine and ^3H-$TPMP^+$ were incubated at room temperature for 5 min before irradiation with laser-flashes of energies indicated below; 1) 2 mJ, 2 flashes, 2) 8 mJ 2o flashes

structural changes of the receptor protein during receptor activation and regulation. Since $TPMP^+$ blocks ion flow through the AChR-ion channel it seems especially useful for investigating the molecular events in connection with channel gating and for localising the participating molecular areas within the quaternary and possibly the primary structure of the AChR.

A prerequisite for this type of investigation would be to adapt the time scale of the labeling experiment to that of the physiological events, i.e. AChR activation (channel opening) and desensitization, which take place in the msec and 1oo msec range. Fig.4 shows that pulse labeling with laser light is feasible. Two flashes of laser light (λ = 266 nm, 2 nsec/pulse (1o Hz)) were sufficient to incorporate enough radioactivity for further analysis. Experiments to combine this laser flash labeling with a stopped flow apparatus are under way.

Acknowledgements: This work was supported by the Deutsche Forschungsgemeinschaft (grant Hu 146/8-2), the Fonds der Chemischen Industrie and the Land Berlin.

References

1) Changeux,J.-P. (1981) : Harvey Lect.75, 85-254.

2) Neubig,R. and Cohen,J.B. (198o): Biochemistry 19, 277o-2779.

3) Lauffer,L. and Hucho,F. (1982): Proc.Natl.Acad.Sci.USA 79, 24o6-24o9.

4) Muhn,P. and Hucho,F. (1983): Biochemistry 22, 421-425.

5) Albuquerque,E.X., Barnard,E.A., Chiu,T.H., Lapa,A.J., Dolly, J.O., Jansson,S.E., Daly,J. and Witkop,B. (1973): Proc.Natl. Acad.Sci.USA 7o, 949-953.

6) Albuquerque,E.X., Tsai,M.C., Aronstam,R.S., Witkop,B., Eldefrawi,A.T. and Eldefrawi,M.E. (198o): Proc.Natl.Acad.Sci. USA 77, 1224-1228.

7) Tsai,M.C., Mansour,N.A., Eldefrawi,A.T., Eldefrawi,M.E. and Albuquerque,E.X. (1978): Mol.Pharmacol. 14, 787-8o3.

8) Popot,J.-L., Sugiyama,H. and Changeux,J.-P. (1976): J.Mol. Biol. 1o6, 469-483.

9) Rang,H.P. and Ritter,J.M. (197o): Mol. Pharmacol.6, 383-39o.

1o) Andreasen, T.J. and McNamee,M.G. (1977): Biochem.Biophys. Res.Commun. 79, 958-965.

11) Neubig,R.R., Krodel,E.K., Boyd,N.D. and Cohen,J.B. (1979): Proc.Natl.Acad.Sci.USA 76, 69o-694

α-BUNGAROTOXIN AND STRYCHNINE AS TOOLS TO CHARACTERIZE NEUROTRANSMITTER RECEPTORS OF THE CENTRAL NERVOUS SYSTEM

Heinrich Betz, David Graham, Friedhelm Pfeiffer, Hubert Rehm
Max-Planck-Institute for Psychiatry, Department of Neurochemistry, D-8033 Martinsried, FRG

Introduction

Electrical excitability, neurotransmitter secretion and chemosensitivity are the essential properties of nerve cells upon which information processing in the brain is based. The molecules mediating these functions are largely associated with the neuronal membrane and are often specifically located at the site of interneuronal communication, the synapse. The interest of our group centers upon the biochemistry and regulation of the membrane components involved in neurotransmission at central synapses. In order to identify and purify relevant macromolecules, we have used polypeptide and alkaloid neurotoxins which perturb the function of the motor system. Here, we report the use of α-bungarotoxin and strychnine, paralyzing and convulsive poisons, respectively, as molecular probes for analyzing neuronal postsynaptic membrane proteins.

Toxins as Tools in Neurochemistry

The α-BTX binding protein of the chick visual system: a putative neuronal nicotinic acetylcholine receptor.

The snake venom protein α-BTX binds very selectively to the agonist recognition sites of the nicotinic acetylcholine receptor (AChR) in skeletal muscle and fish electric organ and thus blocks the function of this protein. Because of its extremely high affinity, α-BTX has established itself as a classical ligand in the analysis of this peripheral effector organ receptor (1,2, and this volume). Binding sites for α-BTX exhibiting nicotinic cholinergic specificity are also widely distributed in the central nervous system of vertebrates and invertebrates (3-5). Though often referred to as neuronal nicotinic AChR, the nature of these α-BTX binding sites is still controversial, principally because α-BTX fails to antagonize nicotinic-cholinergic responses in most nerve cell preparations (for review see 5). In the brain of some lower vertebrates, however, α-BTX has been shown to block a nicotinic AChR function and to modify synaptic connectivity (6,7).

The interest of our group has focussed on the α-BTX binding protein of the chick visual system (retina and optic tectum), a brain structure which contains relatively high concentrations of α-BTX binding sites. Different lines of evidence suggest that α-BTX binding in the chick central nervous system may be associated with a neuronal nicotinic AChR (4):

1. ^{125}I-labelled α-BTX binds with high affinity (K_D~0.6 nM) to membrane preparations or cell cultures of chick neural retina (8,9). This binding is inhibited by nicotinic cholinergic ligands as well as by local anaesthetics and the frog alkaloid histrionicotoxin (8-11). All these compounds are active at the peripheral nicotinic AChR.

2. The size properties of the detergent-solubilized chick α-BTX binding protein (9.5 - 11.0 S) are similar, but not identical to that of muscle AChR (8,12).
3. The α-BTX binding site exhibits changes in agonist (but not antagonist) affinity which suggest conformational alterations corresponding to the well documented desensitization of other neurotransmitter receptors (8).
4. Some antisera against muscle or Torpedo AChR have been shown to weakly, but significantly cross-react with the α-BTX binding site of chick retina or optic lobe (8,12).
5. In the adult retina, α-BTX binding is localized to synapses (13,14).
6. The regulation of the retinal α-BTX binding sites *in vivo* and in tissue culture exhibits striking parallels to the extensively investigated metabolism of AChR in skeletal muscle (15).

Although all of these points provide strong arguments for classifying the retinal α-BTX binding protein as a nicotinic AChR, the obstacle has to be resolved that electrophysiological and ion flux studies with different types of chick neurons fail to demonstrate an antagonism of nicotinic-cholinergic responses by α-BTX (reviewed in 3,5). Several hypotheses, including toxin-inaccessible agonist binding sites, non-identity of functional agonist and antagonist binding domains, and overlapping multisite interactions between receptor, toxin and agonist, have been forwarded to solve this discrepancy between the physiological and binding data (8,9,12,16). In fact, the existence of toxin-inaccessible, "cryptic" α-BTX binding sites in retinal membranes has been demonstrated (17). A solid proof of any of these models would, however, require the identification of the membrane polypeptide(s) involved in agonist and α-BTX binding and the demonstration of their association within the same macromolecular assembly mediating acetylcholine-triggered ion transport.

As a first approach towards these goals, we have localized the retinal α-BTX binding site on a polypeptide of an apparent mol. weight (M_r) of 57,000 (18). This polypeptide was identified by crosslinking experiments using ^{125}I-labelled α-BTX as well as by affinity purification and indirect immunoprecipitation from ^{35}S methionine-labelled cell cultures (Table 1). The crosslinking and immunoprecipitation data in addition suggested that two other polypeptides of lower mol. weight (M_r = 35,000 and 25,000) may be involved in toxin binding.

In order to elucidate the physiological relevance of these α-BTX binding polypeptides, we have initiated an immunolo-

Table 1. Identification of α-BTX binding polypeptide(s)

Method	M_r of labelled band(s)		
Crosslinking[a]			
a) glutaraldehyde	64 K,	43 K,	33 K
b) Lomant's II	63 K,		32 K
-8 K[c]:	55-56 K,	35 K,	24-25 K
Toxin column[a]	57 K		
Immunoprecipitation[b]	57 K,	34 K,	25 K
	55-57 K,	34-35K,	24-25 K

[a] Chick retina
[b] PC12
[c] M_r of α-BTX

gical investigation of the α-BTX binding protein. For this purpose, the toxin receptor site was purified from Triton X-100 extracts of chick optic lobe membranes by sequential affinity chromatography on α-BTX-derivatized agarose and wheat germ agglutinin-Sepharose following a protocol originally developed for the isolation of muscle AChR (19). Thus small quantities (5-20 μg of protein) of highly purified toxin receptor preparations (specific activites 0.6 - 1.0 nmol/mg protein) could be obtained. SDS-Polyacrylamide gel electrophoresis revealed that 15-20% of the protein in these preparations was associated with the previously identified M_r = 57,000 polypeptide. Balb/c mice were immunized with μg quantities of this purified receptor, and their sera were tested for antibody activity. One of the mice exhibited a high titer of antibody and its spleen was used for the production of monoclonal antibodies (mAbs) (20). From 18 initially positive wells, 14 stable antibody-secreting hybridoma lines were established. So far, 12 of the mAbs have been characterized in more detail (Betz and Pfeiffer, unpublished). Some of their properties are summarized in Table 2. Particularly interesting is that the antibodies

Table 2. Properties of mAbs to the α-BTX binding protein of chick optic lobe

mAb No.	Titer (pmol/ml)	Inhibition of α-BTX binding	Antibody binding blocked by cholinergic ligands	Antigenic modulation
1a	79	-	N.D.	+
2a	688	-	N.D.	-
3a	400	-	N.D.	-
4b	73	+	-	(+)
5a	186	-	N.D.	+
6a	1	-	N.D.	+
7a	350	+	+	(+)
8a	400	+	+	(+)
9b	283	+	-	(+)
10a	316	+	-	(+)
11b	208	-	N.D.	+
12a	258	+	-	(+)

N.D. Not determined.

distinguish at least 3 different epitopes on the retinal α-BTX binding protein: a) half of the mAbs block α-BTX binding; b) receptor binding of 2 of the latter class of mAbs is sensitive to cholinergic antagonists. Most, but not all, mAbs produce antigenic modulation, i.e. a loss of α-BTX binding sites, in chick retina cultures (H. Betz, unpublished). None of the mAbs crossreacts with muscle or *Torpedo* AChR; some, however, recognize the α-BTX binding protein of the rat phaeochromocytoma cell line PC12.

Preliminary experiments have shown that these mAbs can be used to construct immunoadsorbents for the rapid isolation of the retinal α-BTX binding protein. In addition, the antibodies should be useful to clarify the relationship between α-BTX binding and nicotinic receptor function of the neuronal membrane. Experiments correlating α-BTX binding and cholinergic agonist-induced ion flux in antibody-treated cell cultures are in progress in our laboratory.

Strychnine, a high-affinity ligand and photoaffinity label of the glycine receptor.

Strychnine, a heterocyclic alkaloid from *Strychnos nux-vomica*, has been shown to antagonize postsynaptic inhibition in the spinal cord and other regions of the central nervous system by binding to the postsynaptic glycine receptor (21, 22). This inhibitory neurotransmitter receptor is coupled to a chloride ion channel at which strychnine binding has been proposed to occur (22). High-affinity binding of [^{3}H]strychnine is glycine-displaceable and thus can be used to monitor glycine receptor sites. We have found strychnine to be an extremely potent probe for analyzing and purifying the glycine receptor protein.

As a first step towards the biochemical characterization of the glycine receptor, [^{3}H]strychnine binding sites were solubilized from synaptic membrane fractions of rat spinal cord using the non-ionic detergent Triton X-100 (23). These investigations revealed that the solubilized glycine receptor behaves as a glycoprotein of M_r = 250,000 whose native conformation is dependent on the presence of phospholipids (23). All subsequent experiments with the solubilized glycine receptor were therefore performed in buffers containing phosphatidylcholine.

The further investigation of the glycine receptor was greatly facilitated by our discovery that strychnine is a natural photoaffinity ligand of this receptor protein. Upon UV illumination in the presence of spinal cord membranes, [^{3}H]strychnine became irreversibly incorporated into its high-affinity receptor binding site (24). SDS-Polyacrylamide gel electrophoresis of the illuminated membranes identified a polypeptide of M_r = 48,000 as the strychnine binding subunit of the glycine receptor (24). Meanwhile this photoaffinity labelling reaction has been characterized in more detail. The kinetics and the pharmacology of the UV-induced incorporation of ^{3}H strychnine into membranes all are consistent with the hypothesis that strychnine indeed is crosslinked at or close to the antagonist binding site of the glycine receptor (25). Also, photoaffinity labelling has been used to explore the topology in the membrane of the M_r = 48,000 subunit: incubation of [^{3}H] strychnine-labelled membranes with trypsin led to a degradation of the 48 K band to polypeptides of lower molecular weight (25). Even extensive protease treatment, however, failed to remove the radioactive label from the membranes. Strychnine binding therefore is thought to occur at a receptor domain which is protected by the hydrophobic environment of the neuronal membrane (25). This interpretation is consistent with the previously postulated association of strychnine binding with the ionic channel of the glycine receptor.

For purification of the glycine receptor a physiologically active derivative of strychnine, 2-aminostrychnine, was coupled to agarose beads via a long hydrophilic spacer (26). This affinity resin bound > 90% of the glycine receptor present in detergent extracts of rat spinal cord membranes. Furthermore, about 20% of the stychnine binding activity could be recovered after extensive washing of the resin and biospecific elution with the receptor agonist glycine. This simple protocol has been developed as a standard procedure for isolating the glycine receptor in one step to near homogeneity: upon SDS-polyacrylamide gel electrophoresis, only three polypeptides of M_r = 48,000, 58,000 and 93,000 were detected in the purified receptor preparation (26). The main biochemical properties of the purified glycine receptor are listed in Table 3. From these and other unpublished data we suggest a model of the glycine receptor

Table 3. Biochemical properties of the affinity-purified glycine receptor

1.	K_D for strychnine binding of 9.3 nM.
2.	Strychnine binding inhibited by the agonists glycine, ß-alanine, taurine, and by avermectin B1a.
3.	Sedimentation coefficient of 8.2 S, Stokes radius of 7.7 nm.
4.	M_r of 250,000.
5.	UV illumination incorporates $[^3H]$strychnine into the 48 K subunit.
6.	48 K subunit present in higher amounts than 58 K and 93 K subunits.
7.	48 K and 58 K subunits are glycosylated.

Data from refs. 26, 27; and unpublished experiments.

as depicted in fig. 1. Further work is, however, needed to consolidate this hypothetical view of the glycine-gated chloride channel of the postsynaptic membrane.

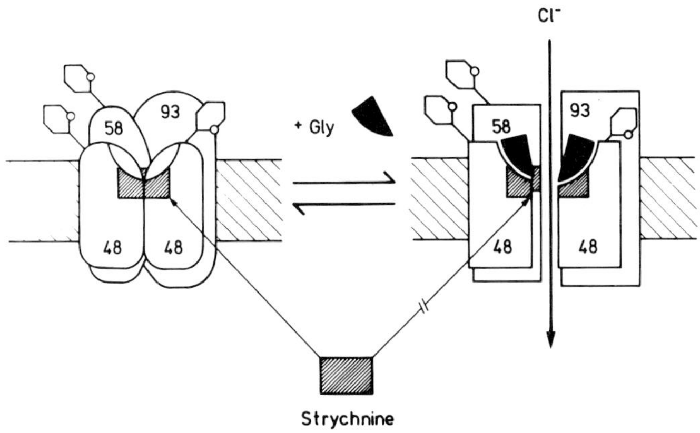

Fig. 1. A hypothetical model of the glycine receptor in its closed and open conformations.

4. Conclusion

Our studies on postsynaptic membrane proteins of the central nervous system, i.e. the putative nicotinic AChR of the chick visual system and the glycine receptor of rat spinal cord, have shown that even for rather low abundancy membrane receptors the classical approach of molecular pharmacology, i.e. ligand binding experiments, can be extended to the level of receptor chemistry provided specific neurotoxins can be used for irreversibly labelling and affinity purifying the protein in question. Also, although only minute quantities of the purified receptors are currently available, the power of the modern techniques of molecular biology (monoclonal antibody production, synthesis of cDNA probes) has drawn previous dreams of analysis to close reality! It therefore seems not too optimistic to predict that within only a few years we should have a very detailed picture of the receptors whose current status of investigation has been presented here.

Acknowledgements. We thank Ms. U. Müller and C. Weyrauch for expert technical assistance and Ms. E. Eichler and C. Bauereiß for help during the preparation of the manuscript. Work in the authors' laboratory has been supported by the Deutsche Forschungsgemeinschaft and the Stiftung Volkswagenwerk. H.B. holds a Heisenberg fellowship of the Deutsche Forschungsgemeinschaft. The provision of laboratory facilities by the Max-Planck-Institute for Psychiatry is gratefully acknowledged.

References

1. Heidmann, T., Changeux, J.P.: Ann. Rev. Biochem. 47, 317-357 (1978).
2. Fambrough, D.: Physiol. Rev. 59, 165-227 (1979).
3. Morley, B.J., Kemp. G.E.: Brain Res. Rev. 3, 81-104 (1981).
4. Betz, H., Rehm, H.: In Neuroreceptors (Hucho, F., ed) pp. 165-171, Walter de Gruyter & Co., Berlin - New York (1982).
5. Oswald, R.E., Freeman, J.A.: Neuroscience 6, 1-14 (1981).
6. Freeman, J.A., Schmidt, J.T., Oswald, R.E.: Neuroscience 5, 929-942 (1980).
7. Freemann, J.A.: Nature 269, 218-222 (1977).
8. Betz, H.: Eur. J. Biochem. 117, 131-139 (1981).
9. Wang, G.K., Molinaro, S., Schmidt, J.: J. Biol. Chem. 253, 8507-8512 (1978).
10. Vogel, Z., Nirenberg, M.: Proc. Natl. Acad. Sci. USA 73, 1806-1810 (1976).
11. Betz, H.: Neurosci. Lett. 33, 153-158 (1982).
12. Norman, R.I., Mehraban, F., Barnard, E.A., Dolly, J.O.: Proc. Natl. Acad. Sci. USA 79, 1321-1325 (1982).
13. Vogel, Z., Maloney, G.J., Ling, A., Daniels, M.P.: Proc. Natl. Acad. Sci. USA 74, 3268-3272 (1977).

14. Daniels, M.P., Vogel, Z.: Brain Res. 201, 45-56 (1981).

15. Betz, H.: J. Neurosci., in press (1983).

16. Lukas, R.J., Morimoto, H., Bennett, E.L.: Biochemistry 18, 2384-2395 (1979).

17. Rehm, H., Betz, H.: Biochem. Biophys. Res. Comm. 102, 1385-1392 (1981).

18. Betz, H., Graham, D., Rehm, H.: J. Biol. Chem. 257, 11390-11394 (1982).

19. Froehner, S.C., Reiness, C.G., Hall, Z.W.: J. Biol. Chem. 252, 8589-8596 (1977).

20. Köhler, G., Milstein, C.: Nature 256, 495-497 (1975).

21. Young, A.B., Snyder, S.H.: Proc. Natl. Acad. Sci. USA 70, 2832-2836 (1973).

22. Young, A.B., Snyder, S.H.: Proc. Natl. Acad. Sci. USA 71, 4002-4005 (1974).

23. Pfeiffer, F., Betz, H.: Brain Res. 226, 273-279 (1981).

24. Graham, D., Pfeiffer, F., Betz, H.: Biochem. Biophys. Res. Comm. 102, 1330-1335 (1981).

25. Graham, D., Pfeiffer, F., Betz, H.: Eur. J. Biochem., in press (1983).

26. Pfeiffer, F., Graham, D., Betz, H.: J. Biol. Chem. 257, 9389-9393 (1982).

27. Graham, D., Pfeiffer, F., Betz, H.: Neurosci. Lett. 29, 173-176 (1982).

IV. Ca-Channels

Axonal Toxins

Presynaptic Toxins

Cardiotoxin

New Venoms

INTERACTION OF TOXINS AND DIVALENT CATIONS IN CALCIUM CHANNELS OF THE NEURONAL MEMBRANE

Platon G. Kostyuk

A.A.Bogomoletz Institute of Physiology, Academy of Sciences of the Ukrainian SSR, Kiev, USSR

Introduction

The development of the technique for intracellular perfusion of isolated nerve cells (1) gave a possibility for reliable separation of all components of the transmembrane ionic current that appears in response to a depolarizing membrane potential shift. In a number of studies performed with this technique it was shown that the somatic membrane of nerve cells of both invertebrate and vertebrate animals generates among others also a considerable inward calcium current which is due to the presence in it of ionic channels selective for divalent cations (2, 3, 4). Since the influx of Ca^{2+} ions through these channels is of primary importance for the coupling between membrane and intracellular processes, a detailed investigation of the mechanisms of their functioning and a search for ways of their effective pharmacological control are the matter of urgent interest.

In the present communication the data obtained in our laboratory about the functioning of calcium channels in the somatic neural membrane and the effect of synthetic toxins on them will be presented.

Toxins as Tools in Neurochemistry

Results

1) Recording of calcium currents. Experiments were performed on nonidentified neurons of the snail, Helix pomatia. The techniques of cell isolation, intracellular perfusion, membrane potential clamping and recording of transmembrane ionic currents did not differ from those described earlier (5). In order to remove outward currents and to record inward calcium currents in a pure form, the cells were perfused with potassium-free Tris-aspartate solution (130 mM), internal pH being adjusted by Tris-HCl. If necessary, sodium currents were removed from the extracellular solution by replacement of Na^+ ions by $Tris^+$.
2) Selectivity of calcium channels under normal conditions. Besides Ca^{2+} ions, inward currents through calcium channels can be effectively carried only by Ba^{2+} and Sr^{2+}. These currents show characteristic slow kinetics that allows their clear - cut discrimination from "fast" inward currents carried through sodium channels. Concentration - dependence of the currents flowing through calcium channels shows saturation and can be approximated by Langmuir's isotherm having potential - dependent K_d. This indicates a binding of penetrating divalent cations to some binding group inside the channel. Such a group can be as well responsible for blocking of calcium channels by other divalent cations (Co^{2+}, Ni^{2+}, Cd^{2+}) having stronger binding ability, as their action can also be approximated by an adsorption isotherm and at the same time is in competitive relation to the concentration of the current - carrying ion.
   A change of fixed charges on the outer membrane surface is also occurring at the increase of the concentra-

tion of divalent cations in the extracellular solution, which leads to a corresponding reduction of near - membrane concentration of carrier ions (usually accumulating at the membrane because of attraction by these charges). This can affect the concentration characteristics of currents flowing through calcium channels. Therefore, a calculation of the changes in near - membrane concentration of carrier ions depending on the changes of fixed charges (the latters were determined by the shift of current-voltage characteristics for the corresponding currents) has been done according to the equation

$$C^s_{out} = C_{out} \exp(-z\varphi \frac{F}{RT}) ,$$

and a correction of corresponding concentration dependences of ionic currents was made (6). After correction for the change in near - membrane concentration the corresponding pK values for ion binding in the calcium channel (pK = $-\log K_d$) are as follows: $pK_{Ba} = 1.0$; $pK_{Sr} = 1.7$ and $pK_{Ca} = 2.0$. This series is similar to that obtained for binding of divalent cations to a single carboxylic group.

3) Pharmacological blocking of calcium channels. No natural toxins which may block specifically calcium channels have been found yet. However, these channels in the somatic membrane, like in other excitable membranes, can be quite effectively blocked by extracellular application of two groups of synthetic organic drugs - the derivatives of verapamil and 1,4 - dihydropyridine. The action of these drugs can be approximated by Langmuir's isotherm with $K_d = 2.6 \times 10^{-5}$M for D-600 (derivative of verapamil) and $K_d = 2.3 \times 10^{-5}$M for niphedipine (derivative of 1,4 - dihydropyridine). The effect of both drugs is completely reversible. Intracellu-

lar application of verapamil also produces some inhibition of the calcium current, however much less effective than the extracellular one. Intracellular application of niphedipine does not produce any detectable changes in the calcium current.

The currents passing through sodium or potassium channels, as well as leakage currents are not affected by both groups of drugs used in the above - mentioned concentrations.

4) Modifications of calcium channel selectivity. A characteristic modification of inward currents through calcium channels takes place if divalent cations are removed from the extracellular medium by adding EDTA to it.
In this case a strong inward current is induced which has the same kinetics as the calcium current but is carried by sodium ions. The region of activation of this current is shifted on the current - voltage curve by 60–70 mV in the hyperpolarizing direction reflecting a complete demasking of fixed charges at the external side of the cell membrane in Ca-free medium. The selectivity of the modified channels carrying this current becomes close to that of original sodium channels. Such current can also be induced by removal of Ca^{2+} ions from the extracellular solution using other Ca - binding substances (EGTA, GMDTA, HTA, citric and orthophosphate acids). In this case the amplitude of the induced current correlates well with the affinity of these substances to Ca^{2+} ions (determined from the level of free Ca^{2+} in the extracellular solution with the help of metallochromic indicator Arsenazo III). The increase in extracellular Ca^{2+} content (or addition of other divalent cations) returns the selectivity of the calcium channels to its initial state. This process can be approximated by Langmuir's

isotherm with pK values $pK_{Ca} = 6.6$; $pK_{Sr} = 5.5$; $pK_{Ba} = 4.8$ and $pK_{Mg} = 4.2$; the effect is not potential-dependent. The obtained series is similar to that for binding of divalent cations by chelates and different Ca-binding proteins in which a high affinity to these cations is due to the presence in the active center of several carboxylic groups forming complex coordinative relations with the cation. Therefore, it can be suggested that the calcium channel, apart from the inner binding group, has an outer binding group which is capable of taking Ca^{2+} ions from the extracellular solution with very high affinity. In the bound state this group prevents monovalent cations from their entering inside the channel. Introduction of chelating agents into the extracellular solution resulting in a decrease of Ca^{2+} concentration below 10^{-7}M makes this group unbound. In this case the calcium channel requires the ability to pass effectively monovalent cations demonstrating selectivity characteristic for the sodium channel. The scheme below summarizes the above-stated suggestions on the functional structure of the calcium channel.

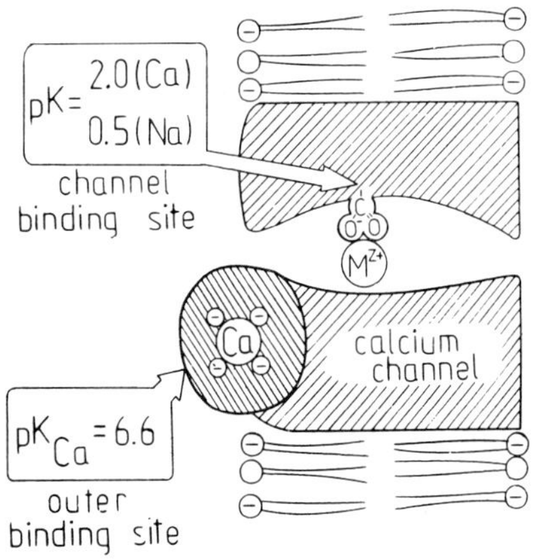

Fig. 1 Scheme illustrating the functional organization of the calcium channel.

5) Pharmacological blocking of modified calcium channels. Calcium channels transformed into sodium ones, as described above, not only preserve their ability to be blocked by specific blockers of the calcium conductance but become 2 - 3 times more sensitive to the action of these agents. Such a change in the sensitivity of calcium channels might be due to both possible competitive interaction between the blocker and Ca^{2+} ions for a common binding site, and to an increase in the near - membrane concentration of the blocker owing to the change in fixed negative surface charges on the membrane surface (at the pH values used these blockers are mainly in cationic form and may accumulate near the surface). In order to solve this question we have estimated the possible amount of such accumulation, using the results about the changes in surface charges following the removal of divalent cations from the extracellular solution. Taking the values of membrane surface potential equal to -7 mV and -25 mV obtained correspondingly in the presence of 30 mM Ca^{2+} in the external solution and after their complete removal from the latter, a corrected constant for blocking the sodium current through the modified calcium channels became 2.8×10^{-5}M for D-600 and 2.25×10^{-5}M for niphedipine, respectively. These values differ only slightly from the similar values obtained for blocking the calcium current through nonmodified channels. The results are in favour of the latter suggestion excluding competitive action of toxins and Ca^{2+} ions for common binding sites in the calcium channel.

One should, therefore, accept that the action of specific organic blockers on the calcium channel is not related to their interaction with one of the groups determining the passage of inorganic cations through the channel.

It seems to be determined by the presence in the calcium channels of some additional molecular structures having high affinity to these toxins. Recent studies of the action of verapamil derivatives containing tertiary or quaternary nitrogen on the calcium channels in the membrane of cardiac muscle fibres (7) led to the suggestion that these derivatives block the channel from the inner side of the membrane previously penetrating the membrane lipids in an uncharged form. However, our findings about much lower effectiveness of the drugs, if applied intracellularly, makes this suggestion doubtful, at least for calcium channels in the neuronal membrane. Obviously, further studies on the site of action of calcium channel toxins are of great importance for both the understanding of the structure of these channels and for the finding of more effective ways for regulation of their function.

References

1. Kostyuk, P. G., Krishtal, O. A., Pidoplichko, V. I.: Nature 257, 691-693 (1975).
2. Kostyuk, P. G., Krishtal, O. A.: J. Physiol. (London) 270, 569-580 (1977).
3. Akaike, N., Lee, K. S., Brown, A. M.: J. Gen. Physiol. 71, 509-531 (1978).
4. Kostyuk, P. G., Veselovsky, N. S., Fedulova, S. A.: Neuroscience 6, 2431-2437 (1981).
5. Kostyuk, P. G., Krishtal, O. A., Pidoplichko, V. I.: J. Physiol. (London) 310, 403-421 (1981).
6. Kostyuk, P. G., Mironov, S. L., Doroshenko, P. A.: J. Membrane Biol. 70, 181-189 (1982).
7. Hescheler, J., Pelzer, D., Trube, G., Trautwein, W.: Pflügers Arch. 393, 287-291 (1982).

TOXINS ISOLATED FROM THE VENOM OF THE SCORPION *CENTRUROIDES SCULPTURATUS*: CHEMICAL STRUCTURE AND ELECTROPHYSIOLOGICAL EFFECTS

Hans Meves

I. Physiologisches Institut der Universität des Saarlandes,
D - 6650 Homburg/Saar

Dean D. Watt

Department of Biochemistry, Creighton University,
Omaha, Nebraska 68178, U.S.A.

Introduction

Scorpion toxins have attracted the attention of electrophysiologists because they strikingly modify the function of the sodium channels. We describe in this article the chemical characteristics and the electrophysiological effects of toxins isolated from the American scorpion *Centruroides sculpturatus*.

Chemical Characteristics of Isolated Toxins

Isolation of individual toxins. Whole venom is a complex mixture of mucous, polynucleotides, small organic molecules and salts, several non-toxic acidic proteins and many basic proteins, the neurotoxins (1,2,3). Resolution of whole venom and purification of the toxins is accomplished by a series of column chromatographic separations with ammonium acetate buffers as developers. The initial separation of whole venom is by chromatography on carboxymethylcellulose, C-11 (CMC). A typical elution profile is shown in Fig. 1. Zones 4-12 con-

Toxins as Tools in Neurochemistry

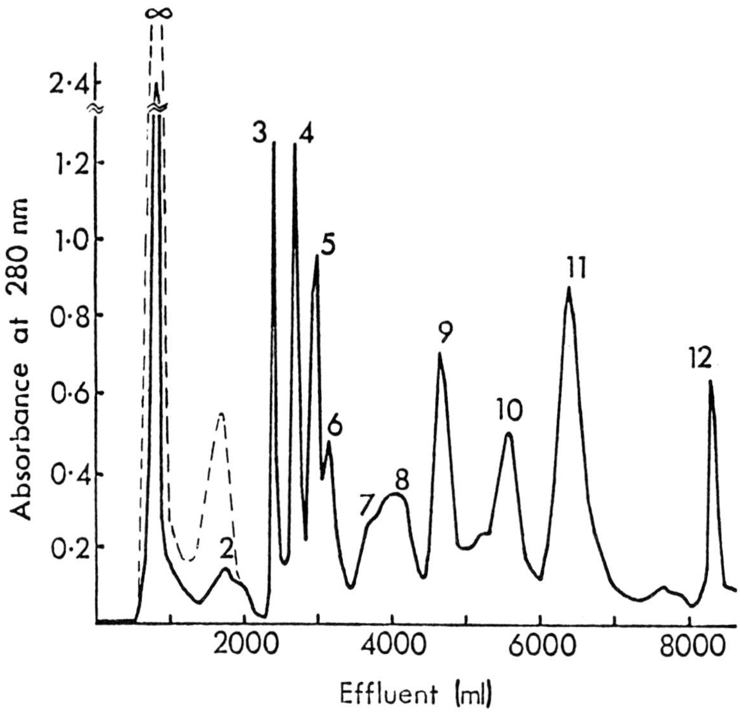

Fig. 1. Elution profile from chromatography of venom from C. sculpturatus. Carboxymethylcellulose column, CM-11 (CMC); NH_4Ac buffers. Solid line O.D. at 280 nm; dashed line at 260 nm. Elution program see (2).

tain the many toxins. Basicity of the toxins increases from left to right across the profile as does the toxicity up to and including Zone 11 which contains the most potent toxins III and IV. The toxins studied thus far vary greatly in toxicity and are characterized by frequent omission of amino acids (4).

Amino acid sequences. Primary structures of seven toxins from C. sculpturatus have been determined (2,3,5). Tabulated in Table I are amino acid sequences of selected toxins from several species of scorpions including four of the toxins from C. sculpturatus. The sequences are arranged, leaving gaps as necessary, so that recurring amino acid residues are aligned

TABLE I AMINO ACID SEQUENCES OF SELECTED SCORPION TOXINS.

CsE v3	-KEGYLVKKSDGCKYGCLKLGENEGCDTECKAKNQGGSYGYCYAF-----ACWC-EGLPESTPTYPLPNKSC
CsE v1	-KEGYLVKKSDGCKYDCFWLGKNEHCNTECKAKNQGGSYGYCYAF-----ACWC-EGLPESTPTYPLPNK-CS
CsE v2	-KEGYLVNKSTGCKYGCLKLGENEGCNKECKAKNQGGSYGYCYAF-----ACWC-EGLPESTPTYPLPNK-CSS
CsE I	-KDGYLVEK-TGCKKTCYKLGENDFCNRECKWKHIGGSYGYCYGF-----GCYC-EGLPDSTQTWPLPNK-CT
Css II	-KEGYLVSKSTGCKYECLKLGDNDYCLRECKQQYGKSSGGYCYAF-----ACWC-THLYEQAVVWPLPNKTCN
Css III	-KEGYLVSKSTGCKYECLKLGDNDYCLRECKQQYGKSSGGYCYAF-----ACWC-EALPDHTQVW-VPNK-CT
AaH I	KRDGYIVYPNN-CVYHCVPP-----CDGLCKKN-GGSSGSSCFLV-PSGLACWC-KDLPDNVPIKDTSRK-CT
AaH II	VKDGYIVDDVN-CTYFCGR---NAYCNEECTKL-KGES-GYCQWASPYGNACYCYK-LPDHVRTKG-PGR-CH
AaH III	VRDGYIVNSKN-CVYHCVPP-----CDGLCKKN-GAKS-GSCGFLIPSGLACWC-VALPDNVPIKDPSYK-CHS
Amm V	LKDGYIIDDLN-CTFFCGR---NAYCDDECKKK-GGES-GYCQWASPYGNACWCYK-LPDRVSIKE-KGR-CN
Be M_{10}	VRDGYIADDKD-CAYFCGR---NAYCDEECKK--GAES-GKCWYAGQYGNACWCYK-LPDWVPIKQKVSGKCN
Bot I	GRDAYIAQPEN-CVYECAQ---NSYCNDLCTKQ-GATS-GYCDWLGKYGNACWC-KDLPDNVPIKI-PGK-CHF
Bot II	GRDAYIAQPEN-CVYECAK---NSYCNDLCTKN-GAKS-GYCQWLGRWGNACYC-IDLPDKVPIRIE-GK-CHF
Bot XI	LKDGYIVDDRN-CTFFCGR---NAYCNDECKKK-GGES-GYCQWASPYGNACWCYK-LPDIVRTVQA-GR-CN
Lqq V	LKDGYIVDDKN-CTFFCGR---NAYCNDECKKK-GGES-GYCQWASPYGNACWCYK-LPDRVSIKEK-GR-CN

Notes, Table I. Numbering is every 5 residues of CsE-v3 excluding gaps. Paired Greek letters identify cysteine residues that form the disulfide bridges. Abbreviations: CsE, Centruroides sculpturatus Ewing, (2,3,5); Css, C. suffusus suffusus (7,11); AaH, Androctonus australis Hector (7); Amm, A. mauretanicus mauretanicus (7); Be, Buthus eupeus (8); Bot, B. occitanus tunetanus (7); Lqq, Leiurus quinquestriatus quinquestriatus (7).

with the sequence of Variant 3, the only scorpion toxin whose tertiary structure is known. The sequences reveal the following characteristics:

(1) Several sequence segments are conserved (boxed columns). (a) The sequence residues 1-6 contains conservative substitutions of amino acids in all cases. (b) The sequences -CXXXC- at positions 12-16 and 25-29 are spatially conserved although there are radical substitutions of amino acids in the sequence 12-16. (c) The sequence, residues 46-48, is invariant being either -Cys-Tyr-Cys- or -Cys-Trp-Cys-. With only two exceptions alanine precedes cysteine-46. (d) The sequence -Leu-Pro-Glu- (or Asp), 51-53, is homologous in all toxins except Toxin II (Css-II).

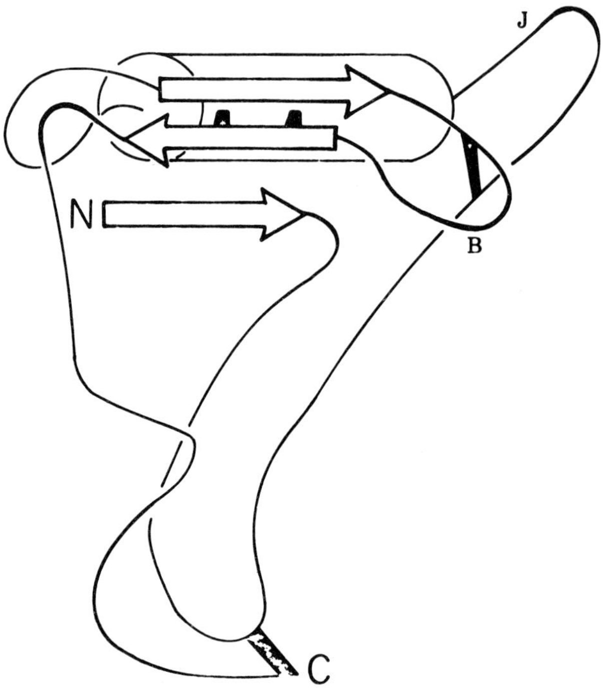

Fig. 2. Diagramatic representation of the structure of Variant 3. The arrows signify three strands of anti-parallel ß-sheet (residues 1-4, 37-41 and 46-50). The cylinder represents the α-helix; the heavy bars the disulfide bridges. The jutting "J" and blunt "B" loops are identified. (Courtesy Dr. C.E. Bugg, University of Alabama/Birmingham).

(2) Two other regions are noteworthy, viz., the loops between Cys-16 and -25 and between Cys-41 and -46, designated the "J" and "B" loops, respectively (Fig. 2). The "J" loop is long in the 'New World' scorpion toxins and short in the 'Old World' toxins. The reverse is true with the "B" loop. Unpublished preliminary data (5) suggest that toxin V more nearly resembles the 'Old World' toxins. The question then is "does Toxin V represent a common link in the phylogenesis of the 'Old' and 'New World' scorpions". A search for other toxins similar in sequence to Toxin V, therefore, seems warranted.

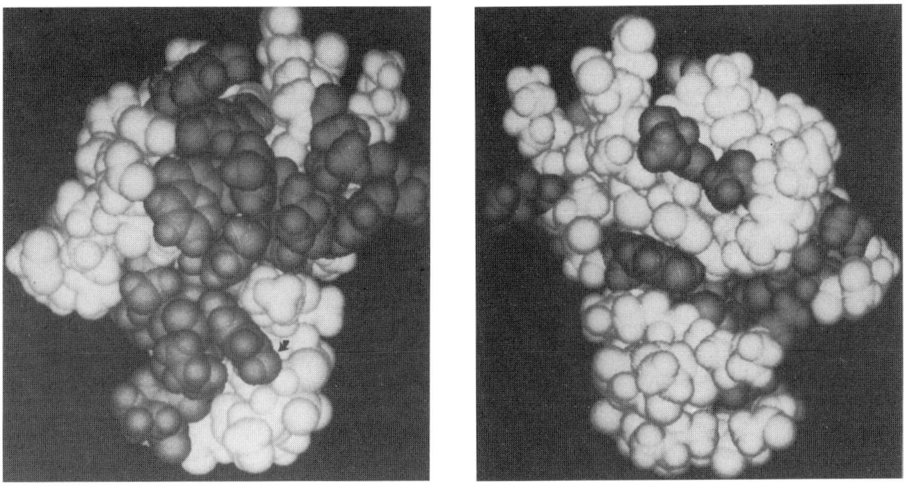

a b

Fig.3. Space filling model of Variant 3 from C. sculpturatus. (a) Hydrophobic residues (darkened) form a continuous patch. Tyr 58 lies on edge, arrow; 7 of the 8 aromatic amino acid sidechains lie in this patch. (b) The model rotated 180° about the vertical axis. Strips of buried hydrophobic residues project through to the hydrophilic side of the molecule. (Courtesy of Dr. R.J. Feldman, National Institutes of Health).

(3) The only strictly invariant residues in the sequences listed are the Cysteines, Tyrosine-5 and Leucine-51. Conservative substitutions, however, are observed with a number of residues.

(4) In general, with the exception of CsE-V, there are greater resemblances among toxins from the 'New World' scorpions than those from 'Old World' species and vice versa. The boxed region comprising the "B" loop is a good example. Similarly, the "C" terminal sequences are quite variable.

Conformation of Variant 3 from C. sculpturatus. The tertiary structure is known for only one toxin, Variant 3 from C. sculpturatus (9,10). Two elements of secondary structure are evident, namely, two and one-half turns of a α-helix (residues 23-32) and a three-strand segment of anti-parallel ß-sheet (residues 1-4, 37-41 and 46-50), (Fig. 2). These secondary structures lie approximately parallel to each other. Three of the four disulfide bridges (dark bars, Fig. 2) are located in the region of this secondary structure. The bridges Cys (25)-Cys(46) and Cys(29)-Cys(48) connect the α-helix and ß-sheet while the third disulfide bridge, Cys(16)-Cys(41), joins the "J" and "B" loops. The fourth bridge, Cys(12)-Cys (65) lies on the opposite end of the molecule and links the N and C termini. The highly contorted section of the main chain at the "C" terminus is likely produced by the four proline residues in this region. Space-filling models(Fig. 3-a,b) highlight the location on the surface of a hydrophobic patch (darkened area). Seven of the eight aromatic amino acid side-chains lie in this patch. Conserved sequence segments, residues 1-4 and 52-53, are contiguous with Tyrosine-58 (arrow). Lysine-13, upper right of Tyrosine-58 (tip of arrow shaft) also lies next to this patch. If the molecule is rotated 180° about its vertical axis, a largely hydrophilic surface is observed (Fig. 3-b). Whether the hydrophobic patch is involved in binding to the membrane is not known. Conserved residues 5,6,45 and 51 are buried in the structure and may be observed as a narrow streak (bottom center), (Fig. 3-b). Definitive statements about structure-function relationships of the scorpion toxins cannot be made on the basis of the structural model for Variant 3, however, because this isolate is the least active biologically of all the toxins isolated from venom of C. sculpturatus.

Electrophysiological Effects of Isolated Toxins

The most potent fractions are toxins I, III, IV and V. The effects of toxins I, III and IV are identical. Toxin V acts differently : it resembles 'Old World' toxins not only in its chemical structure (see above) but also in its electrophysiological effects (12,13).

Toxins I, III and IV. Toxin I (10 μg/ml and toxins III and IV (0.8-3.3 μg/ml) significantly reduce the size of the Na current but have little effect on the kinetics of inactivation. In nodes treated with toxins I,III or IV a depolarizing conditioning pulse shifts the descending branch of the $I_{Na}(E)$ curve to more negative values of membrane potential (Fig. 4A), an effect also seen with the whole venom of *Centruroides sculpturatus* (14). The shift is transient; its amount varies

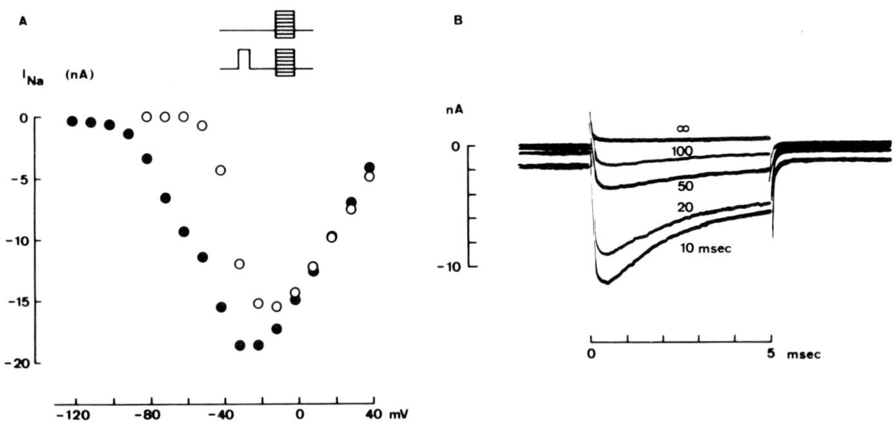

Fig. 4. Node of Ranvier treated with 1.33 μg/ml toxin IV. A: $I_{Na}(E)$ curve measured without (o) and with (●) conditioning pulse (15 ms, 130 mV, followed by 20 ms pause). B: Current produced by 30 mV pulse following the conditioning pulse (15 ms, 130 mV) after a variable pause; duration of pause indicated next to each record. Holding potential -92 mV. From Meves and Watt (unpublished).

between -10 and -60 mV, depending on the height and duration of the conditioning pulse and the pause between conditioning pulse and test pulse (Fig. 4B). The $P_{Na}(E)$ plot in Fig. 5 shows that the conditioning pulse shifts the lower part of the curve to more negative values of membrane potential

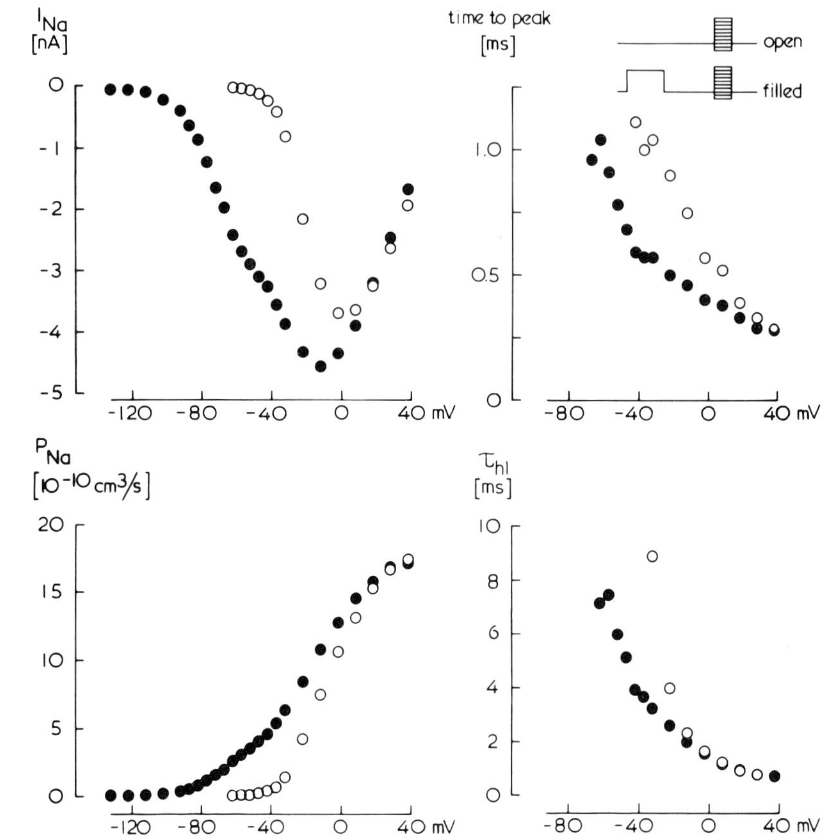

Fig. 5. Node of Ranvier treated with 0.83 µg/ml toxin IV. Peak Na current I_{Na}, peak Na permeability P_{Na}, time to peak and time constant of inactivation (τ_{h1}) as a function of pulse potential, measured without (o) and with (•) conditioning pulse (15 ms, 130 mV, followed by 20 ms pause). From (13).

whereas the maximum P_{Na} reached at E > 40 mV remains unchanged. The voltage dependence of the kinetic parameters (time to peak, time constant of activation (τ_m), time constant of inactivation (τ_{h1})) is also shifted (Fig. 5). The ion selectivity of the Na channels is little altered in nodes treated with toxins I, III or IV and subjected to a depolarizing conditioning pulse.

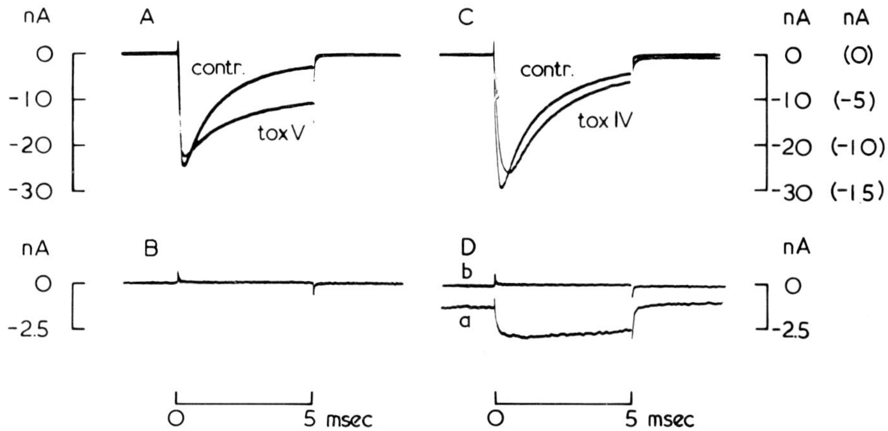

Fig. 6. Comparison of the effect of 3.33 μg/ml toxin V (A,B) and 1.1 μg/ml toxin IV (C,D). A,C: Current produced by 70 mV pulse, B,D: Current produced by 10 mV pulse preceded by conditioning pulse (15 ms, 130 mV; pause 20 ms). Holding potential -92 mV. From (13).

Toxin V and variants 1,2,3. Toxin V (3.3-10 μg/ml) and the variants (50 μg/ml) cause a marked slowing of inactivation, i.e. act like 'Old World' scorpion toxins (15); a strong depolarization is not followed by a large shift of the $I_{Na}(E)$ curve (see Fig. 6 which compares the effects of toxin V and toxin IV). A depolarizing conditioning pulse weakens, however, transiently the effect of toxin V on inactivation, a phenomenon also observed with other 'Old World' scorpions (16).

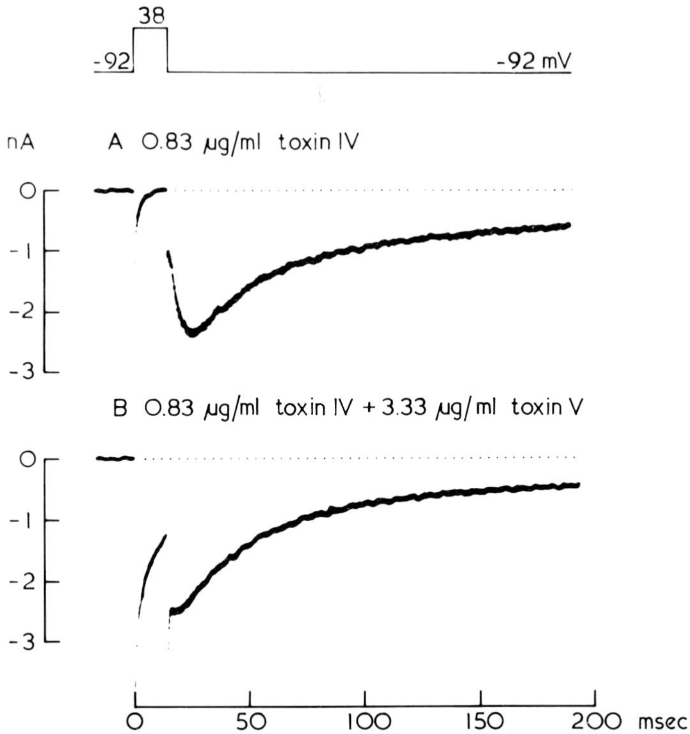

Fig. 7. Simultaneous action of toxin IV and toxin V (B) and of toxin IV alone (A). Record A taken after record B. From (12).

Simultaneous action of toxin IV and toxin V. The effect of toxin IV is not suppressed by a four times higher concentration of toxin V. As shown in Fig. 7, the slowly decaying inward current following a strong depolarizing pulse in a toxin IV-treated node (indicative of a shift of the $P_{Na}(E)$ curve to more negative values of membrane potential) is observed in the simultaneous presence of toxin IV (0.83 µg/ml) and toxin V (3.33 µg/ml). This observation suggests that the two toxins act on different receptors, a conclusion supported by other electrophysiological experiments (17) and by binding studies (18).

All toxins and variants isolated from the venom of Centruroides sculpturatus make inactivation of the Na permeability incomplete.

References

1. Pool, W.R.: Characterization of components of the venom of Centruroides sculpturatus Ewing, Dissertation for Ph.D. Degree, Creighton University, Omaha, Nebraska, USA, 1979.

2. Babin, D.R., Watt, D.D., Goos, S.M., Mlejnek, R.V.: Arch. Biochem. Biophys. 164, 694-706 (1974).

3. Babin, D.R., Watt, D.D., Goos, S.M., Mlejnek, R.V.: Arch. Biochem. Biophys. 166, 125-134 (1975).

4. Watt, D.D., Simard, J.M., Babin, D.R., Mlejnek, R.V.: Toxins: Animal, Plant and microbial, (P. Rosenberg, editor) p. 647-660, Pergamon Press, New York (1978).

5. Mole, J.E., Watt, D.D.: (to be published).

6. Sampieri, F., Habersetzer-Rochat, C.: Biochim. Biphys. Acta 535, 100-109 (1978).

7. Rochat, H., Bernard, P., Couraud, F.: Neurotoxins: Tools in Neurobiology in Advances in Cytopharmacology 3 (B. Ceccarelli, F. Clementi, editors), p. 325-334, Raven Press New York (1979).

8. Grishin, E.V.: Intern. J. Quant. Chem. 19, 291-298 (1981).

9. Fontecilla-Camps, J., Almassy, R.J., Suddath, F.L., Watt, D.D., Bugg, C.E.: Proc. Nat. Acad. Sci. USA 77, 6496-6500 (1980).

10. Fontecilla-Camps, J., Almassy, R.J., Ealick, S.E., Suddath, F.L., Watt, D.D., Feldman, R.J., Bugg, C.E.: TIBS 6, 291-296 (1981).

11. Zlotkin, E., Miranda, F., Rochat, H.: Venoms of Buthinae, Handbook of Experimental Pharmacology, v. 48 Arthropod Venoms (S. Bettini, editor) p. 317-369, Springer-Verlag, New York, (1978).

12. Meves, H., Rubly, N., Watt, D.D.: Pflügers Arch. 393, 56-62 (1982).

13. Hu, S.L., Meves, H., Rubly, N., Watt, D.D.: Pflügers Arch. in press (1983).

14. Cahalan, M.D.: J. Physiol. 244, 511-534 (1975).

15. Koppenhöfer, E., Schmidt, H.: Pflügers Arch. 303, 133-149 (1968).

16. Mozhayeva, G.N., Naumov, A.P., Nosyreva, E.D., Grishin, E.V.: Biochim. Biophys. Acta 597, 587-602 (1980).

17. Wang, G.K., Strichartz, G.: Biophys. J. 40, 175-179 (1982).

18. Wheeler, K.P., Lazdunski, M., Watt, D.D.: Pflügers Arch. in press (1983).

Acknowledgement. Our appreciation to Drs. C.E. Bugg, University of Alabama/Birmingham and R.J. Feldman, National Institutes of Health for models in Figures 2 and 3; to U.S. Public Health Service (NIH Grant NS-12365) and Deutsche Forschungsgemeinschaft (SFB 38) for financial support.

CHEMICAL MODIFICATION OF THE SEA ANEMONE TOXIN II FROM *ANEMONIA SULCATA*: SYNTHESIS OF A TOXIC, PHOSPHORESCENT DERIVATIVE

Michael Rack and Christel Waschow

Institut für Physiologische Chemie
Universität des Saarlandes
D-6650 Homburg/Saar

Introduction

Propagation of the nervous impulse is based on a transient ion flux across the axonal membrane. The descrete membrane components involved in this change of membrane permeability - the ion channels - are well described by the use of electrophysiological techniques. On a molecular level progress has been made by using neurotoxins, which bind to the channels with very high affinity (1,2). Examples are highly radioactive tetrodotoxin (3), comparable analogues (4), photoactivable derivatives of scorpion toxins (5), or fluorescently labeled analogues of tetrodotoxin (6).

The sea anemone toxin II (ATX II) slows down the inactivation of voltage-dependent sodium channels in excitable membranes. Consequently, the duration of the action potential in nerve, heart, and skeletal muscle is increased. Concentrations as small as 10 nM cause a measurable prolongation of the duration of the action potential of a frog myelinated nerve fibre, when potassium channels are blocked by tetraethylammonium ions (7). This high affinity makes ATX II an important tool for the study of sodium channels in excitable membranes.

ATX II is a small protein, consisting of 47 amino acids, and contains no free sulfhydryl groups nor tyrosine (8).

Toxins as Tools in Neurochemistry

ATX II was shown to be accessible to amino group modifying reagents without destruction of its toxicity (9,10,11,12). In the present study, labeling of ATX II with the phosphorescent and fluorescent probe eosinisothiocyanate is described.

Materials and Methods

ATX II was purchased from Ferring (Kiel,FRG). Eosin-5'-isothiocyanate was from Molecular Probes (Plano, TX,USA), Sephadex G-15 and SP-Sephadex A-25 were from Pharmacia (Freiburg,FRG). Gels for isoelectric focusing were from Serva (Heidelberg,FRG). The other chemicals were from Merck (Darmstadt,FRG).

For chemical modification ATX II (2mg = 0,42 µmole) was dissolved in 2 ml sodium phosphate buffer (200 mM) pH 8,0. 4 mg eosinisothiocyanate (5,7 µmole) were dissolved in 0,2 ml dimethylformamide and slowly added to the stirred ATX II solution. After 1,5 h at room temperatur in the dark, labeled ATX II was separated from the unreacted dye by gel filtration over a Sephadex G-15 column (1x15 cm) with 25 mM sodiumacetate, containing 0,1 M NaCl,pH 5,5 as the eluting buffer. ATX II was separated from eosin-labeled ATX II by a passage of the gel filtrate over a SP-Sephadex column (1x10 cm). Concentration and buffer exchange were performed by ultrafiltration using Amicon YM 2 membranes. The yield of modified toxin was 20%.

Isoelectric focusing was performed on gels with a pH range from 3 upto 10. After electrophoresis the proteins in the gels were fixed in 10% trichloroacetic acid for 10 min and then rinsed with destaining solution. Then the gels were bathed for 5 min in an aqueous solution of 0,1% Serva Blau G, 50% methanol, and 10% acetic acid. After destaining for 30 min in an aqueous solution containing 15% methanol and 10% acetic acid the gels were dried at 80°C.

Eosin-labeled ATX II was tested on nodes of Ranvier. Single motor and sensory nerve fibres were dissected from the tibial nerve of the frog *Rana esculenta* and mounted in a perspec cell (13). A node of Ranvier was superfused with Ringer's solution containing 110,5 mM NaCl, 2,5 mM KCl, 2,0 mM MOPS-buffer, pH 7,2. To block the K channels from outside, 12 mM TEA were added to the Ringer's solution. The ends of the fibre were cut in 120 mM CsCl and 12 mM TEA to block K channels from inside.Action potentials and voltage-clamp currents were recorded by standard methods (13,14). The action potentials were measured at the normal resting potential; small deviations of the resting potential which occurred during an experiment were compensated by sending a polarizing current of appropriate size. In the voltage clamp experiments, the fibre was held at a potential -22 mV from the normal resting potential; the normal resting potential was taken as the potential at which 30% of the Na channels were inactivated. The temperature was 15°C.

Results

Isolation of eosin-labeled ATX II. The toxin incubated with eosinisothiocyanate and applied to a Sephadex G-15 column yield a coloured, fluorescent derivative that appeared shortly after the void volume. Excess of eosinisothiocyanate was strongly adsorbed by gel. The fraction appearing shortly after the void volume was applied to a SP-Sephadex-A-25 column, and rinsed with the same buffer as used for gel filtration. The coloured material was eluted promptly. In control experiments ATX II was not eluted under the same conditions.

Attempts to purify eosin-labeled ATX II by anion exchange chromatography were impracticable, because eosin-labeled ATX II once adsorbed by the gel was not to elute by increasing salt concentrations upto 1M NaCl and moderate changes of the pH value of the buffer. Isoelectric focusing of the isolated eosin-labe-

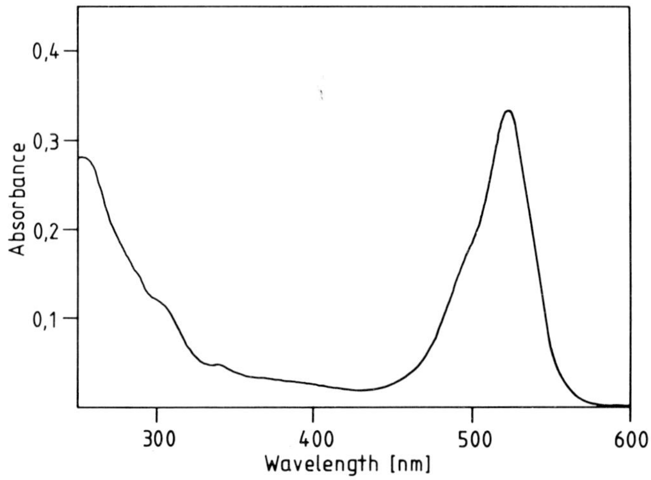

Fig.1 Absorption spectrum of eosin-labeled ATX II
Optical pathlength 1 cm. Reference was Ringer's solution.

led ATX II yielded a single coloured band. This band migrated towards the anode, while ATX II (in control experiments) migrated to the cathode. The isoelectric point of eosin-labeled ATX II is similar to that of bovine serum albumin, while ATX II focused between myoglobin from sperm whale and cytochrome C.

The amount of eosin-ATX II was determined using a value of $8{,}3 \times 10^4\ M^{-1}cm^{-1}$ for the maximum of absorption for eosinisothiocyanate or protein-bound derivatives (15).

Fig.1 shows the absorption spectrum of 4 µM eosin-labeled ATX II in TEA-Ringer's solution, pH 7,2. The spectrum in the visible range is due to the eosin chromophore, while ATX II and eosin contribute to the absorption in the u.v.-range.

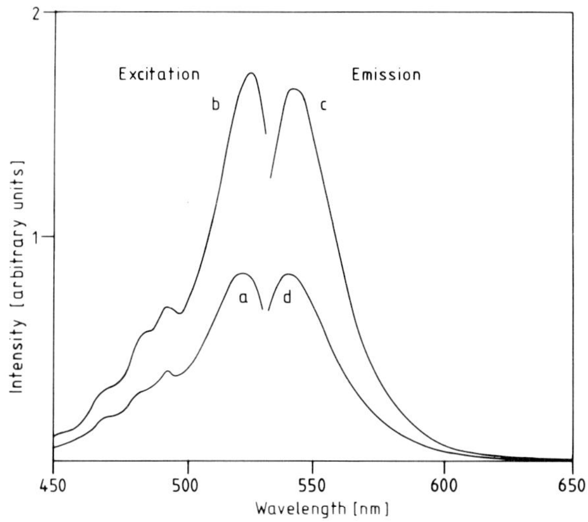

Fig.2 Comparison of fluorescence properties of eosin-labeled ATX II in buffer solution and in the presence of 20% ethanol. Concentration of eosin-ATX II was 500 nM. Spectra were recorded in 25 mM sodiumacetate, containing 100 mM NaCl, pH 5,5 (a,d), and 80% of the same buffer solution with the addition of 20% ethanol (b,c). Excitation spectra were recorded at an emission wavelength of 540 nm (a,b), emission spectra (c,d), excitation, 520 nm.

Fig.2 shows the fluorescence properties of eosin-labeled ATX II and the environment dependence of the eosin fluorescence. When eosin-ATX II was dissolved in 50% ethanol under comparable conditions the fluorescence intensity increased for a factor 5.

Like ATX II the eosin-labeled derivative was still able to provoke a tremendous prolongation of the action potential on the node of Ranvier. The effect was quickly reversible on

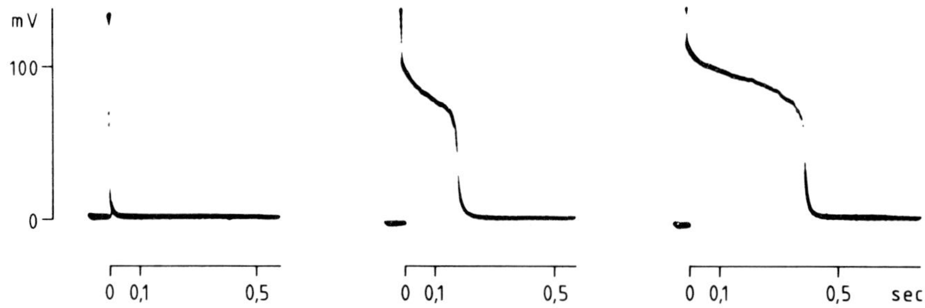

Fig.3 Effect of eosin-labeled ATX II and ATX II on the action potential. Record before the application of a toxin. After application of 1 µM eosin-ATX II, and after application of 1 µM ATX II. Stimulation with depolarizing pulses of 0,15 ms duration.

washing with Ringer's solution. The toxin derivative was not as effective as the same amount of native ATX II on the duration of the action potential.

As shown in Fig.4 eosin-ATX II and ATX II markedly increased the inward current that is left after the end of a 13 ms pulse. The effect of eosin-ATX II was mostly but not completely reversed after 10 min wash with toxin free Ringer. Peak I_{Na} was more reduced by the application of ATX II than with the toxin derivative of the same concentration.
For a quantitative measurement of the toxin effects seen in the voltage clamp experiments the decaying phase of the Na

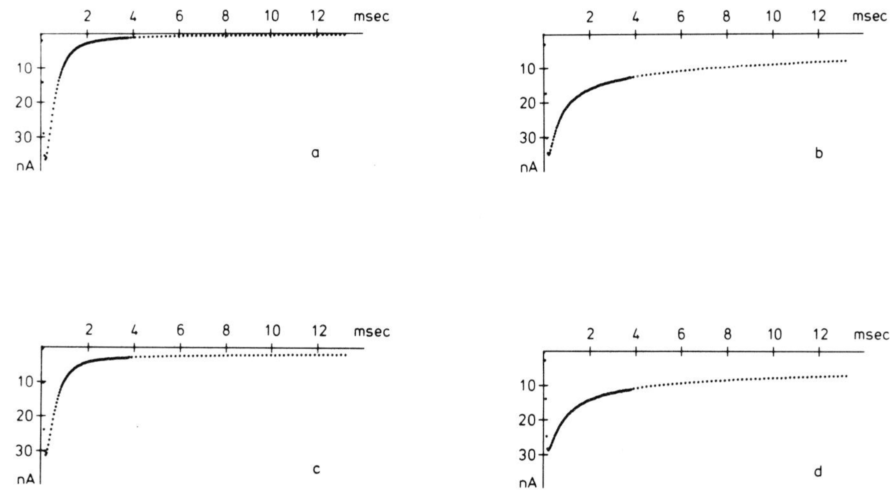

Fig.4 Effects of eosin-ATX II and ATX II on the sodium inward current of the node of Ranvier. (a) control. (b) 4 min in Ringer with 4 µM eosin-ATX II. (c) after 10 min wash with Ringer. (d) 4 min in Ringer with 4 µM ATX II.

currents was fitted by an equation, that describes the Na current as the sum of a non-inactivating, a rapidly inactivating and a slowly inactivating component. The same equation has been used for the Na current of fluoresceinthiocarbamylated ATX II (12) and ATX II treated nodes of Ranvier (16,17). Fig.5 shows computer fits of the sodium currents given in Fig.4.

The non-inactivating component was markedly increased by eosin-ATX II and ATX II to nearly the same extent. The reduction in the size of the rapidly inactivating component was more affected by native ATX II. 2 µM eosin-labeled ATX II reduced the non-inactivating component of the Na-current to 60% of the amount of the reduction achieved by treatment of the nerve fibre with 4 µM eosin-labeled ATX II.

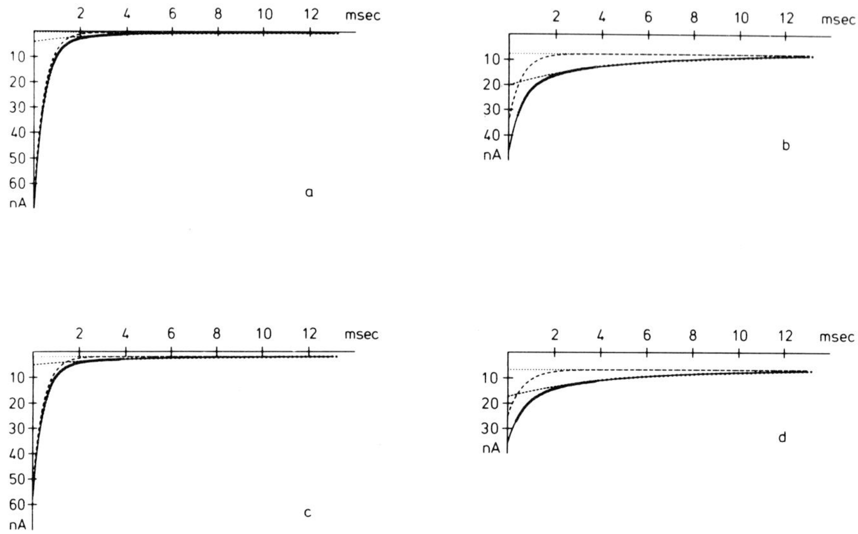

Fig.5 Computer fits of the sodium currents shown in Fig.4 I_{Na} is shown by the uninterrupted curve. The non-inactivating component is given by a dotted line; while the rapidly and the slowly inactivating components are represented by the interrupted curves.

Discussion

The synthesis, some spectroscopic properties, and the electrophysiological characterization of a phosphorescent and fluorescent derivative ATX II is presented. Among the groups present in the ATX II molecule (8) only amino groups are able to form a stable covalent bond with an isothiocyanate group (18,19,12). The eosin chromophore has an anionic character, additionally each amino group modification with isothiocyanate results in the loss of a positive charge. The main aim of the following purification procedure, the removal of unreacted ATX II was therefore achievable by a passage of the gel filtrate through a cation exchange column. Subfractionation of mono-modified eosin-ATX II from poly-modified ATX II by anion exchange chromatography was impractible, because eosin-ATX II once adsorbed on the gel was not to remove by increasing the ionic strength or moderate changes of pH. A similar behaviour of an eosin-labeled small protein was described for a bungarotoxin derivative (20). Using a limited excess of the protein reactive reagent and a slightly alkaline pH a single coloured band was obtained on isoelectric focusing. The toxin derivative provoked a prolongation of the action potential similar to that obtained by treatment of a nerve fibre with ATX II. The duration of the action potential did not reach the same extent when the derivative was used instead of ATX II. In voltage clamp experiments the increase on the non-inactivating component was essentially the same for ATX II and the phosphorescent derivative, while the peak I_{Na} was somewhat more decreased on treatment with ATX II.

These electrophysiological data show that the eosin-modified ATX II is still interacting with sodium channels of the nerve membrane in a way very similar to that of native ATX II.

The availability of a toxic phosphorescent and fluorescent ATX II derivative is of special interest for biophysical and biochemical characterization of the voltage-sensitive sodium channel. Different fluorescent compounds interacting with the voltage-dependent sodium channel will be useful for energy transfer studies. The preparation of e.g. fluorescent tetrodotoxin derivatives was described (6). Appropriately selected chromophores bound to neurotoxin derivatives open the possibility to determine the distance between toxin binding sites on the channel macromolecule.

The duration of the sodium current through the axonal membrane after polarizing impulse is in the order of milliseconds. The lifetime of the triplet state of the eosin molecule is in the ms time range, too. Several membrane proteins have been studied with eosin as the triplet probe (for review see: (21). Treatment of excitable membranes with eosin-labeled ATX II may thus be suitable for studies of the molecular events underlying the excitation mechanism.

Acknowledgements

We wish to thank Prof. H. Meves for his continual support and his advice. The excellent technical assistance of N. Rubly is gratefully acknowledged. This research was financed by the Deutsche Forschungsgemeinschaft, Sonderforschungsbereich 38 "Membranforschung".

References

1. Catterall, W.A.: Ann. Rev. Pharmacol. Toxicol. 20, 15-43 (1980)
2. Lazdunski, M., Balerna, M., Barhanin, J., Chicheportiche, R., Fosset, M., Frelin, C., Jacques, Y., Lombet, A., Pouysségur, J., Renaud, J.F., Romey, G., Schweitz, H., Vincent, J.P.: Ann. N. Y. Acad. Sci. 358, 169-182 (1980)
3. Grünhagen, H.H., Rack, M., Stämpfli, R., Fasold, H., Reiter, P.: Arch. Biochem. Biophys. 206, 198-204 (1981)
4. Chicheportiche, R., Balerna, M., Lombet, A., Romey, G., Lazdunski, M.: Eur. J. Biochem. 104, 617-625 (1980)
5. Beneski, D.A., Catterall, W.A.: Proc. Natl. Acad. Sci. 77, 639-643 (1980)
6. Angelides, K.J.: Biochemistry 20, 4107-4118 (1981)
7. Schmidtmayer, J., Ulbricht, W.: Pflügers Arch. Ges. Physiol. 379, R42 (1979)
8. Wunderer, G., Fritz, H., Wachter, E., Machleidt, W.: Eur. J. Biochem. 68, 193-198 (1976)
9. Barhanin, J., Hugues, M., Schweitz, H., Vincent, J.P., Lazdunski, M.: J. Biol. Chem. 256, 5764-5769 (1981)
10. Stengelin, S., Rathmayer, W., Wunderer, G., Béress, L., Hucho, F.: Anal. Biochem. 113, 227-285 (1981)

11. Rack, M., Meves, H., Béress, L., Grünhagen, H.H.: Hoppe-Seyler's Z. Physiol. Chem. (Abstract) 363, 993 (1982)

12. Rack, M., Meves, H., Béress, L., Grünhagen, H.H.: Toxicon in press

13. Stämpfli, R., Hille, B.: Frog Neurobiology, (Llinas, R., Precht, W., Eds.) Springer-Verlag Berlin, 3-32 (1976)

14. Nonner, W.: Pflügers Arch. Ges. Physiol. 309, 176-196 (1969)

15. Cherry, R.J., Cogoli, M., Oppliger, M., Schneider, G., Semenza, G.: Biochemistry 15, 3653-3656 (1976)

16. Neumcke,B., Schwarz, W., Stämpfli, R.: Biochem. Biophys. Acta 600, 456-466 (1980)

17. Schmidtmayer, J., Stoye-Herzog, M., Ulbricht, W.: Pflügers Arch. Ges. Physiol. 392, R32 (1982)

18. Stark, G.R.: Biochemistry 4, 588-595 (1965)

19. Stark, G.R.: Methods Enzymol. 25, 579-584 (1972)

20. Bartholdi, M., Barrantes, F.J., Jovin, T.M.: Eur. J. Biochem. 120, 389-397 (1981)

21. Cherry, R.J.: Methods Enzymol. 54, 47-61 (1978)

NMR CONFORMATIONAL STUDY: POLYPEPTIDE NEUROTOXINS - HONEY-BEE APAMIN AND SCORPION INSECTOTOXIN I_5A.

Vladimir F. Bystrov, Alexandr S.Arseniev, Vladimir I.Kondakov, Vladimir N.Maiorov, Viktor V.Okhanov and Yuri A.Ovchinnikov
Shemyakin Institute of Bioorganic Chemistry, USSR Academy of Sciences
117988 GSP-1 Moscow V-334, USSR

Introduction

Polypeptide neurotoxins isolated from bee and scorpion venoms capture attention not only because they posses highly specific biological activity and are useful as essential tools in neurobiological research, but also because they are convenient models for correlation of protein biological function with chemical and spatial structure of the molecules. As soon as molecular conformation represents a crucial factor in the functioning of physiologically active compounds, the nuclear magnetic resonance (NMR) is the choice among various spectroscopic techniques available for solution study (1). Out from many NMR techniques specialy designed for this task the most powerful at the moment are two-dimensional (2-D) spectroscopy (2), interpretation of vicinal coupling constants (3) and perturbation by specific intra- and intermolecular probes (4,5).
This paper is a brief account of the results obtained in the Shemyakin Institute of Bioorganic Chemistry on structure of two neurotoxins . Previously it was shown that proper combination of NMR data with results of other physical techniques (circular dichroism, laser Raman, fluorescence and electron spin resonance spectroscopy) supported by extensive chemical studies is capable to provide detailed information on conformation of polypeptides in solution (4,5). In this study it was revealed

Toxins as Tools in Neurochemistry

that honey-bee apamin and scorpion insectotoxin I_5A posess stable molecular conformation as gross of backbone folding is concerned. For molecules with definite dominant conformation the usual treatment is to present the picture of three-dimensional structure as if it were a rigid frame, although it is obvious that in solution the molecules are subject to some degree of intramolecular motion (6).

Apamin

Apamin, an active principle of the honey-bee *Apis mellifera* venom, is an octadecapeptide with two disulfide bonds:

C-N-C-K-A-P-E-T-A-L-C-A-R-R-C-Q-Q-H-NH_2

It is unique in being the smallest known peptide neurotoxin with exitatory effect on the central nervous system and the only known polypeptide that passes the blood-brain barrier (7). Apamin becomes essential tool in the study of mechanism of nerve pulse transmission.

The overall conformational stability of apamin molecule directly follows from circular dichroism (CD) spectra study (8), which demonstrated that the structure highly abundant with right-handed α-helical component is persistent against pH change of aqueous solution in wide range, presence of organic solvents and denaturating agent (6M quanidinium chloride), and even chemical modification. Proton and ^{13}C-NMR spectra also demonstrate the relatively rigid backbone folding of apamin and were employed as tools for three-dimensional structure determination (4,5,9,10).

Signals in 300 and 500 MHz proton, and 75 MHz ^{13}C-NMR spectra of apamin have been identified by two levels of assignment in the one-dimensional spectroscopy. On the first level the proton signals are interconnected by homonuclear proton-proton spin decoupling and assigned to a particular spin multiplet system according to the specific NMR types of amino acid residues (11).

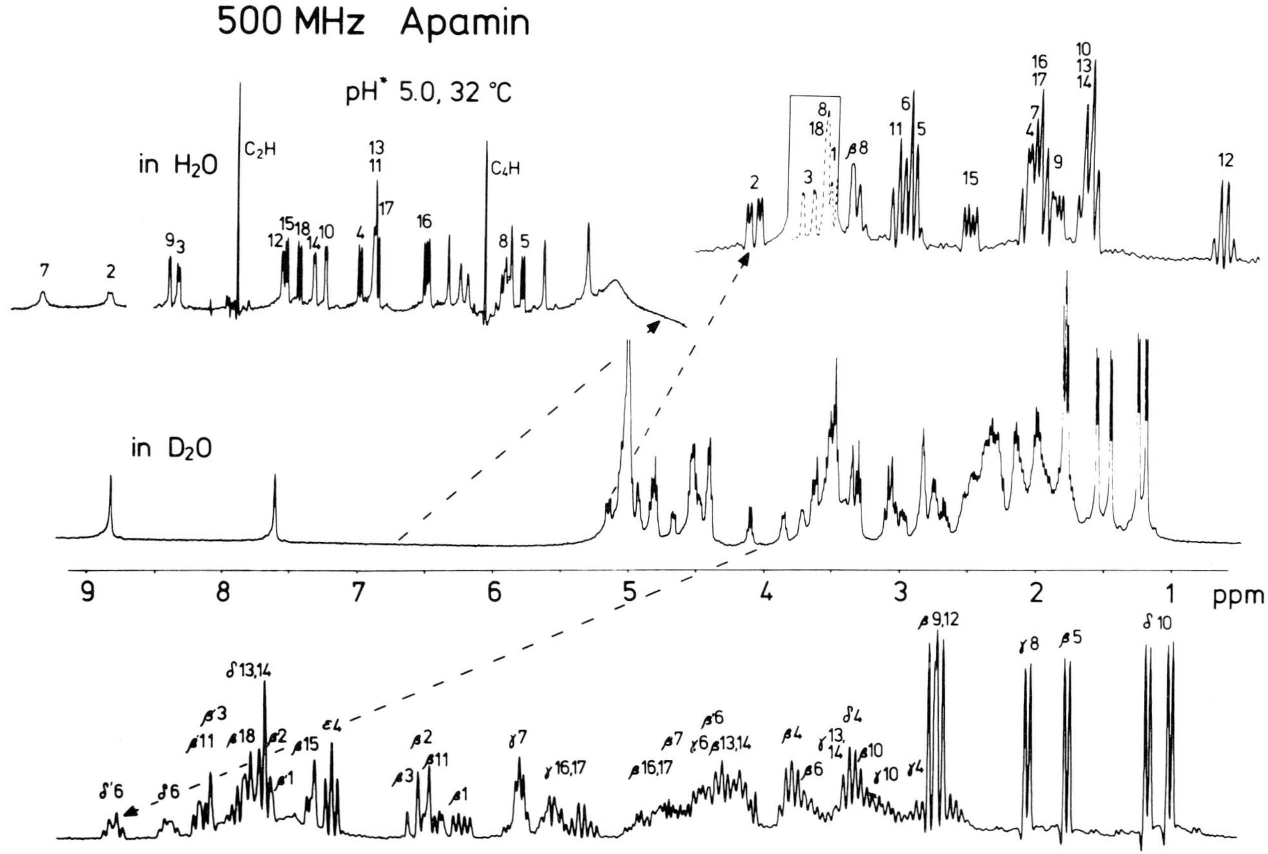

Fig. 1. 500 MHz proton NMR spectra of *Apis mellifera* apamin 10 mM solution recorded in H_2O for the NH proton region (upper left corner) and in D_2O for other regions, pH 5.0 at 32°C.

The second level implies assingment of the observed spin system to a specific position of the amino acid residue in the apamin primary structure by heteronuclear $^{13}C\{^{1}H\}$ selective decoupling of carbonyl signals (12). The results of resonance assignment are shown in Fig. 1, and the identification of proton signal parameters with particular amino acid residues is given in Table 1. The presented signal assignment are mainly based on the results of (4,5,9-12) taking into account the recent reassignment of Leu^{10}, Arg^{13}, Gln^{16} and Gln^{17} resonances made by Wemmer and Kallenbach (13) by use of 360 MHz 2-D spectroscopy.

Table 1. Apamin proton NMR parameters

Apamin residues	$^{3}J_{HNC^{\alpha}H}$ Hz	$^{3}J_{HC^{\alpha}C^{\beta}H}$ Hz	$t_{1/2}$* hours
Cys-1	-	7.8,7.8	-
Asn-2	9.5	11.2,4.2	<0.1
Cys-3	5.2	11.9,2,8	0.2
Lys-4	7.6		4
Ala-5	6.8	6.1	11
Pro-6	-	6.2,6.2	-
Glu-7	6.9	9.6,4.4	0.1
Thr-8	8.5	1.9	4
Ala-9	2.7	6.1	<0.1
Leu-10	4.2		>100
Cys-11	5.9	7.6,7.6	14
Ala-12	3.0	6.1	26
Arg-13	5.5		20
Arg-14	3.5		2.7
Cys-15	5.4	9.6,4.2	26
Gln-16	6.8		11
Gln-17	7.1		0.5
His-18	7.9	8.6,6.1	<0.1

* At pH 2.9 and 14°C

NMR parameters listed in Table 1 are most important for the apamin conformational study. Backbone $H\text{-}NC^{\alpha}\text{-}H$ and side chain $H\text{-}C^{\alpha}C^{\beta}\text{-}H$ vicinal proton-proton coupling constants provide information on the torsional angles ϕ and χ^{1} of the $N\text{-}C^{\alpha}$ and $C^{\alpha}\text{-}C^{\beta}$ bonds, respectively (3). Half-time of deuterium exchange ($t_{1/2}$) discriminates intramolecular hydrogen bonded backbone NH groups (14). In addition, chemical shift pH-dependence studied in (4,5,9,10) revealed the microenvironment of ionogenic groups: N-terminal α-amino (pK 6.7), ε-amino of Lys^{4} (pK 10.8), γ-carbonyl of Glu^{7} (pK 3.6), and

imidazole ring of His^{18} (pK 6.6).
When examining the NMR parameters (Table 1) attention is captured by seven residues in succession (Ala^{9}-Cys^{15}) with low values of H-NC^{α}-H proton coupling (2.7-5.9 Hz) as well as by six residues in the seven residue fragment Leu^{10}-Glu^{16} with very slow deuterium exchange ($t_{1/2}$>11 hours). Both of these NMR features are inherent to polypeptide right-handed helical structure (3_{10},α- or π-helix) (14). Thus it could be assumed that the 10-16 fragment is incorporated in a helix and its backbone NH groups are hydrogen bonded to the carbonyls of n-i residues (i=3,4 and 5 for 3_{10}-, α- and π-helix, respectively). By model building it was found that only the α-helix could be constructed due to steric hindrances introduced by 1-11 and 3-15 disulfide links. The presence of the right-handed α-helical region in apamin is in accord with CD (8) and laser Raman (15) spectra, as well as with secondary structure prediction (16) based on Chou and Fasman method (17).
The model building around the α-helical fragment 6-16 permits the formation of β-turn 2-5 for slow exchanging NH group of Ala^{5} residue. In the proposed apamin spatial structure shown in Fig. 2 the following main features of the NMR study (4,10) were also taken into account: (i) carboxyl group of Glu^{7} residue is at ~ 0.5 nm from the N-terminal α-amino group and is also proximate to the methyl groups of Ala^{5} and Thr^{8} residue, (ii) the side chain of at least one arginine residues is in the vicinity of Leu^{10} methyl group, (iii) the ε-amino group of Lys^{4} residue is distant to the other residues, (iv) the side chains of His^{18} and of both arginine residues are far from the other ionogenic residues, (v) the backbone NH groups of Lys^{4} and Thr^{8} residues are capable of forming weaker intramolecular hydrogen bonds, possibly with side chain functions of the molecule, (vi) the torsional angles ∅ must be in conformity with the measured vicinal coupling constants $^{3}J(H\text{-}NC^{\alpha}\text{-}H)$ and $^{3}J(^{13}C'\text{-}NC^{\alpha}\text{-}H)$.
In addition, the laser Raman spectra study (15) demonstrates the similarity in conformational arrangement of both disulfide bridges 1-11 and 3-15. On the basis of correlation (18) the

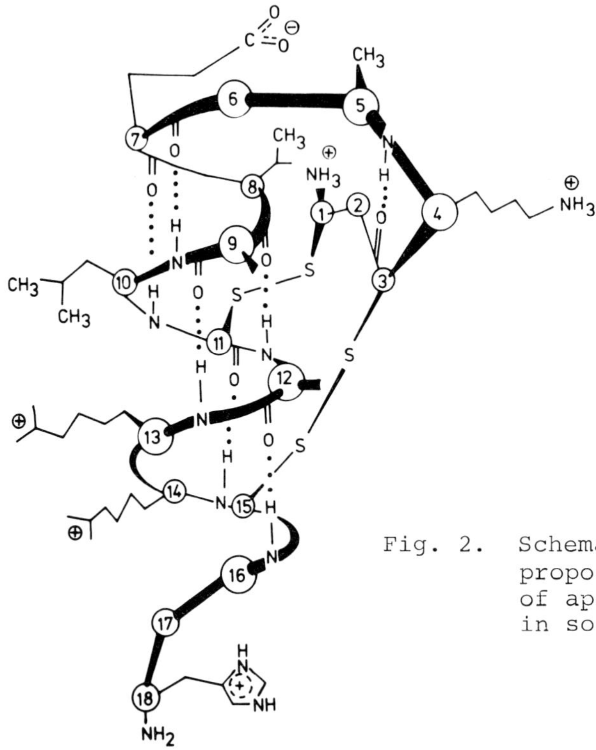

Fig. 2. Schematic presentation of proposed spatial structure of apamin *Apis mellifera* in solution.

trans-gauche-gauche-gauche-trans orientation for the C^{α}-C^{β}-S-S-C^{β}-C^{α} fragment was proposed.
The backbone torsional angles ϕ for the solution spatial structure (Fig. 2) were evaluated from the $^{3}J(H\text{-}NC^{\alpha}\text{-}H)$ coupling constants and the ambiguity in the ϕ values was partially put aside by the $^{3}J(^{13}C'\text{-}NC^{\alpha}\text{-}H)$ coupling (9-12). The Ramachandran ϕ-ψ energy map indicated that the apamin residue points fall within the allowed regions. The rigidity of the apamin conformation is achieved by relative orientation of the α-helical unit 6-16, β-turn 2-5 and the disulfide bridges Cys^{1}-Cys^{11} and Cys^{3}-Cys^{15}. Additional, but weaker intramolecular hydrogen bonds can be formed by the backbone NH of Lys^{4} with the side chain carbonyl group of Asn^{2} residue and by the backbone NH of Thr^{8} with the itself hydroxyl oxygen, or backbone carbonyl of Ala^{5} residue. The C-terminal Glu^{17} and His^{18} are the only residues not incorporated

into the apamin tertiary structure. It must be mentioned that the apamin conformation shown in Fig. 2 differs from the previously proposed structure in (5,10) by elongation of the α-helical part up to 16-th residue.
The solution spatial structure of apamin presented in Fig. 2 explains the spectroscopic results (4,8,9,13,15) much better than structure proposed earlier on the basis of conformational energy calculations using the atomic (19) and residual representations (20) as well as secondary structure prediction (16, 21). It could be considered as similar to the conformation adopted on a receptor because of its persistence against media conditions and chemical modification as well as the fact that synthetic apamin possesses full biological activity and the same CD and NMR spectra (22) which is considered (16) as evidence that the information for the correct folding of the peptide chain is contained in the chain and not in an apamin precursor. However the overall rigid spatial structure is not the only feature that is important for biological action. The chemical structure of arginine side chain (8,23) is also crucial as follows from a dramatic loss of neurotoxic activity when both Arg^{13} and Arg^{14} are substituted either by lysine (24,25) or by ornithine (8, 26,) residues. Further progeress in structure - activity relationship study of apamin depends on localization and isolation of receptor system, and studying of its interaction with apamin and its derivatives.

Insectotoxin I_5A

From the venom of Caucasian scorpion *Buthus eupeus* obtained by electrical stimulation of telsons were isolated 12 different neurotoxins (27). One of them is insectotoxin I_5 posessing paralysing activity for insects presumably by acting on glutamate receptor of postsynaptic membrane (28). This insectotoxin and its isotoxin I_5A (29) belong to the homolog family of "short" insectotoxins (30) consisting of 35-36 amino acid re-

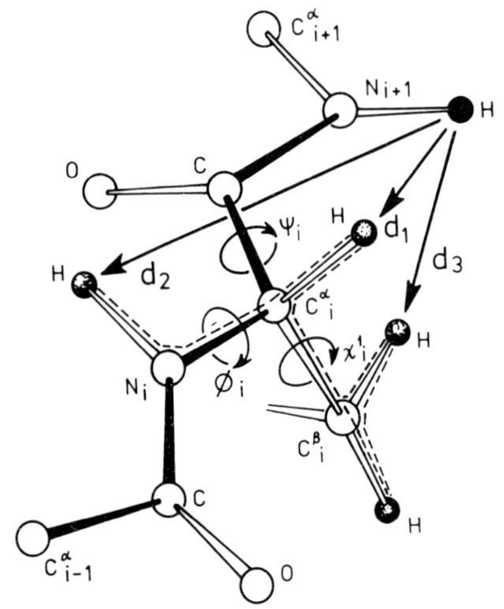

Fig. 3. NMR conformational parameters of peptide fragment. Arrows indicate interresidue pathways of dipolar interproton interactions of amide NH_{i+1} proton with $C^{\alpha}H_1$ (d_1), amide NH_i (d_2) and $C^{\beta}H$ (d_3) protons. Dashed lines show intraresidue vicinal spin-spin scalar coupling ^3J(H-NC$^{\alpha}$-H) and ^3J(H-C$^{\alpha}$C$^{\beta}$-H).

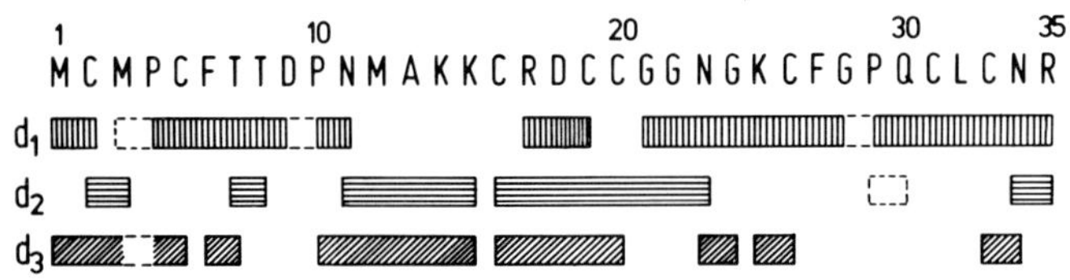

Fig. 4. Primary structure of the *Buthus eupeus* insectotoxin I_5A (29) and survey of the sequential NOE connectivities d_1, d_2 and d_3 (see Fig. 3) by which the individual resonance were assigned. The fragments isolated by dashed lines correspond to specific cases of proline residues for which the NOE interactions of Pro $C^{\delta}H_2$ protons with NH_i and $C^{\alpha}H_i$ protons of the preceding residue are observed.

sidues with four disulfide bonds. It should be stressed that the disulfide bridge positions were not established.

The solution spatial structure of insectotoxin I_5A is estimated by extensive use of 2-D NMR spectroscopy. By using scalar coupling correlated spectra in COSY and SECSY arrangements and especialy dipolar coupling correlated NOESY spectra practically every signal was assigned to particular amino acid residue in the primary structure (29, 31). The first NOESY set of interproton dipolar interactions d_1, d_2 and d_3 (Fig. 3) for neighboring residues as revealed by nuclear Overhauser effect (NOE) is shown in Fig. 4 along with the amino acid sequence of insectotoxin I_5A. In addition more than 100 dipolar NOE interactions were observed for protons located in unneighbouring amino acid residues, that while remote in the primary structure becomes close to one another owing to folding of peptide backbone (31).

The conformational analysis of insectotoxin I_5A proceeds in three steps - configuration of X-Pro amide bonds, secondary structure and three-dimensional structure.

Table 2. The interproton distances for X^i-Pro^{i+1} fragments depending on configuration of the amide bond and the ψ_i angle rotation.

X-Pro	$H^{\alpha}_i \cdots H^{\alpha}_{i+1}$	$H^{\alpha}_i \cdots H^{\delta}_{i+1}$
trans	0.43-0.48 nm	0.20-0.39 nm
cis	0.18-0.38 nm	0.43-0.50 nm

There are three prolyl residues in the insectotoxin molecule, that are in the fragments Met^3-Pro^4, Asp^9-Pro^{10} and Gly^{28}-Pro^{29}. By general structural consideration the interproton distances shown in Table 2 are estimated from refined molecular geometry data (32) depending on the *trans* or *cis* configuration of the X^i-Pro^{i+1} bond and on the ψ angle rotation in the i-th residue. It follows that one could distinguish *cis/trans* configuration by determination of the closest distance between the corresponding protons.

Taking into account that under employed NOESY experimental condition(mixing time 100 ms) we have observed NOE cross-peaks for protons separated by less than 0.30-0.35 nm, it is clear, for instance, that for the *trans* X-Pro bond we expect to observe only $H_i^{\alpha}\cdots H_{i+1}^{\delta}$ dipolar interaction. Indeed, as shown in Fig. 5 only NOE cross-peaks for the $C_i^{\alpha}H$ and $C_{i+1}^{\delta}H_2$ protons were observed, but no $C_i^{\alpha}H\cdots C_{i+1}^{\alpha}H$ peaks were detected. Thus it becomes obvious that all three X-Pro bonds in insectotoxin I_5A are in the *trans* configuration.

Table 3. Extreme d_1, d_2 and d_3 distances (in nm) in the peptide fragment (Fig. 3) for regular protein secondary structure components and unregular structure.

Distance	α_R	β_P	β_A	Unregular
d_1	0.35-0.36*	0.22-0.23	0.22-0.24	0.22-0.36**
d_2	0.24-0.33	0.38-0.46	0.39-0.45	0.14-0.47
d_3	0.21-0.41	0.32-0.46	0.27-0.44	0.13-0.47

* For glycine residue the d_1 range for the α-helical structure is 0.26-0.30 nm.

** For glycine residue the maximum d_1 distance is 0.33 nm.

The secondary structure evaluation could be grounded on consideration of extreme values of d_1, d_2 and d_3 distances (Fig. 3) for principal components of the regular protein secondary structure (33). Table 3 shows these values for right-handed α-helix, as well as for parallel and antiparallel β-structures. Calculation was performed by changing in the region of $\pm 20^{\circ}$ the standard values of the ϕ and ψ torsional angles taken correspondingly as -56°, -44° for right handed α-helix -114°, 114° for parallel and -136°, 134° for antiparallel β-structures (34). In the d_3 calculation the χ_i^1 angle was allowed to rotate freely from -180° to $+180^{\circ}$. From Table 3 it follows that for α-helical

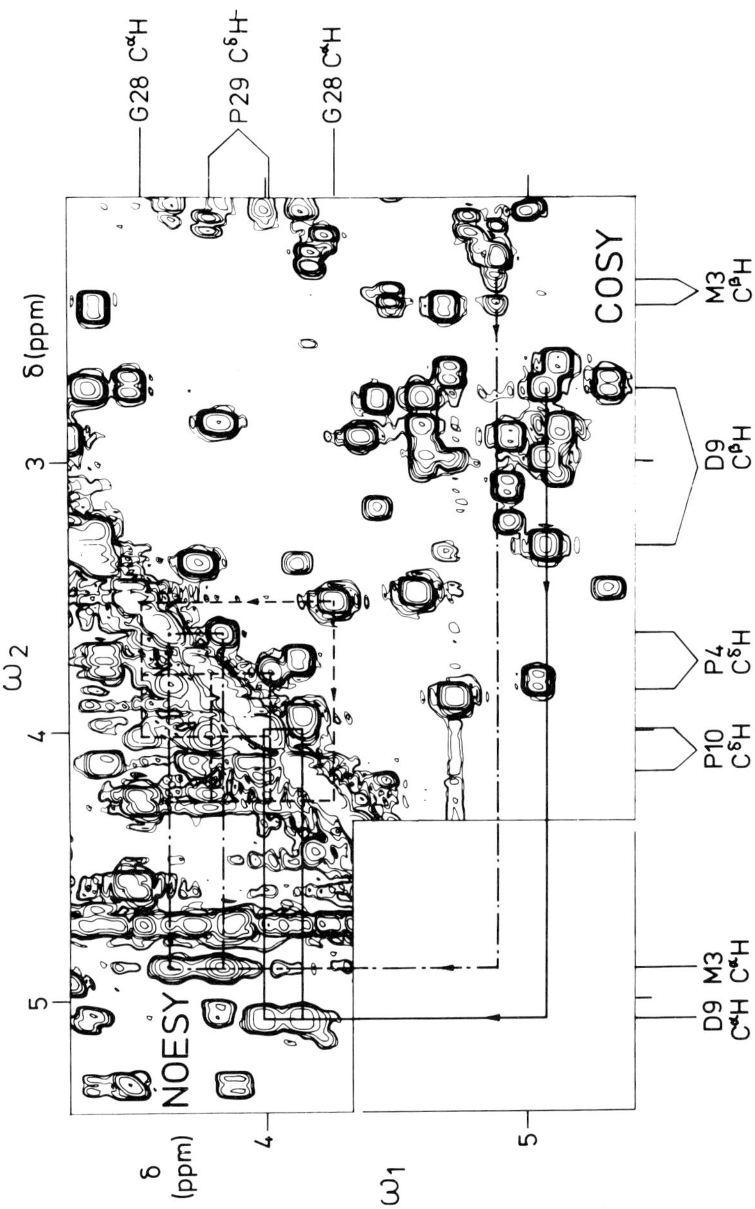
G28 CαH
P29 CδH
G28 CαH
δ (ppm)
ω2
3
4
5
COSY
NOESY
M3 CβH
D9 CβH
P4 CδH
P10 CδH
D9 CαH
M3 CαH
δ (ppm)
ω1
4
5

parts of polypeptide chain we are expected to observe the sequential NOE d_2 cross-peaks without any d_1 peaks (except of glycine residues). Contrary, for β-structures it is inherent the presence of the d_1 peaks and absence of d_2 interactions. The d_3 peaks have to be absent in the case of parallel β-structure. Now, if we analyze from this point of view the observed set of d-interactions (Fig. 4) it becomes obvious that the α-helical part extends from the Asn^{11} to at least Cys^{19} residue of insectotoxin I_5A. The β-structure is accomplish in the Asn^{23}-Phe^{27} and Gln^{30}-Asn^{34} fragments. Because these two fragments are located nearby in the sequence one could assume that they are arranged in the antiparallel β-structure with β-bend constituted by Gly^{28} and Pro^{29} residues. The CD and laser Raman spectra are consistent with right-handed α-helix and β-structure content of the insectotoxin I_5A secondary structure. Confirmation of the above proposed regular secondary structure and construction of three-dimensional structure for corresponding fragments are based on examination of additional NMR data on hydrogen-bonded backbone amide NH groups, torsional ∅ angles estimated from the vicinal spin-spin couplings, and unneighbor residue interproton NOE interactions (31). The vicinal ^3J (H-$C^{\alpha}C^{\beta}$-H) couplings, for instance, were extracted from the NMR spectrum by use of proton 2-D J-resolved experiment shown in Fig. 6. The partial results of NMR conformational analysis are shown in Figs. 7 and 8. The insectotoxin fragment Asn^{23}-Asn^{34} is presented by the antiparallel β-sheet with β-bend in the 26-30 positions (Fig. 7). In addition to the shown backbone

Fig. 5. Combined 500 MHz COSY-NOESY countor plot (ω_1 3.3-5.4 ppm, ω_2 2.0-5.4 ppm) for 0.011 M solution of insectotoxin I_5A in D_2O, pH 2.9 at 30°C. The mixing time τ_m in the NOESY experiment was chosen equal to 0.1 sec. Connections indicate diagonal and cross peaks assigned to the $C_i^{\alpha}H$ and $C_{i+1}^{\delta}H_2$ proton in the Met^3-Pro^4 (-·-), Asp^9 -Pro^{10} (——) and Gly^{28}-Pro^{29} (- - -) fragments. Their chemical shifts are shown on the diagram margins. The assignment as indicated by arrows was started from the J-peaks in the COSY diagram for $C^{\alpha}H$ and $C^{\beta}H_2$ of the i-th residue in case of Met^3 and Asp^9.

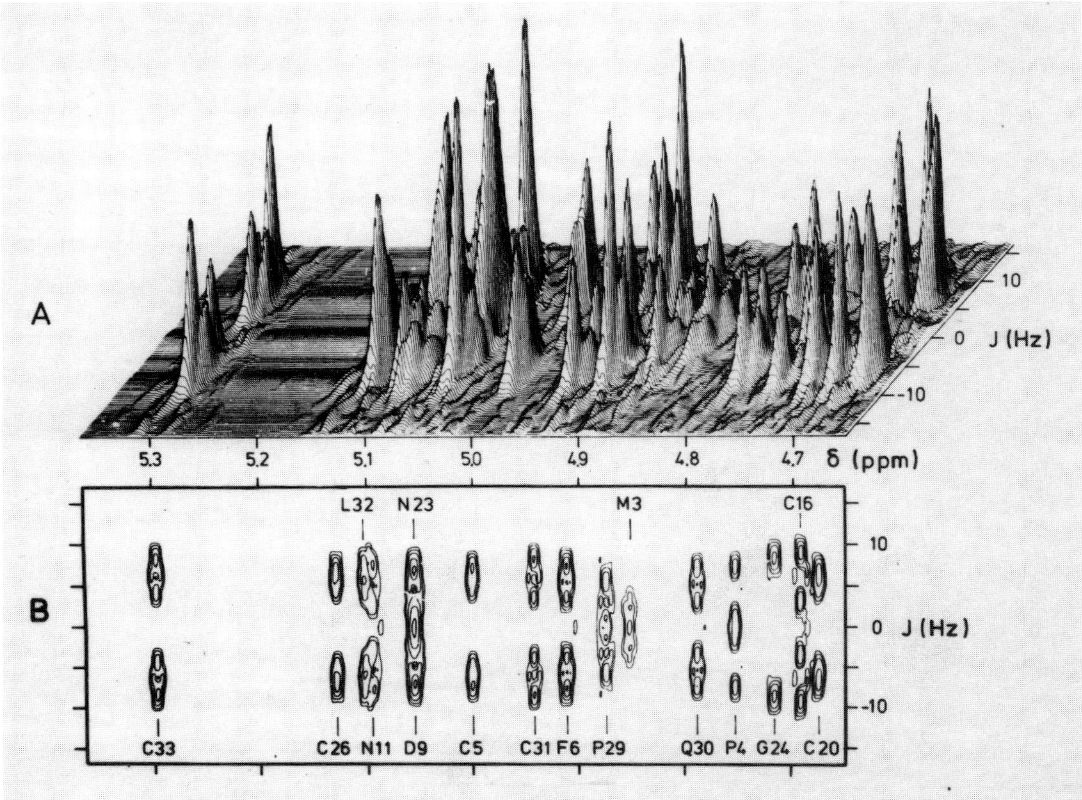

Fig. 6. 2-D J-resolved NMR spectrum of $C^{\alpha}H$ protons of insectotoxin I_5A 11 mM solution in D_2O, pH 2.9 at 50°C. Stacking (A) and contour (B) plot presentation.

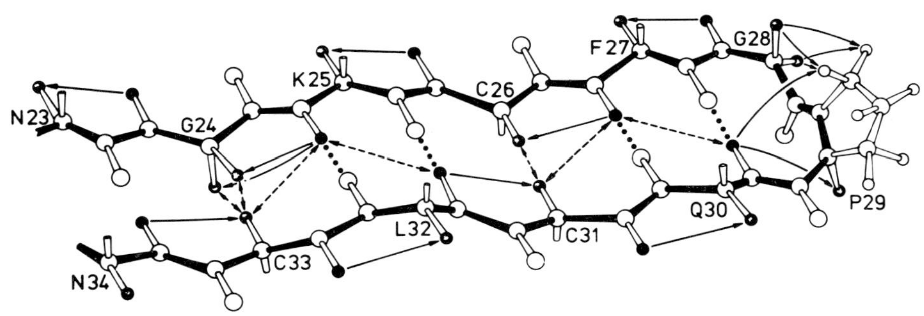

Fig. 7. The antiparallel β-sheet conformation for the Asn^{23}-Asn^{34} fragment of the *Buthus eupeus* insectotoxin I_5A in solution. Solid and dashed arrows correspond to neighbour residue (see Figs 3 and 4) and to unneighbour residue proton NOE dipolar coupling, respectively. By rows of four points the inter-residue hydrogen bonds are shown, the corresponding NH groups being in very slow exchange with D_2O solvent ($t_{1/2}$ > 100 hours at pH 2.7, 32°C).

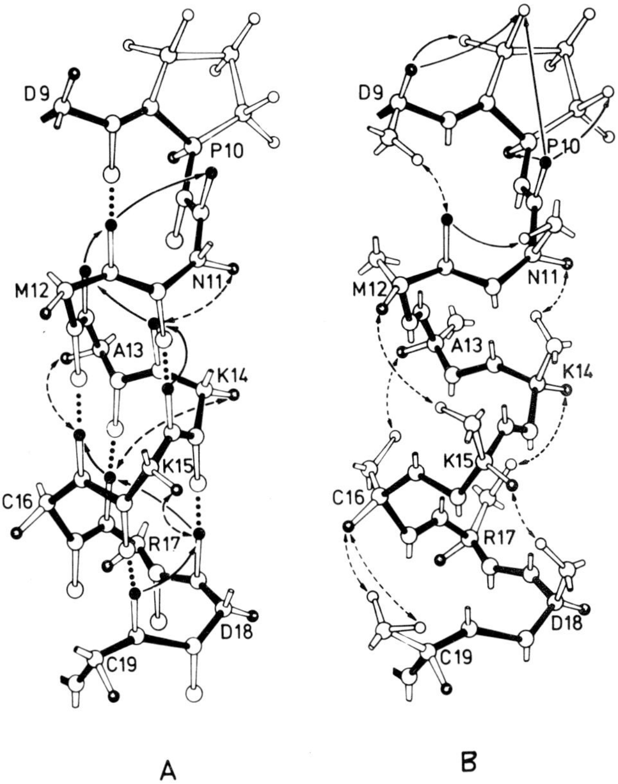

Fig. 8. The α-helical conformation of the Asp^{9}-Cys^{19} fragment of the *Buthus eupeus* insectotoxin I_5A in solution. Arrows indicate interproton NOE connectivites between backbone NH and $C^{\alpha}H$ protons (A) and between backbone and side chain protons (B).

connectivities there are several NOE interactions between side chain protons inherent for antiparallel β-structure, for instance, the aromatic Phe^{27} protons are dipole-dipole coupled with the side chain and backbone NH protons of the Lys^{25} and Leu^{32} residues. It is worth to mention that the $^3J(H\text{-}NC^{\alpha}\text{-}H)$ scalar coupling constants are in agreement with this structure ø angles. Values of 7.0-10.1 Hz for residues 25,26,27,30,31,32 and 33 correspond to ø -140 $\pm 15^{o}$ inherent to β-structure (3).
The insectotoxin fragment Asp^9-Cys^{19} (Fig. 7) contains the combination of the type I β-turn 9-12 and right-handed α-helix 12-19. The $^3J(H\text{-}NC^{\alpha}\text{-}H)$ coupling constants for the Asn^{11}-Cys^{19} fragment are also in agreement with this structure ø angles. For instance, the residues participating in the α-helical conformation have $^3J(H\text{-}NC^{\alpha}\text{-}H)$ values in the range of 2.4-5.0 Hz, which corresponds to ø angles -70 ± 10° (3). The ambiguity of $^3J(H\text{-}NC^{\alpha}\text{-}H)$ and ø angles relationship could be removed by taking into account the NOE results on d_1, d_2 and d_3 distances.
The work now in progress to combine all the NMR experimental data by use of distance geometry algorithm (35) to obtain the complete three-dimentional structure of the insectotoxin I_5A molecule and to evaluate the disulfide bond locations. The preliminary results (31) obtained in pseudoatom approximation reveal general backbone folding of the molecule shown in Fig. 9A.
In conclusion it is exiting to compare the obtained solution structure of "short" insectotoxin I_5A with recently established X-ray crystal structure (Fig. 9B) of the "long" type scorpion toxin v-3 (65 amino acid residues) from *Centruroides sculpturatus* Ewing (36). In spite of absence of any homology in amino acid sequences and different biological action (30), the spatial structures of these scorpion toxins contain many closely related facets. Both have α-helical and antiparallel β-structural fragments with nearly the same number of amino acid residues. Moreover those fragments are similarily packed one to another in space. However toxin v-3 has additional 30 residues located in the N- and C-terminal parts of the molecule. It is assumed (36) that exactly these parts of "long" toxins are responsible for

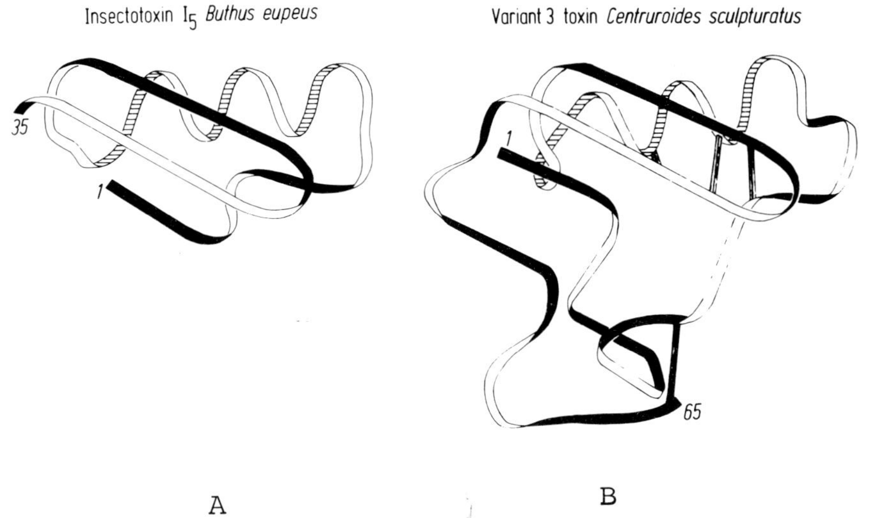

Fig. 9. Comparison of backbone folding of "short" and "long" types of scorpion toxins. A - NMR solution structure of the *Buthus eupeus* insectotoxin I_5A. B - X-Ray crystal structure of *Centruroides sculpturatus* Ewing toxin v-3 as redrawn from (36).

biological action. Thus it is likely that the structural similarity of "short and "long" scorpion toxins at least not directly connected with their mechanism of action. Presumably these types of scorpion toxins posses the same predecessor but are on different stages of evolution.

Conclusion

The proposed spatial structures of two polypeptide neurotoxins are considered as a first approximation. Further investigation is necessary to elucidate the relative flexibility of side chains and possible low amplitude motion of backbone, if any. But, of course, the main road of progress in structure-function relationship research depends on localization and isolation of corresponding receptor systems and studying of it interaction with the toxins.

References

1. Jardetzky, O., Roberts, G.C.K.: NMR in Molecular Biology, Academic Press, New York 1981

2. Bax, A.: Two-Dimensional Nuclear Magnetic Resonance in Liquids, D.Reidel Publ. Co., Dordrecht 1982.

3. Bystrov, V.F.: Progr. NMR Spectr. 10, 41-81 (1976).

4. Bystrov, V.F., Arseniev, A.S., Gavrilov, Yu.D.: J.Magn.Res. 30, 151-184 (1978).

5. Bystrov, V.F., Ivanov, V.T., Okhanov, V.V., Miroshnikov, A.I., Arseniev, A.S., Tsetlin, V.I., Pashkov, V.S., Karlsson, E.: in Advances in Solution Chemistry (Bertini, I., Lunazzi, L., Dei, A., eds) pp.231-251, Plenum Press, New York 1981.

6. Williams, R.J.P.: FEBS Lett. 146, 229 (1982).

7. Habermann, E., Fisher, K.: in Advances in Cytopharmacology (Ceccarelli, B., Clementi, E., eds) vol.3, pp.387-393, Raven Press, New York 1979.

8. Miroshnikov, A.I., Elyakova, E.G., Kudelin, A.B., Senyavina, L.B.:Bioorgan. Khim. 4, 1022-1028 (1978).

9. Okhanov, V.V., Afanasiev, V.A., Gurevich, A.Z., Elyakova, E.G., Miroshnikov, A.I., Bystrov, V.F., Ovchinnikov, Yu.A.: Bioorgan. Khim. 6, 840-864 (1980).

10. Bystrov, V.F., Okhanov, V.V., Miroshnikov, A.I., Ovchinnikov, Yu.A.: FEBS Lett. 119, 113-117.

11. Bystrov , V.F., Okhanov, V.V., Arseniev, A.S., Afanasiev, V.A. Gurevich, A.Z.: Bioorgan. Khim. 6, 386-392 (1980).

12. Okhanov, V.V., Afanasiev, V.A., Bystrov, V.F.: J.Magn.Res. 40, 191-195 (1980).

13. Wemmer, D., Kallenbach, N.R.: Xth International Conference on Magnetic Resonance in Biological Systems, Abstracts, p.I87, August 29-September 3, 1982.

14. Wyssbrod, H.R., Gibbons, W.A.: Surv. Progr. Chem. 6, 209-325 (1973).

15. Nurkhamedov, A.H., Elyakova, E.G., Efremov, E.S., Miroshnikov, A.I.: Bioorgan. Khim. 7, 16-24 (1981).

16. Hider, R.C., Ragnarsson, U.: FEBS Lett. 111, 189-193 (1980).

17. Chou, P.Y., Fasman, G.D.: Adv. Enzymol. 47, 45-148. (1978).

18. Van Wart, H.E., Scheraga, H.A.: J.Phys.Chem. 80, 1823-1831 (1976).

19. Popov, E.M., Melnikov, P.N.: Bioorgan.Khim. 5, 828-847, 1011-1024, 1471-1493 (1979); 6, 21-30 (1980).

20. Busetta,B.: FEBS Lett. 112, 138-142 (1980).

21. Hider, R.C., Ragnarsson, U.: Biochim.Biophys. Acta 667, 197-208 (1981).

22. Nuriddinov, A.R., Okhanov, V.V., Tsetlin, V.I.,Ivanov, V.T.: Bioorgan.Khim. 7, 5-15 (1981).

23. Vincent, J.P., Schweitz, H., Lazdunski, M.: Biochemistry 14, 2521-2525 (1975).

24. Sandberg, B.E.B.: Int.J.Pept.Prot.Res. 13, 327-333 (1978).

25. Granier, G., van Rietschoten, J.: in Natural Toxins (Eaker D., Wadstrom, eds) pp.481-486, Pergamon Press, Oxford 1980.

26. Sandberg, B.E.B., Ragnarsson, U.: Int.J.Pept. Prot. Res. 11, 238-245 (1978).

27. Grishin, E.V.: in Frontiers of Bioorganic Chemistry and Molecular Biology (Ananchenko, S.N., ed.) pp.93-96, Pergamon Press, Oxford 1980.

28. Grishin, E.V., Volkova, T.M., Soldatova, L.N.: Bioorgan. Khim. 8, 155-164 (1982).

29. Arseniev, A.S., Kondakov, V.I., Maiorov, V.N., Volkova, T.M., Grishin, E.V., Bystrov, V.F., Ovchinnikov, Yu.A.: Bioorgan. Khim., in press.

30. Rohat, H., Bernard, P., Courand, F.: in Advances in Cytopharmacology (Ceccarelli, B., Clementi, F., eds.) vol.3, pp.325-334 (1979).

31. Arseniev, A.S., Kondakov, V.I., Maiorov, V.N., Bystrov, V.F., Ovchinnikov, Yu.A.: Bioorgan.Khim. in press.

32. Momany, F.A., McGuire, R.F., Burgess, A.W., Sheraga, H.A.: J.Phys.Chem. 79, 2361-2381 (1975).

33. Billeter, M., Braun, W., Wüthrich, K.:J.Molec.Biol. 155, 321-346 (1982).

34. Mandel, N., Mandel, G., Trus, B.L., Rosenberg, J., Carlson, G., Dickerson, R.E.: J.Biol.Chem. 252, 4619-4636 (1977).

35. Crippen, G.M.: Distance Geometry and Conformational Calculations, Research Studies Press, Chichester 1981.

36. Fontecilla-Camps, J.C., Almassy, R.J., Ealick, S.E., Suddath, F.L., Watts, D.D., Feldmann, R.J., Bugg, C.E.: Trends in Biochem Sci. 6, 291-296 (1981).

EFFECTS OF DIFFERENT SPIDER VENOMS ON ARTIFICIAL AND BIOLOGICAL MEMBRANES

B.A.Tashmukhamedov, P.B.Usmanov, I.Kazakov, D.Kalikulov, L.Ya.Yukelson, B.U.Atakuziev

Institute of Biochemistry, Uzbek Academy of Sciences and State University of Tashkent, USSR

Introduction

Spider venoms show a wide variety of the mechanism of action. The venom of spiders of three different families have been studied. It has been found that spider venoms of the family Theridiidae contained presynaptic neurotoxins forming channels for Ca^{2+}, K^{+}, Na^{+} in bilayer phospholipid membranes (BLM).

These neurotoxins are composed of high molecular weight components (mol.wt.more than 1oo ooo daltons), which are responsible for specificity for presynaptic membranes of vertebrates or insects and of a low molecular weight channel forming component (mol.wt. below 5 ooo daltons). The latter has been found in the crude venom as free separate peptide as well.

Postsynaptic neurotoxins blocking the acetylcholine receptor of vertebrates and the glutamate receptor of insects were demonstrated in the spider venom of the family Araneidae.

A neurotoxin blocking axonal conduction due to blocking of membrane sodium channels was found in the spider venom of the family Segestriidae. Some results concerning the actions of the venoms on BLM and neuromuscular synapses of frog and locust are given.

Toxins as Tools in Neurochemistry

Results

Spider venom of the family Theridiidae. The presynaptic venom effect of black widow spider Latrodectus is well known (1).

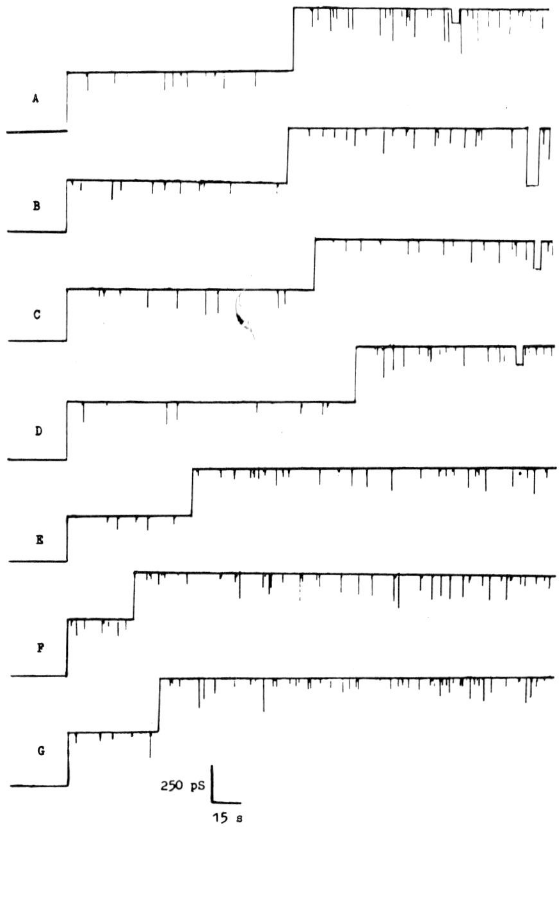

Fig.1 Channel conductance of bilayer membranes treated with spider venoms. A - crude venom of Steatoda paykulliana; C,D,E,F - fractions I,II,III,IV, respectively, of Steatoda paykulliana venom; G - low molecular peptide obtained after rechromatography of fraction I. Membrane was formed in 1oo mM KCL, 5 mM Tris (pH 7.5). The membrane potential was clamped at 1oo mV.

It has been shown earlier that black widow venom contains selective neurotoxins for vertebrates and insects with molecular weight around 13o ooo daltons (2).

In preliminary experiments with BLM it was shown that the crude venom of Latrodectus mactans tredecimguttatus despite of its very complex toxin composition forms a uniform population of cation channels with respect to its selectivity and amplitude (Fig.1 A). Analysis of the data obtained allowes to suggest that in the crude venom there was only one protein forming monotypic channels. The same results were obtained with the venom of the spider Steatoda (Lithyphantes) paykulliana. The channel amplitudes were 35o or 5o pS under standard conditions with 1oo mM KCL or 1oo mM $CaCl_2$, respectively (Fig. 1 B). The spider venoms were fractionated on Sephadex G-1oo in order to identify this channel forming component (Fig.2 A). Crude venoms were divided into four fractions containing protein components of the following molecular masses: fraction I - 1oo ooo daltons and higher, fraction II - 6oooo-7oooo, fraction III - 3o ooo-4o ooo and fraction IV - 5 ooo daltons and less.

Experiments with BLM indicate that there are at least four fractions in every venom (I,II,III and IV) that cause an increase of the integral BLM conductance. All these fractions under comparable conditions form channels of identical selectivity and amplitude (Fig.1 C,D,E,F). An uniform population of channels is unlikely to be formed by proteins with such different molecular weights. Therefore we have suggested that the only channel former in these venoms is the small protein of the latter fraction with a molecular weight of about 5 ooo daltons. This protein is found as a complex component of larger neurotoxins. This suggestion is strenghened by our finding that by repetitive chromatography of fraction I, the component with a molecular weight of 5 ooo $\pm$ 5oo was separated by several cycles of gel filtration (Fig.2 B). This component forms channels (Fig.1 G) which bear similarities to the channels formed by crude venom and by its high molecular weight fractions I,II and IV. The same peptide with a molecular weight of 5 ooo was separated from fraction II and III by rechromatography (Fig.2 C,D).

Experiments with neuromuscular junctions of the frog Rana temporaria and locust Locusta migratoria indicate that fraction I of spider venoms contains a presynaptic toxin for vertebrates and fraction III contains a toxin for insects (Fig. 3 A,B).

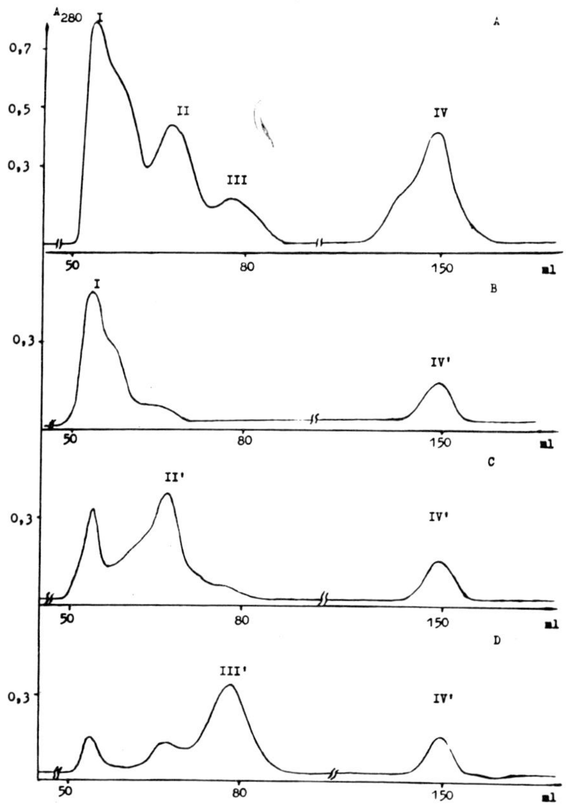

Fig.2 Gel filtration of crude venom of Steatoda paykulliana (A) and its fractions I (B), II (C) and III (D) on Sephadex G-1oo (1,5 x 8o cm column), eluted with o,o1 M Phospate buffer, pH 8.o containing o.o1 M NaCl.

Fraction II increases miniature potential frequency of both frog and locust. For the study of the biological role of the low molecular weight fraction IV and also of the low molecular weight peptide separated from fraction I their influence

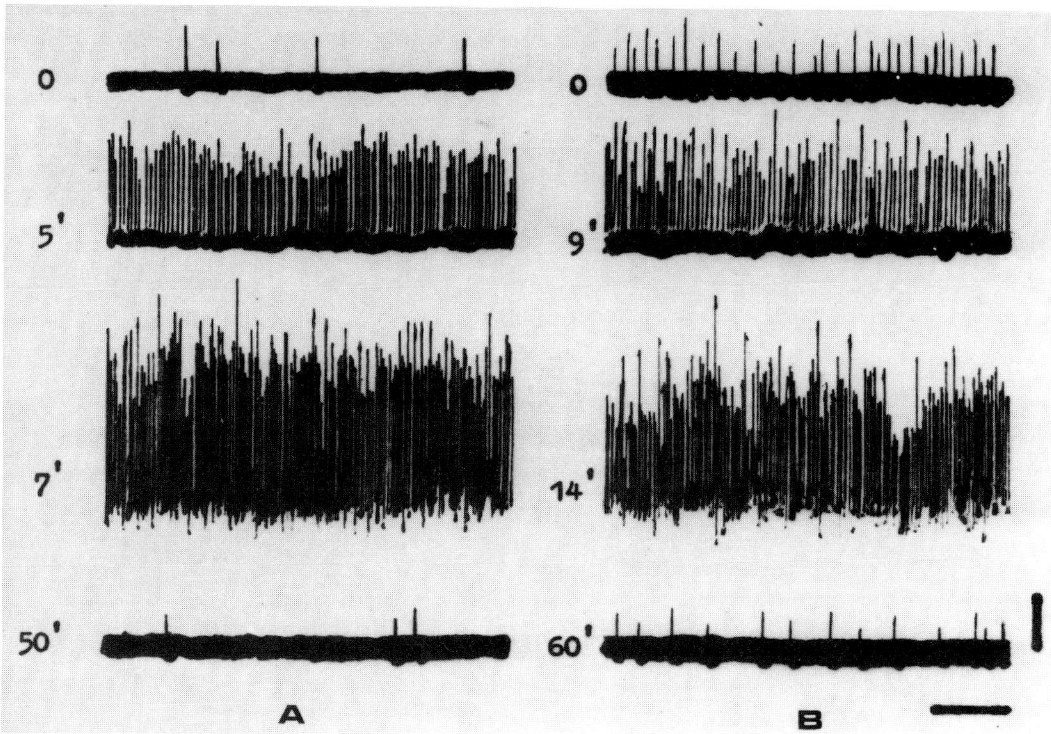

Fig.3 Increase in MEPPs frequency at frog (A) and MEPSPs at locust (B) neuromuscular junctions induced by the fraction I (A) and fraction III (B) of Steatoda paykulliana venom. Each trace shows a record obtained at the indicated time following the application of the fraction. Calibration marks: o.5 mV, 1 sec.

on neuromuscular synapses of frog and locust was investigated. It was shown that these peptides induced peculiar volleys of miniature potentials after long latent periods of 4o - 6o min (Fig.4) These volleys appear to be nerve terminal reactions to discrete increases of presynaptic membrane permeability for Ca^{2+} induced by a channel forming peptide. The long latent period, the lack of sëlectivity and different actions of the low molecular peptide are the result of absence of its recognition structure.

It may be concluded that presynaptic neurotoxins of these spider venoms have an oligomerous structure. Highly selective actions of the toxins appear due to a larger protein molecule, the carrier molecule. These molecules provide the recognition and the subsequent toxin interaction with the presynaptic membranes of vertebrates or insects. In this case, the low molecular weight subunit - the channel former - penetrates the lipid membrane and forms ionic channels for Ca^{2+} and therefore increases miniature potential frequencies.

Spider venom of the family Araneidae. It was shown that the venom of Argiope lobata had no effect on BLM but blocked synaptic transmission. The experiments have been performed with neuromuscular junctions of the frog Rana temporaria and the locust Locusta migratoria. The spontaneous miniature potential amplitudes of frog and locust synapses decreased and then disappeared after application of spider venom (Fig.5 A, 6 A).

The resting potential as well as the miniature endplate potentials (MEPPs) of frog and the miniature excitatory post-synaptic potential (MEPSPs) frequency of locusts were not changed. The frog endplate potentials (EPPs) and the locüst excitatory postsynaptic potentials (EPSPs) were suppressed similarly (Fig.5 B, 6 B). The MEPPs and EPPs of the frog neuromuscular junctions but not locust MEPSPs and EPSPs were recovered after rinsing them with standard Ringer's and Hoyle's solution. In

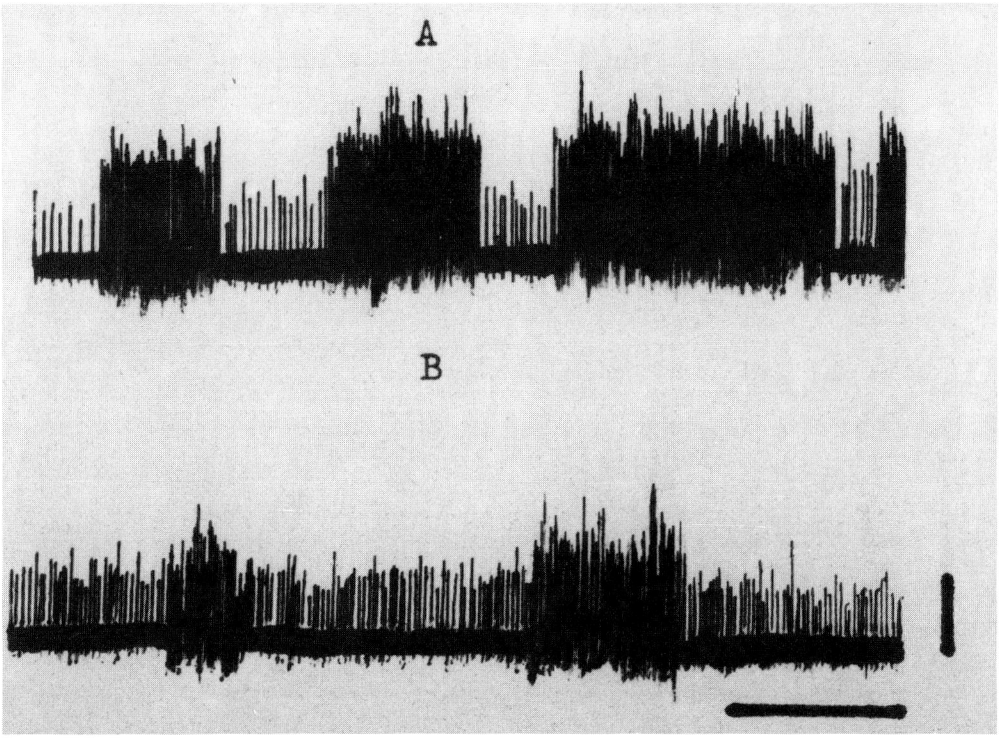

Fig.4 Volleys of MEPPs of frog (A) and MEPSPs of locust (B) neuromuscular junctions induced by Fraction IV of Steatoda paykulliana venom. Calibration marks: o.5 mV, 2o sec.

addition it was found that the venom blocked the potentials which had been evoked by the iontophoretic application of acetylcholine (Fig.7 A) and glutamate (Fig.7 B). It follows that the complex of insect neurotoxin with the postsynaptic glutamate receptor of locusts has a high stability constant.

Then crude spider venom was separated on Sephadex G-75 (Fig. 8 A). It was shown that active neurotoxine peptides are pre-

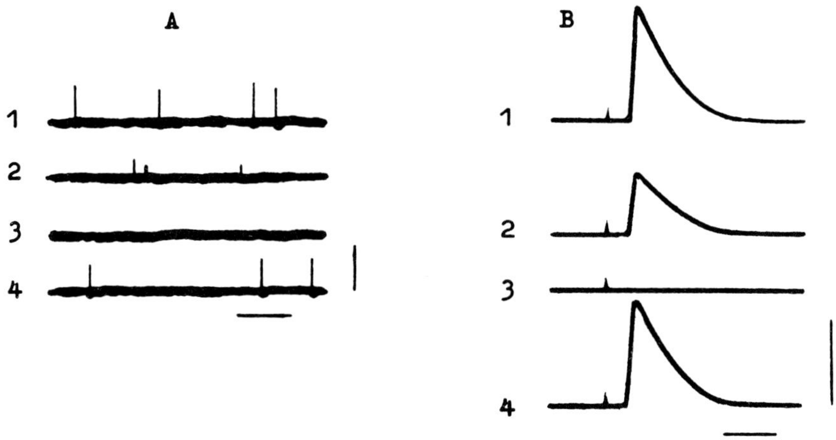

Fig.5 Effect of the crude venom of Argiope lobata on the frog MEPPs (A) and EPPs (B). A_1-B_1-control records, A_2-B_2-5 min and A_3-B_3-1o min after application of the venom, A_4-B_4- after rinsing. Calibration marks: A-0,5 mV, 1 sec B- 5 mV, 4o msec.

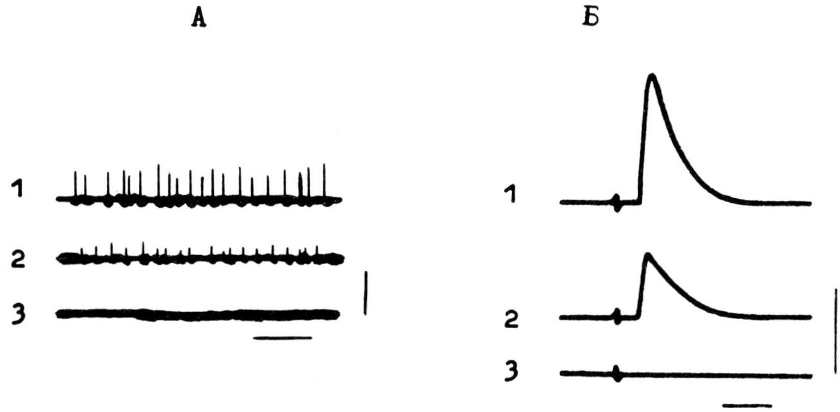

Fig.6 Effect of the crude venom of Argiope lobata on the locust MEPSPs (A) and EPSPs (B). A_1-B_1- control records, A_2-B_2- 6 min and A_3-B_3- 12 min, after application of the venom (o.o5 mg/ml). - Calibration marks: A-0.5 mV, 1 sec; B- 5 mV, 1o msec.

sent in a fraction corresponding to a molecular weight of 6 ooo - 7 ooo daltons. After separation of this fraction on Bio-Gel P 4 eight fractions were obtained (Fig.8 B). One of them (No.3) blocked postsyanptic membranes of both frog and locust, and the other three (No.5,6,8) blocked only locust synapses.

Previously snake postsynaptic neurotoxin s have been used for the isolation and identification of cholinoreceptor protein of vertebrate and invertebrate synapses (3). Similarly the insect neurotoxin described here can be used for the isolation of glutamate receptors of insect synapses as well.

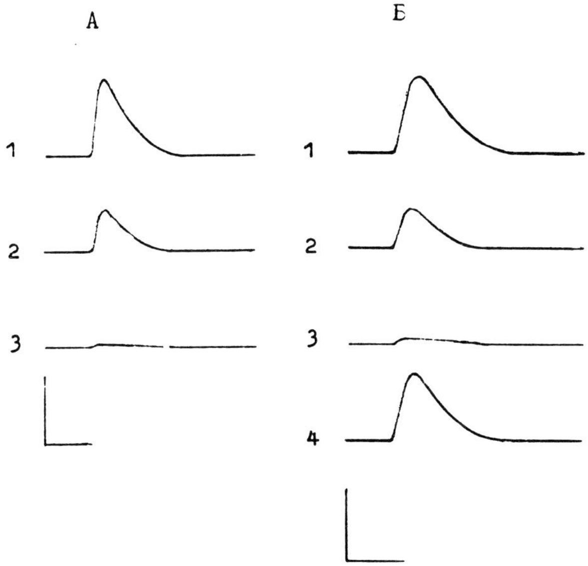

Fig.7 Effect of the crude venom of Argiope lobata on iontophoretically applied acetylcholine (A) and glutamate (B) potentials. A_1-B_1- control records, A_2-B_2- 7 min and A_3-B_3- 15 min after application of the venom, A_4- after rinsing. - Calibration marks: A- 5 mV, 6o msec; B- 5 mV, 2oo msec.

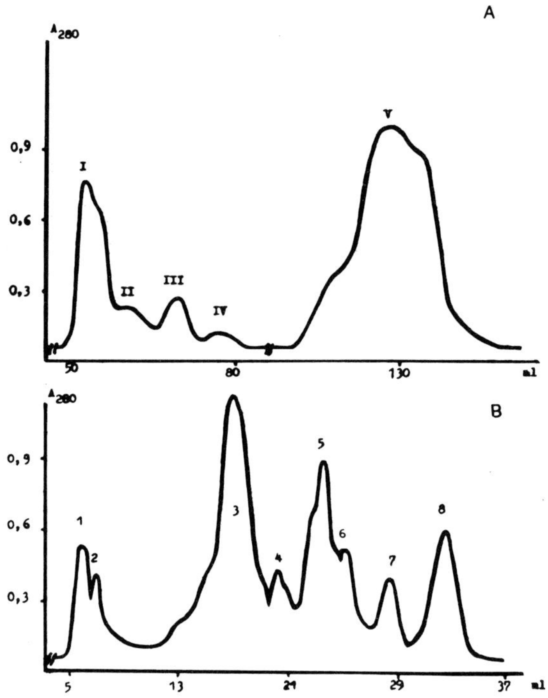

Fig.8 Fractionation of the venom of Argiope lobata on Sephadex G-75 (A) and fraction V of this venom on Bio-Gel P 4 (B). A - gel filtration on Sephadex G-75 (2.5 x 9o column), eluted with o.o1 M phosphate buffer, pH 7.9, containing o.o1 M NaCl; B - gel filtration on Bio-Gel P 4 (o.9 x 55 column), eluted with the same buffer.

In 1982 a neurotoxin was separated from the venom of the spider Nephila clavata (family Araneidae, subfamily Nephilinae) which can block glutamate receptors of lobster neuromuscular synapses (4). Pobably the common characteristic of the venoms of the family Araneidae is that they all contain postsynaptic neurotoxin s which block glutamatergic synapses.

Spider venom of the family Segestriidae. It was shown that crude Segestria florentina spider venom induced spontaneous twitching of frog muscular fibres. Repetitive activity after a single shock was initiated in the motor nerve by application of the venom. A prolongation of action potentials mainly due to increased duration of falling phase was observed after addition of the venom to the isolated sciatic nerve trunk (Fig. 9 A). It was shown that application of the venom to the ventral nerve cord of Periplaneta americana resulted in a burst of spontaneous activity and a gradual decrease of the induced responses (Fig.1o A).

Analysis of the data indicated that the effect of the spider venom is similar to that of scorpion venom. For comparison similar studies have been performed with scorpion venom (Fig. 9 B, 1o B). It was shown that the effects of spider and scorpion venoms were similar. In addition it was shown that the

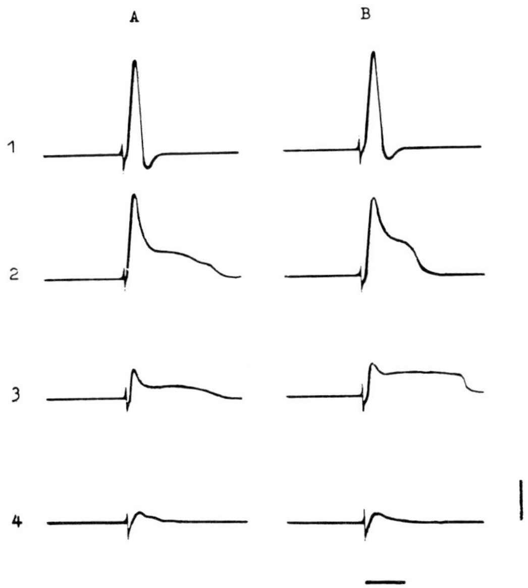

Fig.9 Effect of the crude venoms of Segestria florentina (A) and Orthochirus scrobiculosus (B) on the frog sciatic nerve trunk. - A_1-B_1- control action potentials, A_2-A_3-A_4 in 1o,3o,6o min and B_2-B_3-B_4 in 1,3,5 min respectively, after application of the spider (2o mg/ml) or scorpion (1 mg/ml) venoms. - Calibration marks: 2 mV, 5 msec.

spider venom did not contain any depolarizing agents such as phospholipase or cobra cytotoxin. Spider venom did not induce any conductivity in Phospholipid membranes.

From these data it is clear that Segestria florentina spider venom contains neurotoxins specifically interacting with sodium channels of electroexcitable membranes of insects and vertebrate species.

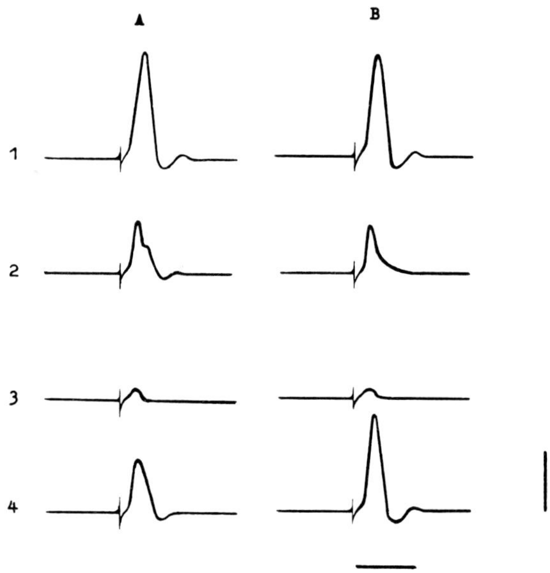

Fig 1o Effect of the crude venoms of Segestria florentina (A) and Orthochirus scrobiculosus (B) on the cockroach ventral nerve cord. - A_1-B_1- control action potentials, A_2-B_2- 15 min and A_3-B_3-2o min after application of the spider (1o mg/ml) and scorpion (5 mg/ml) venoms. Calibration marks: 2 mV, 5 msec.

Conclusion

Analysis of our data leads to the conclusion that spider venoms of different families show a wide variety of mechanisms of action. One of these affects presynaptic membranes,

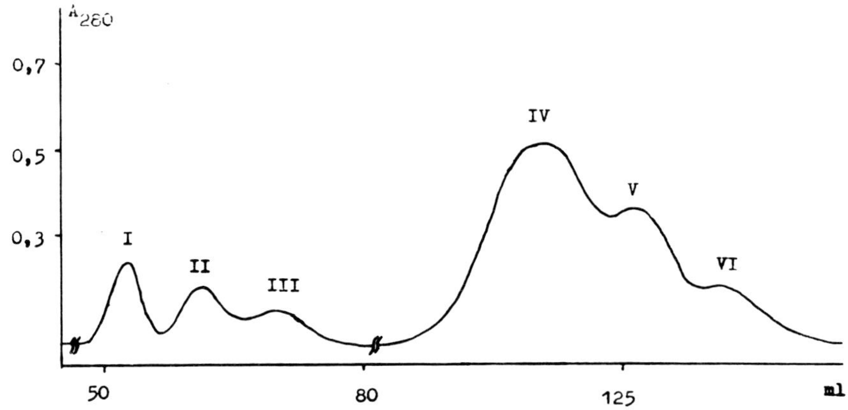

Fig.11 Fractionation of the venom of Segestria florentina by Gel filtration on Sephadex G-1oo (1.5 x 9o cm column), eluted with o.o1 phosphate buffer, pH 7.9, containing o.o1 M NaCl. - Crude venom was fractionated by Sephadex G-1oo . An active neurotoxic factor was found in fraction III.

another different postsynaptic neuroreceptor membranes, a third acts on ionic channels of excitable membranes. Probably all of them like the toxins of snakes, scorpions and insects can be used for various neurobiological investigations.

References

1) Longenecker, H.E., Hurlbut, W.P., Mauro,A., Clark, A.W.: Nature 225 , 7o1-7o3 (197o)

2) Bettini,S., Maroli,M. in: Arthropod Venoms (Ed.S Bettini), Springer-Verlag, Berlin-Heidelberg-New York 48, 149-185 (1978)

3) Changeux,J.P., Kasai,M., Lee,C.V.: Proc.Natl.Acad.Sci.USA 67, 1241-1247 (197o)

4) Kawai,N., Niwa,A., Abe,T.: Brain.Res. 247, 169-171 (1982)

STRUCTURAL PECULIARITIES AND MECHANISM OF ACTION OF PRESYNAPTICALLY ACTING NEUROTOXINS FROM THE VENOMS OF LATRODECTUS AND LITHYPHANTES SPIDERS

Sh. Salikhov, M. Adylbekov, T. Slavnova, M. Tashmukhamedov, J. Abdurakhmanova, A. Korneyev, A. Sadykov

Institute of Bioorganic Chemistry, Tashkent, USSR

The particular attention given to neurotoxins may be explained by their usage as "tools" to investigate the molecular mechanisms of neuro-muscular transmission and to isolate the structural complexes of biological membranes responsible for life-giving processes in a cell. This ability made it possible to successfully use postsynaptically acting neurotoxins (α-bungarotoxin et al.) for the isolation and study of structure of postsynaptic membrane acetylcholine receptors (1-3). Regretfully not much research has been done concerning the presynaptic membrane. This can be explained by the fact that neurotoxins affecting it rarely occur in nature and they are not well studied. This is why the isolation of new presynaptic neurotoxins, the study of their structure and mechanism of action and their usage as specific markers for the investigation of the structural organization of the presynaptic membrane will provide more information for understanding the molecular mechanisms of neuromuscular transmission.

According to the mechanism of action and the effect produced presynaptically acting neurotoxins may be divided into the following groups:

1) Neurotoxins isolated from the venoms of Elapidae snakes (notoxin, taipoxin, β-bungarotoxin, crotoxin). They are considered to be proteins of basic nature possessing phospholipase activity.

Toxins as Tools in Neurochemistry

2) Neurotoxins produced by the microorganisms Clostridium botulinum (botulinic toxin) and Clostridium tetani (tetanic toxin).

3) Neurotoxins isolated from the venom of Latrodectus spiders.

Neurotoxic phospholipases and toxins produced by Clostridium microorganisms selectively act on cholinergic systems by blocking acetylcholine release. These toxins can be widely used in electro-physiological and biophysical studies; however they can not be used as much for the isolation of presynaptic membrane components because the neurotoxic phospholipases possess enzyme activity while the toxins produced by microorganisms go through the presynaptic membrane into the nerve terminal. This is why the neurotoxins from Latrodectus spiders venom appear to be most prospective "tools" to study the presynaptic membrane structure. These neurotoxins do not act as enzymes and they bind irreversibly to the presynaptic membrane. The main effect of Latrodectus spider toxins is to induce massive transmitter release. On the basis of data from the literature it can be concluded that the venom of Latrodectus spiders contains at least three toxic components (m.w. 13o. ooo, 7o ooo and 5ooo respectively), each of which specifically influences transmitter release from various nerve terminals (4-8).

Regretfully at present there is not enough information to discuss the relationship between the difference in structure and function of these toxins.

To study the structure and mechanism of the action of Latrodectus venom neurotoxins the authors isolated and identified the toxins from the venom of Latrodectus tredecimguttatus spiders widespread in Central Asia and from the rare species Latrodectus pallidus pavlovski Charit and from the venom of spiders Latrodectus Dahli Levi bred from cocoons in an insec-

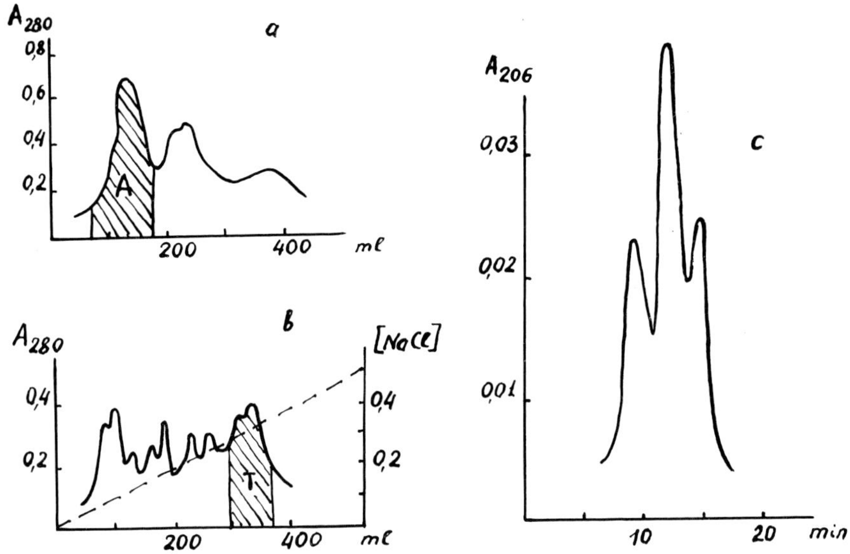

Fig.1 a) Gel filtration of Latrodectus tredecimguttatus venom on a column (2.5 x 9o cm) with Sephadex G-5o (fine) in o.o5 M Tris HCl, pH 8.2; b) Chromatography of fraction A on a column (2.5 x 2o cm) with cellulose DE-52 in Tris NaCl gradient, pH 8.2; c) Rechromatography of fraction T on a TSK-4ooo-column.

tarium. For comparison the authors also ioslated the neurotoxins from another genus, Lithyphantes, whose venom is very similar to that of Latrodectus in its physiological effect. The fractionation technique was the same for all species. It included gel and ion-exchange chromatography. The procedures resulted in 1o fractions from the venom of each species. The fractions possessing maximal activity eluated at o.28 - o.3oM NaCl from the column containing cellulose DE-52 (Fig.1). Rechromatography of the active fraction has been carried out on a TSK-4ooo column (LKB, Sweden).

As a result two neurotoxins have been isolated from each species. The first one probably corresponds to α-latrotoxin, isolated earlier (4) and has m.w. 13o ooo. Its amino acid composition is presented in Table 1. Ile appears to be the N-ter-

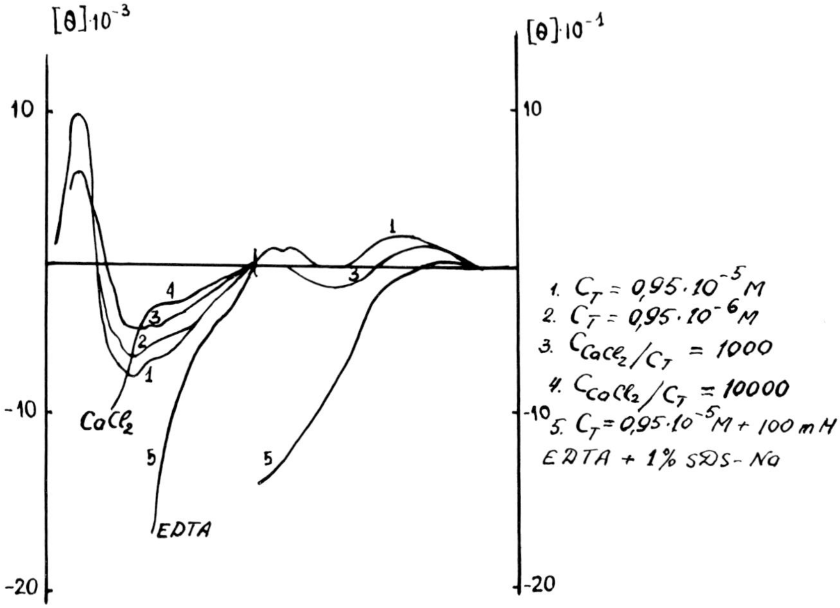

Fig.2 CD-spectra of β-latrotoxin dependent on Ca^{2+} concentration and EDTA

minal amino acid. The second type of neurotoxin, β-latrotoxin with the N-terminal blocked, has m.w. 75 000.

The data obtained in the process of isolation and purification of these toxins hint at a quaternary structure. At neutral and weak alkaline pH they are present as biologically active dimers and tetramers being cleavable into non-active monomers at low pH values and in the presence of chelate-forming agents and detergents (9). The presence of quaternary structure is proved by studies of CD-spectra of β-latrotoxin under various conditions (Fig.2).

Calculation of the amount of regular structures in the toxin protein molecule according to (1o) gives 12.6 % of α-helix, 4o % of β-structure and 4o % of non-organized fragments. Addition of SDS and EDTA to the solution turns the complete structure into the state of a random coil.

The amino acid composition shows marked similarities in protein molecules of various spider species (Table 1).

Table 1 Amino acid composition of neurotoxins from venoms of Latrodectus and Lithyphantes spiders

Amino Acid	L. tredecim-guttatus α-latro-toxin	L. tredecim-guttatus β-latro-toxin	L. Dahli Levi	L. pallidus pav-lovski	Lithyphantes paykulliana
Asp	15o	8o	96	95	92
Thr	6o	29	37	31	34
Ser	68	33	32	29	38
Glu	154	67	86	82	86
Pro	48	34	28	28	28
Gly	66	38	37	36	4o
Ala	74	4o	38	36	41
Val	56	49	38	39	39
Met	1o	12	11	12	12
Ile	54	38	4o	4o	41
Leu	1oo	59	65	65	66
Tyr	38	25	23	24	22
Phe	48	34	31	32	3o
His	3o	23	22	25	23
Lys	72	36	42	38	42
Arg	52	37	34	38	35
Trp		4			

In spite of the considerable similarities in functional activity and in physico-chemical parameters a final conclusion may be drawn only after finding homologies in the amino acid sequences. At present the authors are analysing the primary structure of β-latrotoxin. The composition of tryptic hydrolysates has been established and all the peptides obtained as a result of CNBr-cleavage and cleavage by glutamine protease have been isolated and their partial structure has been determined. As a result of lyophilization and the period following it, the neurotoxin partially denatures, which is due to labile Pro-Asp bonds found in tryptic peptides. During high-pressure rechromatography on the TSK-4ooo column (Fig.3) a peptide with the following amino acid composition has been found: Asp 2o, Thr 4, Ser 4, Glu 22, Pro 5, Gly 5, Ala 6, Cys 3, Val 5, Ile 3, Leu 9, Tyr 2, Phe 7, His 2, Lys 9, Arg 1.

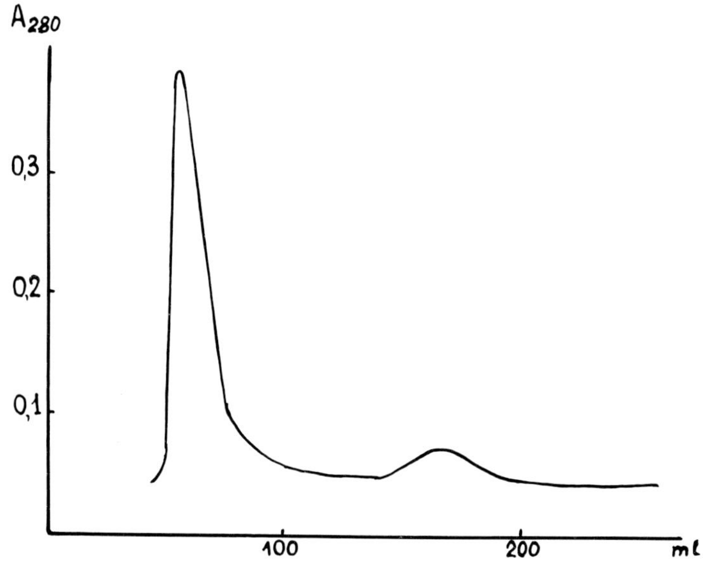

Fig.3 Gel filtration of β-latrotoxin after lyophilization on a column (1.5 x 1oo cm) with Sephadex G-1oo (fine)

To study the mechanism of action of β-latrotoxin, electrophysiological investigations with neuromuscular frog preparations have been carried out. If the toxin at a concentration of 5 x $1o^{-7}$ - 1 x $1o^{-5}$g/ml is added to a Ringer solution containing Ca^{2+} and Mg^{2+}, an increase in M.E.P.P. frequency may be observed after some latent period, the duration of the latter depending on the toxin concentration. The frequency increased 1oo-2oo times when the toxin showed its maximal effect. The removal of the toxin after 1o minutes contact with the preparation did not prevent the development of its effect. This means that its effect is irreversible. Addition of toxin to the medium with the Ca^{2+} concentration maintained at $1o^{-8}$ M by adding 1 mM EGTA in the absence of other bivalent ions did not induce the typical effect expressed in M.E.P.P. frequency increase. However, exchanging the medium for one containing Ca^{2+}, Mg^{2+} or Sr^{2+} ions immediately resulted in a rise of M.E.P.P. frequency. This toxin effect can be eliminated by changing the medium for the initial solution without bivalent cations. Thus, the effect of the toxin can be controlled by adding or removing the bivalent cations from the medium. Threshold concentrations for Ca^{2+} and Mg^{2+} are o.5 mM, but the maximal effect has been observed only in the presence of 1 mM Mg^{2+} or Ca^{2+} or 4 mM Sr^{2+} (Fig.4).

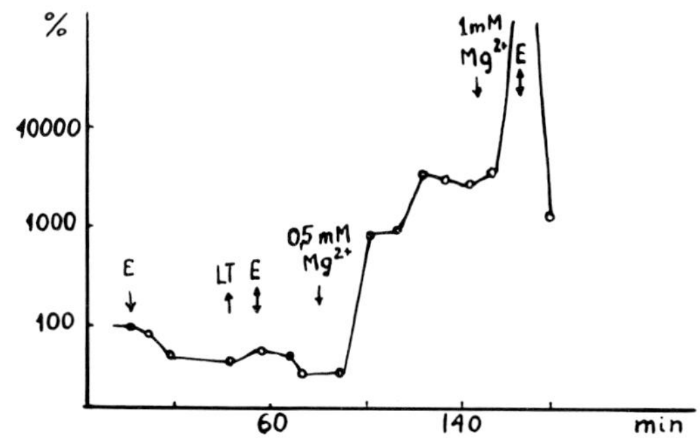

Fig.4 Activation of β-latrotoxin effect by Mg^{2+} ions

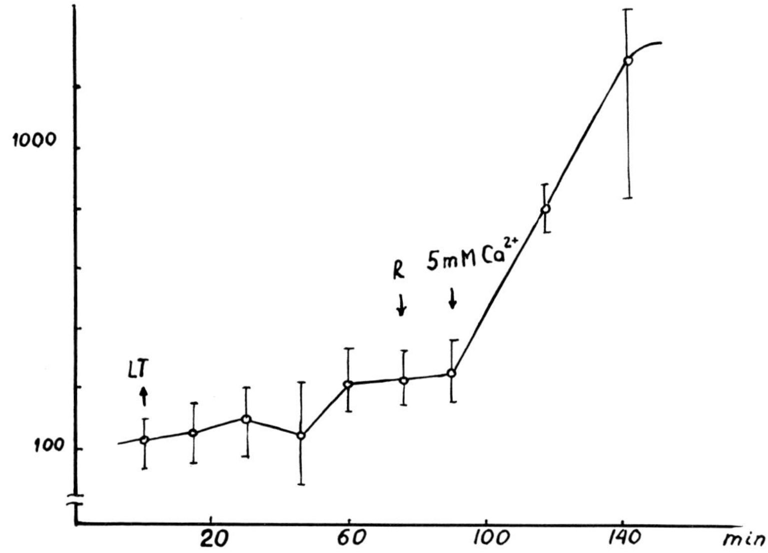

Fig. 5 Reactivating influence of increased Ca^{2+} concentration on the effect of 1 x 10^{-5} g/ml of β-latrotoxin previously treated with 5 mM EGTA

As has been already shown, β-latrotoxin is a monomer at a pH lower than 5.o.This is why these experiments have been carried out at a pH as low as 4.8. The addition of the toxin to the bath with normal Ringer (Ca^{2+} = 1.8 mM, pH 7.9) resulted in a M.E.P.P. frequency rise within 5 - 1o minutes. The exchange of the solution for a similar one but with pH 4.8 rapidly decreased M.E.P.P. frequency.This effect can be reversed by raising the pH to 7.4. The efficiency of β-latrotoxin action on bilayer phospholipid membranes depends also on the pH of the medium: It increases considerably in the pH range 5.o - 8.o. Thus the effect of the toxin that is already bound to the presynaptic membrane can be controlled by the pH. Several points of view exist concerning the molecular nature of the activating effect of bivalent cations. The venom of Latrodectus spider and its neurotoxin have been shown to increase model phospholipid membrane permeability for K^+, Na^+ and Ca^{2+} ions by forming the conductivity channels (11,12). If the toxin can form additional ion channels in a presynaptic membrane, then the penetration of bivalent cations into the axoplasma may result in an increased discharge of transmitter quanta. However, the typical effect of the toxin can be observed also in the absence of bivalent cations by increasing the tonicity of the bathing solution by adding 12o mM saccharose or by raising the temperature of the medium. These findings may be considered as evidence for the necessity for the presynaptic membrane of the toxin molecule itself to be in a certain state of order to ensure the proper effect. Actually, preincubation of the toxin for 2 hours with 1 - 1o mM EGTA or 1o mM EDTA appeared to prevent the development of the typical effect in spite of the fact that the toxin incubated first in Ca^{2+}-free medium was then added to the medium containing 1.8 mM Ca^{2+}. If native toxin is added to the medium together with an equal amount of toxin previously treated with EGTA, the M.E.P.P. frequency immediately increases a few dozen times. The loss of activity as a result of the effect of the chelate-forming agents may be explained by the necessity of having bivalent cations for maintaining the active form of the toxin molecule, i.e. its dimeric or tetrameric structure. This agrees with studies of the CD-spectrum of the toxin and is evident by experiments restoring the toxin activity lost after treatment with EGTA. This can be achieved by a 15 min addition to the bath of the toxin previously incubated with EGTA and then washing it with normal Ringer containing 1.8 mM Ca^{2+}. As in previous experiments no increase in M.E.P.P. frequency has been observed. But if the Ca^{2+} concentration is raised to 5 mM, M.E.P.P. frequency will be markedly increased (Fig.5). This indicates that the changes in the toxin molecule induced by incubation in the presence of EGTA and EDTA may be eliminated by the usage of higher Ca^{2+} concentration even in case the non-active toxin has been already strongly absorbed on a presynaptic membrane.

The ability of β-latrotoxin to form conductivity channels in phospholipid bilayers of various composition hints at its selective presynaptic effect. Toxin receptors in the presynaptic membrane appear to specifically bind and concentrate toxin up to its effective channel-forming concentration. The findings given above show that conditions leading to transition of the toxin to its monomeric form deprive it of the capability to act specifically on a presynaptic membrane, but they probably do not influence its irreversible binding. Certain conditions (pH, bivalent cations) favouring aggregated forms (such as dimers, tetramers) are necessary for the effect of β-latrotoxin.

Thus the ability of β-latrotoxin to bind specifically and irreversibly to a presynaptic membrane makes it an important "tool" for the isolation of functionally important components of a presynaptic membrane and for the study of their structure and function.

References

1) Heidmann,T., Changeux,J.P.: Ann.Rev.Biochem.47, 317-357, (1978)

2) Lindstrom,J., Einarson,B., Merlie,J.: Proc.Natl.Acad.Sci. USA, 75, 765-773 (1978)

3) Fentrell,C.M.S. in: "Commentaries in the neurosciences", Oxford, Pergamon Press, 255-28o (198o)

4) Frontali,N., Cecarelli,B., Gorio,A., Mauro,A., Siekevitz, P., Mu-Chin Tzeng, Hurlbut,W.P.: J.Cell.Biol. 68, 462-479, (1976)

5) Grasso,A.; Biochem. Biophys. Acta 439, 4o6-412 (1976)

6) Ornberg,R.L., Smith,T.Jr., Benton, O.W.; Toxicon 14, 329-333 (1976)

7) Frontali,N., Grasso,A.: Arch.Biochem.Biophys. 1o6, 213-218 (1964)

8) McCrone,Y.D., Hatala,R.J. in: "Animal Toxins", Oxford N-Y, Pergamon (1967)

9) Salikhov, Sh., Tashmukhamedov,M., Adylbekov,M., Korneyev, A., Sadykov, A.; DAN SSSR, 262, 2, 485-488 (1982)

1o) Greenfield,N., Fasman,G.D.: Biochemistry, 8, 41o8 (1969)

11) Finkelstein,A., Rubin,L.L., Tzeng, M.C.: Science 193, 1oo9-1o11 (1976)

12) Krasilnikov, O.V., Ternovsky, V.I., Tashmukhamedov,B.A.: Biofizika 27, 72-75 (1982)

MYOTOXIC AND NEUROTOXIC PHOSPHOLIPASES A

Dietrich Mebs

Zentrum der Rechtsmedizin, University of Frankfurt
D-6000 Frankfurt/Main- 70

Some of the most potent snake venom toxins such as taipoxin from Oxyuranus scutellatus and notexin from Notechis scutatus venom have phospholipase A activity and act at the presynaptic site of the nerve endplate by inhibiting the evoked release of acetylcholine. Several studies indicate that the catalytic site of the enzyme is involved in the toxic action (1). Moreover, considerable myotoxicity, i.e. destructive effects to skeletal muscle, is associated with these toxins and some other snake venom phospholipases A. In human envenomation sea snake (Enhydrina schistosa) as well as several Australian elapid snakes cause general muscle pain and myoglobinuria indicating extensive muscular damage (2).

From sea snake venom Fohlman and Eaker (3) isolated a lethal phospholipase A which produced myoglobinuria in mice. We fractionated lo Australian snake venoms (Oxyuranus s.scutellatus, Notechis s.scutatus, N.ater serventyi, N.a.humphreysi, Acanthophis antarcticus, Austrelaps superbus, Pseudonaja textilis, Pseudechis australis, P.porphyriacus, P.colletti guttatus)by column chromatography on CM-Sephadex C-25 and tested the fractions obtained in mice using myoglobinuria as indication for muscle damaging effect (4). Only venom fractions from Austrelaps superbus, Pseudechis australis, P.porphyriacus and P.colletti guttatus produced this effect. On the other hand, Harris et al. (5) had demonstrated that notexin and taipoxin cause considerable histological alterations in skeletal muscle. Due to the high neurotoxicity of these two toxins mice may die before myoglobinuria can develop. The factors from the four venoms were characterized to be neutral or basic phospholipases A containing 120 to 129 amino acid residues exhibiting many similarities to other myotoxic phospholipases A such as myotoxin from sea snake venom (3) and notexin. They possess a relatively low lethality (4.3 to 7.7 mg/kg, s.c. injection compared to 0.006 mg/kg for taipoxin), but higher enzymatic activity when tested on egg yolk suspension (about 100 times higher than taipoxin).

Using the increase of serum creatine kinase level as a parameter of muscle damage the following results were obtained : After intramuscular injection of 0.125 mg/kg (Notechis s.scutatus) and 0.5 mg/kg (other venoms) respectively, only the venoms of Pseudechis australis and P.colletti guttatus produced a steep increase of enzyme activity in the serum of rats after 4 and 16 hr exceeding 30 to 70 times the normal value, but returning to almost normal values after 48 hr (fig.1).

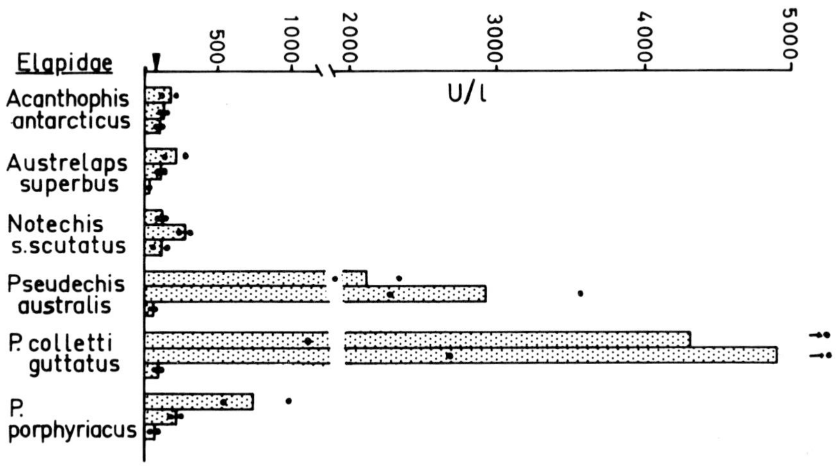

Fig.1. Serum creatine kinase level of rats 4, 16 and 48 hr after the injection of Australian snake venoms : Notechis s.scutatus, 0.125 mg/kg, others 0.5 mg/kg, i.m.

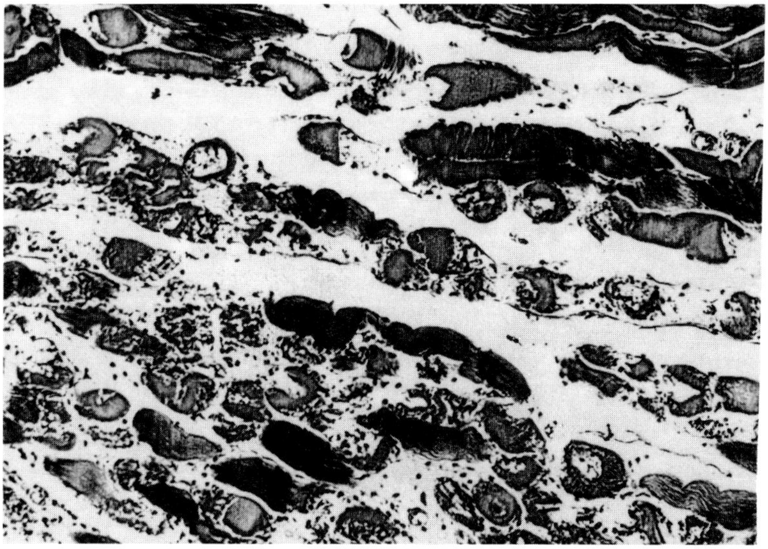

Fig.2. Necrosis of hind leg muscle of rat 4 hr after the injection of tiger snake venom (Notechis s.scutatus, 0.125 mg/kg), about 100 x

Histological examinations of the muscle tissue at the site of injection indicate that slight local swelling and severe degenerative necrosis of muscle cells developed 4 hr after i.m. injection. This is most evident in the case of Notechis s.scutatus venom (fig.2), although its effect on creatine kinase level is rather insignificant. The muscle fibres exhibit various stages of destruction; early changes are characterized by increasing vacuolation, followed by necrosis with complete loss of cellular structure. Although the plasma membrane is disrupted releasing a mass of disintegrated muscle fibre material the basal lamina seem to remain intact.

This feature of a rapidly established, severe degeneration of skeletal muscle strongly favors the suggestion that an enzymatic action of phospholipase A mediates these effects, i.e. hydrolysis and solubilization of membrane lipids leading to the perturbation of the membrane. It might also be speculated that the molecular target of the highly neurotoxic as well as of moderately toxic, rather myotoxic phospholipases A is a structure common to both nerve terminals and muscle (6). The action of taipoxin and notexin at the neuromuscular junction interfering with transmitter release at extremely low concentration may be due to specific binding mechanisms not associated with phospholipases A of relatively low toxicity.

Following the exposure of rat skeletal muscle to notexin or taipoxin Harris et al. (5,7) observed that the regeneration of the muscle fibres begins by 2 to 3 days; reconstruction of the neuromuscular junction takes place provided the basal lamina are preserved. Innervation is completed by 5 to 7 days. The rapid and specific degeneration of skeletal muscle structures by myotoxic phospholipases A may be useful in an animal model where repair and regeneration processes in muscle and nervous system are studied.

References

1. Howard, B.D., Gundersen, C.B.: Annu.Rev.Pharmacol.Toxicol.20, 307-336 (1980).
2. Marsden, A.T.H., Reid, H.A.: Br.med.J. 1, 1290-1293 (1961).
3. Fohlman, J., Eaker, D.: Toxicon 15, 385-393 (1977).
4. Mebs, D., Samejima, Y.: Toxicon 18, 443-454 (1980).
5. Harris, J.B., Johnson, M.A., MacDonell, C.A.: In: Natural Toxins, D.Eaker, T.Wadström (eds.), Pergamon Press, Oxford & New York, pp. 569-578 (1980).
6. Howard, B.D.: Trends.Pharmacol.Sci. 3, 167-169 (1982).
7. Harris, J.B.: Toxicon 20, suppl. 1, 23 (1982).

ß-BUNGAROTOXIN; NEW ASPECTS OF AN OLD APPROACH TO THE PRE-SYNAPTIC MEMBRANE

Hubert Rehm, Heinrich Betz
Max-Planck-Institut für Psychiatrie, Abt. Neurochemie
D-8033 Martinsried bei München

Introduction

The exchange of information between neurons is the most important activity of the brain. Right in the center of this activity sits the synapse. It is not astonishing therefore, that changes in the metabolism and the structure of the synapse are thought to underly the expression of various mental diseases (depression, schizophrenia, Alzheimer's disease) as well as vital brain functions such as memory establishment and plasticity. The presynaptic part of the synapse, the nerve terminal, is presumed to play the more important role in the expression of these phenomena. The key to the understanding of prominent mental diseases as well as to the working of memory therefore seems to be buried in the biochemistry of the presynaptic terminal.

However little is known about the biochemical machinery of this specialized neuronal structure. This is despite the existence of neurotoxins which attack exclusively the presynaptic part of the synapse. We are rather ignorant about the mechanisms of action of these presynaptic toxins and the molecules they interact with, despite the fact that presynaptic toxins are in general very potent. Two of them (botulinum toxin, tetanus toxin) are the most poisonous

Toxins as Tools in Neurochemistry

substances known. This potency certainly indicates a vital function for the binding sites of presynaptic toxins in the nerve terminal membrane. A promise which seduced us to try presynaptic toxins as tools to elucidate presynaptic mechanisms.

The toxin of our choice was ß-Bungarotoxin (ß-BTX), a member of the family of presynaptic snake venom neurotoxins with phospholipase A_2 (PLA_2) activity. ß-BTX (fig. 1) is a basic protein component of the venom of Bungarus multicinctus, the banded krait. The toxin has a MW of 21,000 and consists of two subunits: subunit A with MW 13,500 and subunit B with MW 7,000 (1).

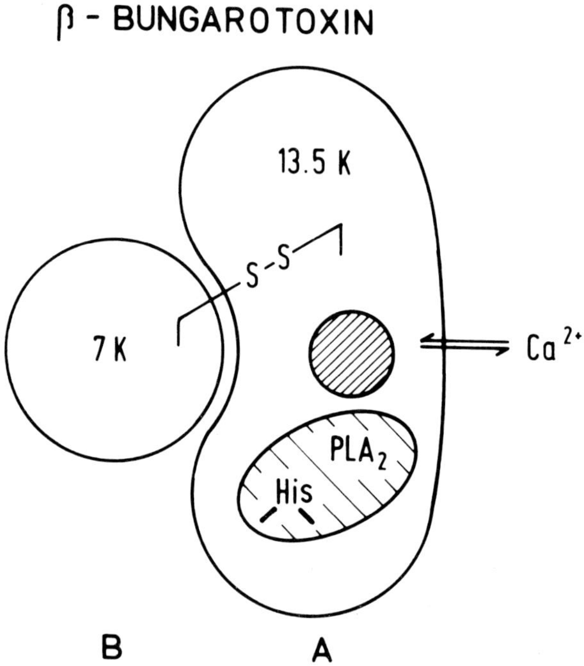

Subunit A bears a Ca^{2+}-dependent PLA_2-activity (1-3) whereas subunit B shows sequence homology with protease inhibitors (1, 4-8). Both subunits are linked by S-S bridges (4).

At the neuromuscular junction, and probably in the central nervous system as well, ß-BTX blocks spontaneous and evoked neurotransmitter release in a complex triphasic manner (9,10). Upon prolonged incubation with the toxin, the presynaptic terminal is destroyed (11).

Results

To elucidate the molecular chain of events leading to the above described symptoms it is logical to start with the first interaction of toxin and neuron: the binding of ß-BTX to a membrane component. Unfortunately, ß-BTX is, due to its basicity, a very sticky molecule which interacts with next to any surface. Furthermore the number of specific binding sites in the neuronal membrane is very low. It is not astonishing therefore, that several workers (12,13,14) failed to detect a physiologically relevant binding of ß-BTX. We found it absolutely necessary to control our binding data by experiments giving direct information about the physiological properties of the toxin:
As our binding experiments were done with chick brain membranes, we in parallel investigated the cytotoxic effects of ß-BTX on a convenient brain structure, the chick retina explant (15). The retinae were kept in rotating Erlenmeyer flasks at 37°C in glucose-enriched medium. Different concentrations of ß-BTX and/or other agents were added. After 24 h the activities of various marker enzymes in the medium and in the tissue explants were measured. The distribution of these marker enzymes between medium and tissue indicated the cytotoxic activity of ß-BTX at the chosen conditions.

The results of these experiments are shown in fig. 2. Choline acetyltransferase and glutamate decarboxylase, which are marker enzymes for cholinergic and GABAergic neurons, are lost from the explants at very low toxin concentrations already (0.2 nM). The contents of tyrosine hydroxylase, a marker for dopaminergic neurons, and glutamine synthetase, a marker for glial cells (16), are affected

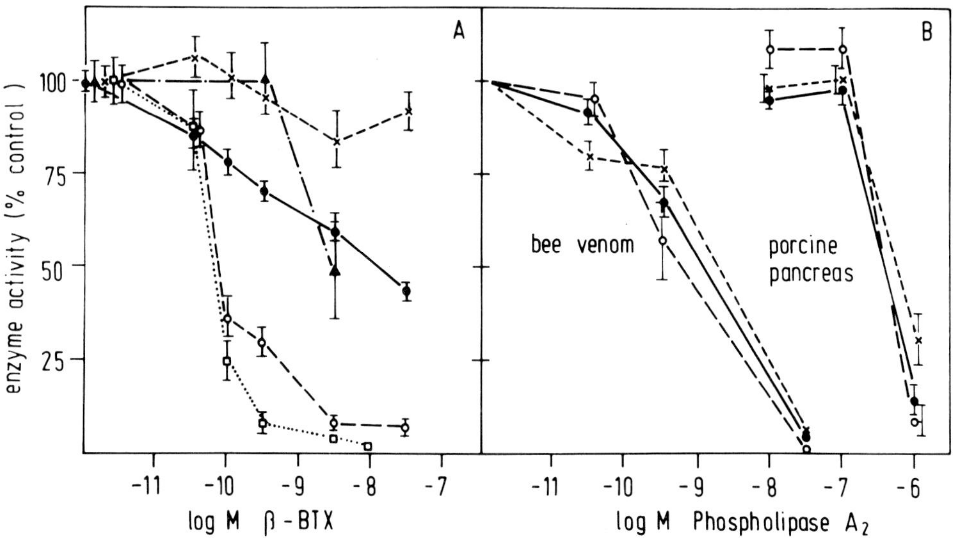

Fig. 2: Effects of ß-BTX and phospholipases A_2 on the enzyme activities of retina explants. Retina from newly hatched chickens were maintained in organ culture for 24 h in the presence or absence of ß-BTX. Then the enzyme activities remaining in the tissue were determined (15). The data are expressed in % of control, which is the total activity of each enzyme in two retinae after 24 h of culture in medium with no toxin added. The values represent the means ± SEM of at least 3 determinations.

lactate dehydrogenase; glutamine synthetase; tyrosine hydroxylase; choline acetyltransferase; glutamate decarboxylase.

Reproduced with permission from reference (15).

only at higher concentrations of ß-BTX, or not at all. This suggests that ß-BTX induces a selective death of cholinergic and GABAergic neurons in the chick retina explant.

Despite a much greater enzymatic activity (9), PLA_2 from bee venom or porcine pancreas are less toxic than ß-BTX and do not show selective cytotoxicity. Obviously the cytotoxic activity of ß-BTX is not an unspecific PLA_2 effect. This point is further strengthened by the finding that ß-BTX is not toxic for phaeochromocytoma (PC12) cells, or for ciliary ganglion cells of 12 day-old chicken embryos (H. Rohrer, personal communication). Furthermore, in the development of the chick embryo retina, ß-BTX displays its cytotoxicity only from the 7th - 8th day onwards with a steep increase at day 12-13 of embryonic development (H. Rehm, unpublished observations). A straightforward explanation for these findings is the existence of a molecule which has to be expressed by the cell before ß-BTX can display its cytotoxic activity.

In an effort to identify this molecule in chick brain membranes the afore-mentioned binding experiments were performed. The toxin was iodinated with the lactoperoxidase method (17). This iodinated toxin was still neurotoxic and its binding properties were unchanged compared to native ß-BTX (17). To determine the amount of ^{125}I-ß-BTX specifically bound, we routinely used a filtration assay, although in some cases a quantitative centrifugation assay was employed (17). A high affinity binding site with a K_d of 0.2 - 0.4 nM and a (major) low affinity binding site with a K_d around 10 nM were detected. The low affinity binding was blocked by basic proteins like cytochrome c, or nerve growth factor. As the cytotoxicity of ß-BTX could not be suppressed by these proteins, it was concluded that only the high affinity binding site is relevant for the physiological action of ß-

BTX. Fortunately, the low affinity binding component could largely be eliminated by treating the membrane fractions with DNAse and low ionic strength buffer. The remaining high affinity toxin binding then gave a linear Scatchard plot with a comparatively low density of binding sites: 50 fmol/mg membrane protein (fig. 3).

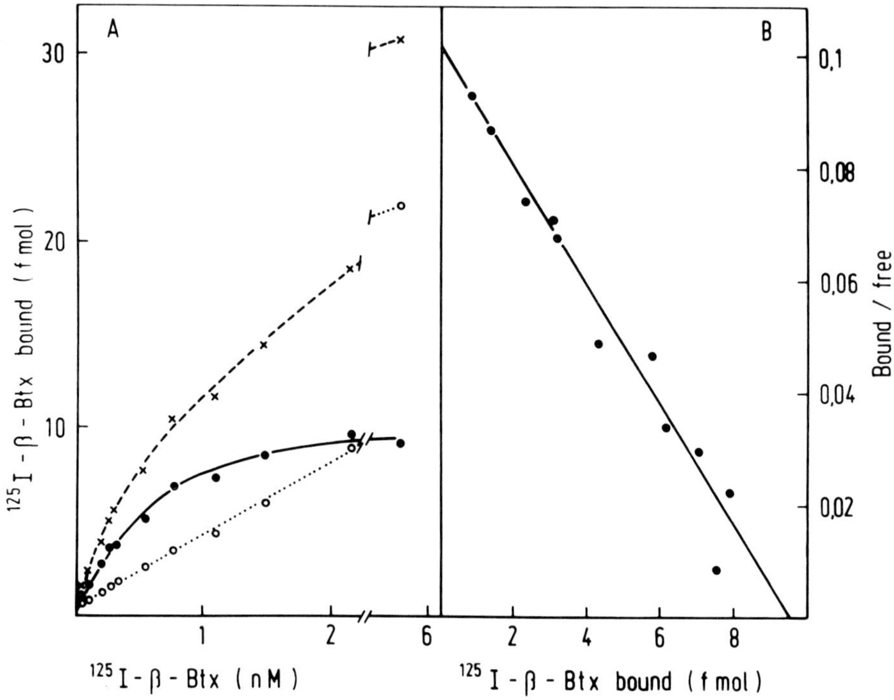

Fig. 3: Binding of ^{125}I-ß-BTX to brain membranes. Membranes (390 µg of protein) from the brains of newly hatched chickens were incubated with different concentrations of ^{125}I-ß-BTX for 15 min, and then the amount of membrane-bound ^{125}I-ß-BTX was determined by a filtration procedure (17). A, specific binding ( ) is the difference between total binding ( ) and unspecific binding as determined in the presence of 238 nM unlabeled ß-BTX ( ). B, Scatchard analysis of specific binding of the data shown in A. The values are the mean of three determinations.
Reproduced with permission from reference (17).

This high affinity binding of ^{125}I-ß-BTX was highly specific: neither ß-BTX modified with p-bromophenacylbromide at a single histidine residue of subunit A, nor crotoxin, another presynaptic PLA_2 of snake venom, inhibited the binding of ^{125}I-ß-BTX. The latter finding suggests that the members of the family of presynaptic toxins from snake venoms have dissimilar targets on the nerve terminal membrane, despite their great similarities in structure and physiology.

The kinetics of ^{125}I-ß-BTX binding are in good agreement with electrophysiological results. The binding of the toxin to membranes is completed within 5 min. At the neuromuscular endplate, the first signs of ß-BTX-induced electrophysiological alterations can be observed 2-3 min after the addition of toxin (9). When ß-BTX has been in short contact with a neuromuscular preparation, a subsequent washout of the toxin has little effect on the time course and the extent of the intoxication (18). From this one should expect a slow dissociation of ß-BTX from its binding site. Indeed in dilution experiments the off-rate was slow ($t_{1/2}$ = 105 min).

The physiological relevance of our binding data is further strengthened by the fact, that a specific binding of ^{125}I-ß-BTX was found only in organs which contain ß-BTX sensitive cells or nerve endings (17).

An important feature of ^{125}I-ß-BTX binding is its Ca^{2+}-dependence. This cation can be replaced by Sr^{2+} or Co^{2+}, but not by Mg^{2+} (17). ß-BTX is known to have a Ca^{2+}-binding site (on subunit A) (9), but of course it is not clear from our data whether ß-BTX needs Ca^{2+} to bind the membrane molecule, or whether the membranes need Ca^{2+} to bind ß-BTX. The results of a recent publication (19) on the binding of tritiated ß-BTX to rat brain membranes, where no Ca^{2+}-dependence of binding was found suggests the second possibility. In this case the ß-BTX binding molecules from chick and rat would be expected to differ in their Ca^{2+}-dependence.

The ß-BTX binding site was sensitive to protease treatment (17). This finding, and our previous experience (20) with crosslinking of protein ligands to protein receptors, suggested a further characterization of the toxin binding molecule through photoaffinity crosslinking methods. For that purpose, ^{125}I-ß-BTX was allowed to bind to the membranes, free toxin was washed away, and the membrane-toxin complex was then reacted with N-succinimidyl-6(4'-azido-2'-nitrophenylamino)-hexanoate, first in the dark, and then with light (21). The crosslinked membranes were then washed, solubilized in sodium dodedylsulfate (SDS) sample buffer and run on SDS-polyacrylamide gels (22). Autoradiographs of the gels showed one labeled band with a MW of 116,000. No labeling of this 116 K band was found when Ca^{2+} was omitted from the binding buffer, or when liver was taken as membrane source or when the membranes were preincubated with an excess of unlabeled ß-BTX (21). Picomolar concentrations of unlabeled ß-BTX already reduced the labeling intensity of the 116 K band, as would be expected from the high affinity binding described above. Crosslinking of ^{125}I-ß-BTX (without membranes) gave no labeled 116 K band.

As the addition of ß-mercaptoethanol to the crosslinked samples had no effect on the apparent molecular weight of the 116 K band, we conclude that the presynaptic membrane binding protein for ß-BTX contains a polypeptide of MW 95,000 (116,000 - 21,000).

Conclusion

In this report we have characterized the cytotoxicity of ß-BTX upon central neurons. It was shown that in the chick retina explant cholinergic and GABAergic neurons are preferentially destroyed by this potent poison. Furthermore, binding experiments with ^{125}I-labeled ß-BTX demonstrated the existence of a single class of high affinity toxin binding sites in chick brain membranes. By comparison with our cytotoxicity assay, evidence could be obtained that this binding site is the long-looked for membrane molecule through which ß-BTX exerts its toxic action upon the nerve terminal. Finally, a component of the ß-BTX binding site was identified by photoaffinity crosslinking to be a polypeptide of MW 95,000.

Certainly these findings have not yet fulfilled our hope of throwing light on the mysteries of the biochemistry of the nerve terminal. Nevertheless we think, that via solubilization and purification of this, without doubt, important presynaptic membrane protein we might be able to get access to an exciting piece of information about the working of the brain.

Acknowledgement. This work was supported by the Deutsche Forschungsgemeinschaft. The technical assistance of Ms. Ute Müller, the help of Ms. E. Eichler and C. Bauereiß during the preparation of the manuscript and the provision of laboratory facilities by the Max-Planck-Gesellschaft are gratefully acknowledged.

References

1. Howard, B.D., Gundersen, C.B.: Ann. Rev. Pharmacol. Toxicol. 20, 307-336 (1980).
2. Strong, P.N., Goerke, J., Oberg, S.G., Kelly, R.B.: Proc. Natl. Acad. Sci. USA 73, 178-182 (1976).
3. Wernicke, J.F., Vanker, A.D., Howard, B.D.: J. Neurochem. 25, 483-496 (1975).
4. Kelly, R.B., Brown, F.R.: J. Neurobiol. 5, 135-150 (1974).
5. Kondo, K., Narita, K., Lee, C.Y.: J. Biochem. (Tokyo) 83, 101-115 (1978).
6. Kondo, K., Toda, H., Narita, K.: J. Biochem. (Tokyo) 84, 1301-1308 (1978).
7. Lee, C.Y.: Advances in Cytopharmacology, Vol. 3, pp. 1-16, Raven Press, New York 1979.
8. Strydom, D.J., Foubert, F.J.: Hoppe-Seyler's Z. physiol. Chem. 362, 1377-1384 (1981).
9. Abe, T., Alemá, S., Miledi, R.: Eur. J. Biochem. 80, 1-12 (1977).
10. Halliwell, J.V., Tse, C.K., Spokes, J.W., Othman, J., Dolly, J.O.: J. Neurochem. 39, 543-550 (1982).
11. Abe, T., Limbrick, A.R., Miledi, R.: Proc. R. Soc. Lond. B Biol. Sci. 194, 545-553 (1976).
12. Mac Dermot, J., Westgaard, R.H., Thompson, E.J.: Biochem. J. 175, 281-288 (1978).
13. Oberg, S.G., Kelly, R.B.: Biochim. Biophys. Acta 433, 662-673 (1976).
14. Esquerda, J.E., Solsona, C., Marsal, J. Neuroscience 3, 751-758 (1982).
15. Rehm, H., Schäfer, T., Betz, H.: Brain Research 250, 309-319 (1982).
16. Linser, P., Moscona, A.A.: Proc. Natl. Acad. Sci. USA 76, 6476-6480 (1979).
17. Rehm, H., Betz, H.: J. Biol. Chem. 257, 10015-10022 (1982).
18. Kelly, R.B., Wedel, R.J., Strong, P.N.: Advances in Cytopharmacology, Vol. 3, pp. 77-85, Raven Press, New York 1979.
19. Othman, I., Spokes, J.W., Dolly, J.O.: Eur. J. Biochem. 128, 267-276 (1982).

20. Betz, H., Graham, D., Rehm, H.: J. Biol. Chem. 257, 11390-11394 (1982).

21. Rehm, H., Betz, H.: EMBO Journal, submitted (1983).

22. Lämmli, U.K.: Nature 227, 680-685 (1970).

NEUROTOXINS AS TOOLS IN CARDIAC EXCITATION CONTRACTION COUPLING

Ursula Ravens

Abteilung Pharmakologie im Zentrum Klinisch-Theoretische Medizin II der
Christian-Albrechts-Universität Kiel, Hospitalstrasse 4-6
D-2300 Kiel

Introduction

The basic sequence of events in cardiac excitation contraction coupling involves: excitation of the myocardial cells by a propagated action potential; increase of the free calcium ion concentration in the cytosol (by release or by entry); and finally activation of the contractile proteins by their interaction with calcium ions. The degree of contractile activation and thus of tension development depends on the free calcium ion concentration in the cytosol. Although sodium ions do not interact directly with the contractile proteins, they do play an indirect role in cardiac excitation contraction coupling. This is particularly evident when studying the effects of some neurotoxins in heart muscle, which influence not only neuronal but also cardiac sodium channels and thereby modify sodium entry into heart muscle cells. The two neurotoxins chosen as experimental tools for the present investigation were a) tetrodotoxin (TTX) because it blocks the sodium channels and thereby diminishes sodium entry (1, 2) and b) the Anemonia sulcata toxin ATX II because in nerves it inhibits sodium channel inactivation (3, 4) giving rise to prolonged sodium entry.

Materials and Methods

Right ventricular papillary muscles of guinea pigs were superfused with oxygenated tyrode solution at 32°C in a small organ bath. Electrical stimulation was applied via the tips of two platinum wires. The force of contraction was registered with an iso-

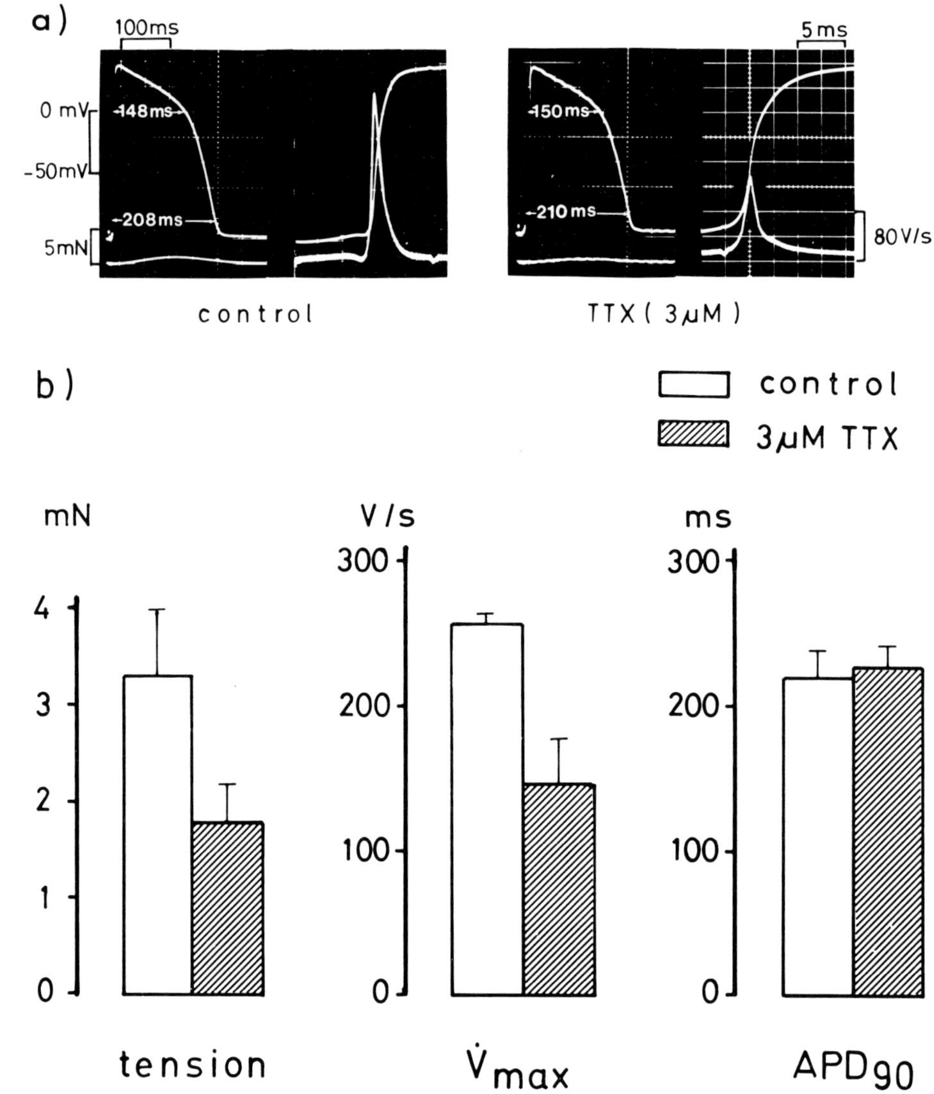

Fig. 1 The effect of 3 µM tetrodotoxin (TTX) on action potential configuration, on maximum depolarization velocity ($\dot{V}_{max}$) and on force of contraction of guinea pig papillary muscle (rate of stimulation $1\ s^{-1}$).
a) Original tracings; b) mean values ± S.E.M. from 5 experiments.
APD_{90} : action potential duration at 90% of repolarization.

metric force transducer and the intracellular action potentials were recorded with conventional glass microelectrodes.
Isolated myocytes were obtained from guinea pig hearts according to the method of Isenberg and Klöckner (5). For voltage clamp experiments a single electrode technique was applied using patch clamp electrodes (6).

Results and Discussion

In guinea pig papillary muscle, 3 μM TTX has little effect on the action potential duration (APD), but reduces the maximum depolarization velocity ($\dot{V}_{max}$) from 270 V/s to 130 V/s and depresses the force of contraction (Fig. 1a).The results from 5 experiments are summarized in Fig. 1b.The negative inotropic effect of TTX may be related directly to the reduced sodium entry which in turn decreases the intracellular calcium available for contraction by augmenting calcium extrusion from the cell via the Na/Ca exchange mechanism (7; for review see 8).However, another possible explanation is an indirect effect on slow inward current: because of the decreased depolarization velocity in the presence of TTX, less slow inward current is activated and force of contraction decreases, but tension development is restored when applying a voltage clamp step which imitates the normal $\dot{V}_{max}$ closely (9).

With the neurotoxin ATX II from Anemonia sulcata as a tool to increase the sodium load, the force of contraction increases and the action potential duration is markedly prolonged (Fig. 1). These effects develop within 30 min and are completely reversible upon washout (10). The ATX II-induced prolongation in action potential duration depends on the rate of stimulation, at 0.1 s^{-1} and with 50 nM ATX II action potentials may last several seconds. The effects of ATX II on force of contraction and APD are abolished when the sodium channels are inactivated by partial depolarization after exposure to elevated extracellular potassium (Fig. 3).

As in myelinated nerve fibres (11), the efficacy of ATX II in heart muscle depends on the extracellular calcium ion concentration (Fig. 4). Over the whole concentration

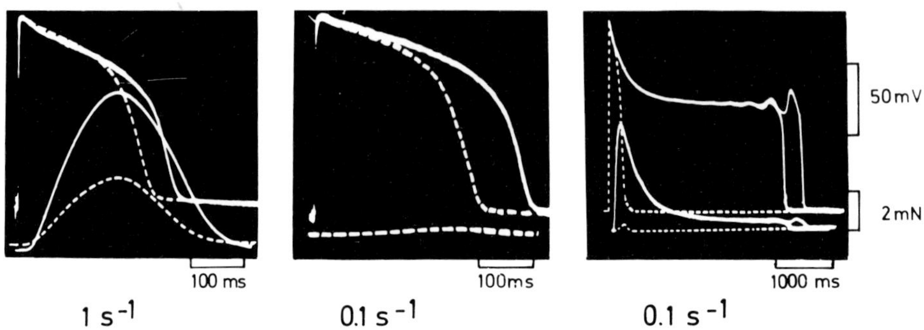

Fig. 2 Original recordings of action potentials and force of contraction after exposure to 10 and 50 nM ATX II, respectively, at two different rates of stimulation. Dashed and continuous tracings are the control records and after 30 min of exposure to ATX II, respectively. Calibrations as indicated by the bars.

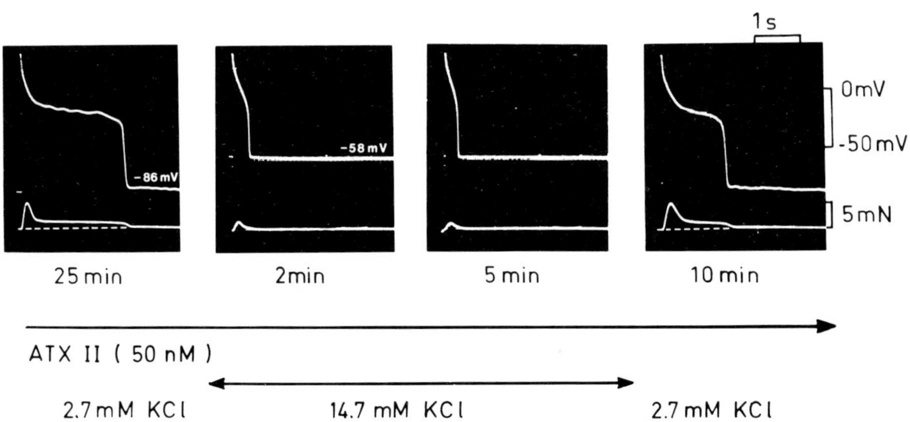

Fig. 3 The influence of partial depolarization in high potassium solution on the action potential prolonging and the positive inotropic effect of 50 nM ATX II in guinea pig papillary muscle stimulated at $0.1\ s^{-1}$.

range studied ATX II is more effective in prolonging the action potential duration at 0.9 and 1.8 mM $[Ca^{++}]_o$, the maximal attainable contractile augmentation, however, is similar at all three calcium ion concentrations studied. This influence of calcium ions can also be demonstrated after the full effect has already been established. Fig. 5 shows the result of an experiment in which the rate of stimulation was decreased from 1 s^{-1} to 0.1 s^{-1} 30 min after addition of 50 nM ATX II because the APD-prolonging

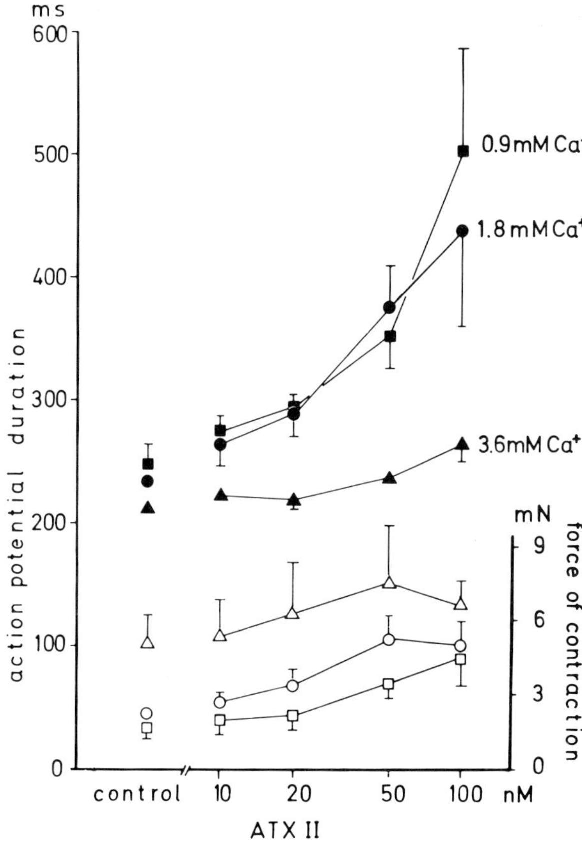

Fig. 4 Cumulative dose-response curves of the effects of ATX II on action potential duration at 90% of repolarization (closed symbols) and force of contraction (open symbols) at various extracellular calcium ion concentrations (mean values ± S.E.M. from 5 - 7 experiments). Rate of stimulation 1 s^{-1} .

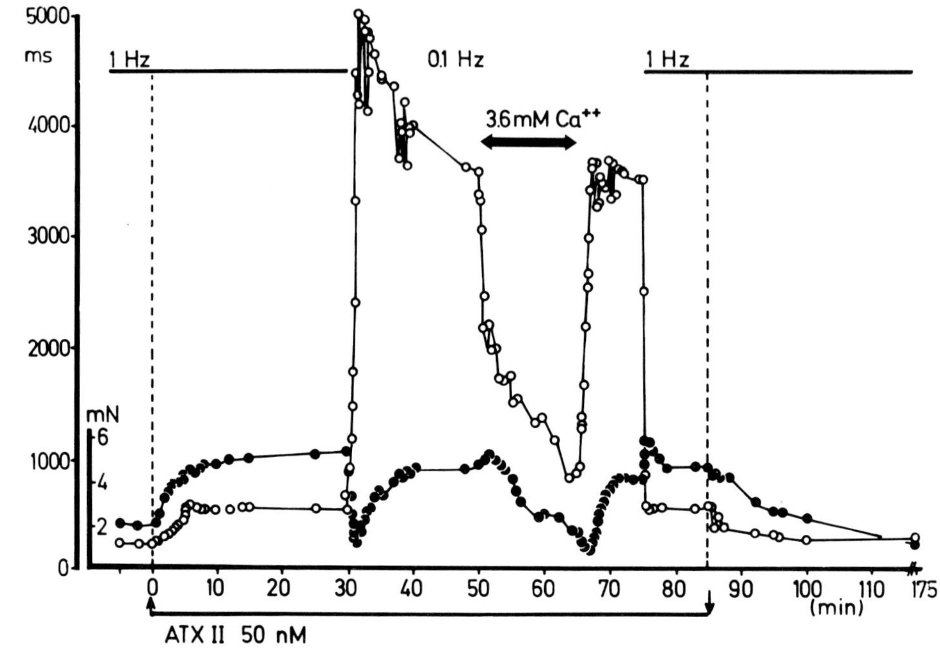

Fig. 5 Attenuation of the effects of ATX II on action potential duration (open symbols) and force of contraction (closed symbols) by increasing the extracellular calcium ion concentration from 1.8 mM to 3.6 mM(horizontal arrow).

effect is very pronounced at this low frequency (12). Increasing the extracellular calcium concentration from 1.8 to 3.6 mM reduces the APD and diminishes the positive inotropic effect. Both ATX II effects are fully re-established when returning to 1.8 mM $[Ca^{++}]_o$. Thus calcium ions seem to inhibit the electrical response to ATX II, the reason for the lack of influence of calcium ions on the extent of the ATX II-induced contractile augmentation may be that the direct enhancement in force of contraction by elevated $[Ca^{++}]_o$ may just balance the smaller effect of ATX II at high $[Ca^{++}]_o$.

Several possible changes in membrane current could be responsible for the ATX II-induced prolongation in action potential duration, i.e. prolonged inward current through sodium and/or slow inward channels, or a reduction in outward potassium current. Since ATX II affects sodium channels in nerve membranes (3, 4) it seems most likely that these channels are involved also in heart muscle. More direct evidence was obtained from voltage clamp studies in isolated heart muscle cells (13). Fig. 6 shows net membrane currents in response to long-lasting clamp pulses from -85 to -27 mV. Two min after the exposure to 20 nM ATX II a slowly decaying inward current appears which is reversible in the presence of 60 μM TTX. With ATX II, the net membrane current remains inward for approximately 500 ms after the initiation of the potential step and can therefore easily explain the observed prolongation in action potential duration. From the complete current-voltage relations under control conditions and in the presence of 20 nM ATX II, the ATX II-induced current can be obtained by simple substraction of the two currents. It has a threshold at -60 mV, a maximum at -30 mV and a reversal potential arround +40 mV. These potential characteristics together with

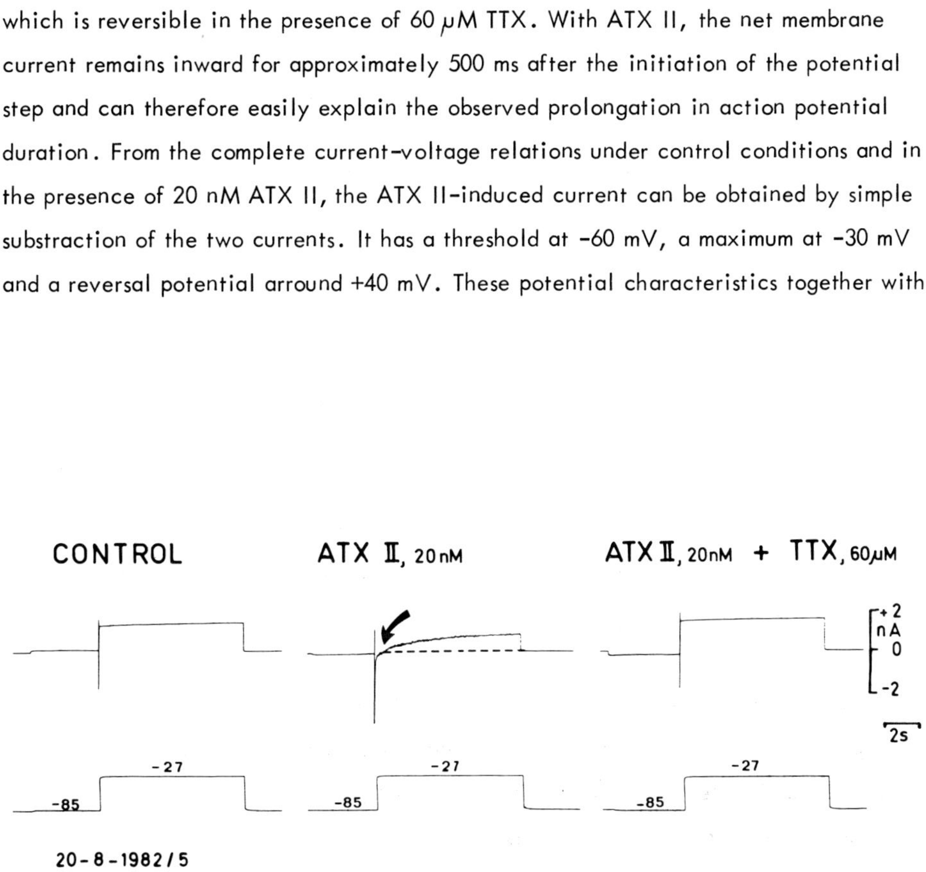

Fig. 6 Brush-pen recordings of net membrane currents (upper tracing) and membrane potential (lower tracings) during a voltage clamp experiment with an isolated heart muscle cell from guinea pig ventricle. Note that the initial downward deflection is not a faithful record of the rapid sodium inward current because of the inertia of the recording system. In the presence of ATX II (20 nM) net membrane current remains inward for approximately 0.5 s (arrow).

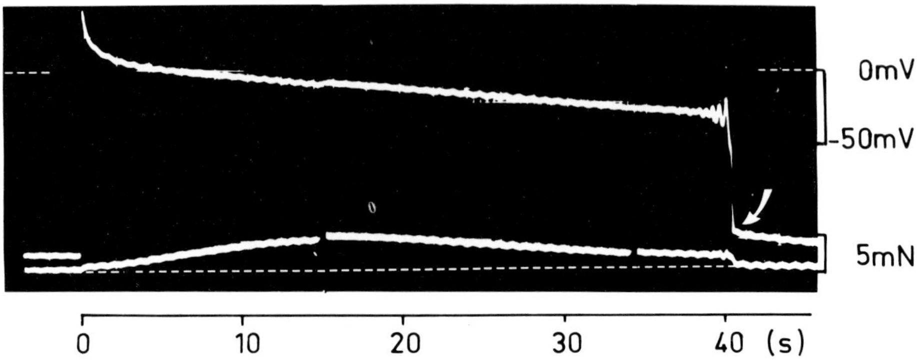

Fig. 7 First action potential after preincubating a guinea pig papillary muscle with 50 nM ATX II during 30 min of rest. Note that during the long lasting action potential tonic tension develops. The arrow points at some residual depolarization after termination of the action potential, which may be the result of extracellular ion accumulation during this extremely long action potential.

the reversibility by TTX and a lack of ATX II on slow inward current (when sodium channels are inactivated by clamping the membrane potential to -45 mV) lead to the conclusion that the ATX II-induced current flows through modified sodium channels also in heart muscle. The prolonged sodium current in the presence of ATX II imposes a marked sodium load onto the cells which may be related to the positive inotropic effect via the Na/Ca exchange mechanism.

Some evidence that the Na/Ca exchange may be involved can be taken from the development of tonic tension during a very long action potential in guinea pig papillary muscle obtained after a 30 min period of rest (Fig. 7). During the plateau phase calcium ions could enter the cells via the Na/Ca exchange carrier (8) since the slow inward channels should be largely inactivated. However from this experiment the possibility cannot be excluded that some calcium ions enter the cells via non-inactivating slow inward channels (14).

In conclusion, TTX and ATX II provide valuable information about the excitation-contraction coupling in heart muscle in spite of the fact that they were primarily discovered as neurotoxins.

References

1. Narahashi, T.: Physiol. Rev. 54, 813-889 (1974).
2. Deitmer, J.W., Ellis, D.: J. Physiol. (Lond.) 300, 269-282 (1980).
3. Bergman, C., Dubois, J.M., Rojas, E., Rathmayer, W.: Biochim. Biophys. Acta 455, 173-184 (1976).
4. Romey, G., Abita, J.-P., Schweitz, H., Wunderer, G., Lazdunski, M.: Proc. Natl. Acad. Sci. U.S.A. 73, 4055-4059 (1976).
5. Isenberg, G., Klöckner, U.: Pflügers Arch. 395, 6-18 (1982).
6. Hamill, O.P., Marty, A., Neher, E., Sakmann, B., Sigworth, F.J.: Pflügers Arch. 391, 85-100 (1981).
7. Bhattacharyya, M.L., Vassalle, M.: J. Electrocardiology 15, 351-360 (1982).
8. Mullins, L.J.: Ion transport in heart, Raven Press, New York 1981.
9. Isenberg, G. (personal communication).
10. Ravens, U.: Naunyn-Schmiedeberg's Arch. Pharmacol. 296, 73-78 (1976).
11. Ulbricht, W., Schmidtmayer, J.: J. Physiol. (Paris) 77, 1103-1111 (1981).
12. Beress, L., Ritter, R., Ravens, U.: Europ.J. Pharmacol. 79, 265-272 (1982).
13. Isenberg, G., Ravens, U.: Naunyn-Schmiedeberg's Arch. Pharmacol. 322, R 15 (1983).
14. Reuter, H.: Ann. Rev. Physiol. 41, 413-424 (1979).

Index

(authors in italics)